Artificial Intelligence for Ocean Remote Sensing

Artificial Intelligence for Ocean Remote Sensing

Guest Editors

Hua Su

Wenfang Lu

Xiao-Hai Yan

MDPI

Basel • Beijing • Wuhan • Barcelona • Belgrade • Novi Sad • Cluj • Manchester

Guest Editors

Hua Su
The Academy of Digital China
Fuzhou University
Fuzhou
China

Wenfang Lu
School of Marine Sciences
Sun Yat-Sen University
Guangzhou
China

Xiao-Hai Yan
Center for Remote Sensing
College of Earth, Ocean
and Environment
University of Delaware
Newark, DE
USA

Editorial Office

MDPI AG
Grosspeteranlage 5
4052 Basel, Switzerland

This is a reprint of the Special Issue, published open access by the journal *Remote Sensing* (ISSN 2072-4292), freely accessible at: https://www.mdpi.com/journal/remotesensing/special_issues/4BKYD16M06.

For citation purposes, cite each article independently as indicated on the article page online and as indicated below:

Lastname, A.A.; Lastname, B.B. Article Title. *Journal Name* **Year**, *Volume Number*, Page Range.

ISBN 978-3-7258-3609-3 (Hbk)
ISBN 978-3-7258-3610-9 (PDF)
https://doi.org/10.3390/books978-3-7258-3610-9

Contents

remote sensing

MDPI

Article

Gas Plume Target Detection in Multibeam Water Column Image Using Deep Residual Aggregation Structure and Attention Mechanism

Wenguang Chen [1], Xiao Wang [1,*], Binglong Yan [2], Junjie Chen [1], Tingchen Jiang [1] and Jialong Sun [1]

1 School of Marine Technology and Geomatics, Jiangsu Ocean University, Lianyungang 222005, China; 2021220403@jou.edu.cn (W.C.); 2022220207@jou.edu.cn (J.C.); jiangtc@jou.edu.cn (T.J.); 2006000076@jou.edu.cn (J.S.)

2 Lianyungang Water Conservancy Planning and Design Institute Co., Ltd., Lianyungang 222006, China; lygslsjy@163.com

* Correspondence: wangxiao@jou.edu.cn; Tel.: +86-137-7548-9830

Abstract: A multibeam water column image (WCI) can provide detailed seabed information and is an important means of underwater target detection. However, gas plume targets in an image have no obvious contour information and are susceptible to the influence of underwater environments, equipment noises, and other factors, resulting in varied shapes and sizes. Compared with traditional detection methods, this paper proposes an improved YOLOv7 (You Only Look Once vision 7) network structure for detecting gas plume targets in a WCI. Firstly, Fused-MBConv is used to replace all convolutional blocks in the ELAN (Efficient Layer Aggregation Networks) module to form the ELAN-F (ELAN based on the Fused-MBConv block) module, which accelerates model convergence. Additionally, based on the ELAN-F module, MBConv is used to replace the 3×3 convolutional blocks to form the ELAN-M (ELAN based on the MBConv block) module, which reduces the number of model parameters. Both ELAN-F and ELAN-M modules are deep residual aggregation structures used to fuse multilevel features and enhance information expression. Furthermore, the ELAN-F1M3 (ELAN based on one Fused-MBConv block and three MBConv blocks) backbone network structure is designed to fully leverage the efficiency of the ELAN-F and ELAN-M modules. Finally, the SimAM attention block is added into the neck network to guide the network to pay more attention to the feature information related to the gas plume target at different scales and to improve model robustness. Experimental results show that this method can accurately detect gas plume targets in a complex WCI and has greatly improved performance compared to the baseline.

Keywords: multibeam water column image; gas plume; target detection; YOLOv7; deep residual aggregation structure; SimAM block

Citation: Chen, W.; Wang, X.; Yan, B.; Chen, J.; Jiang, T.; Sun, J. Gas Plume Target Detection in Multibeam Water Column Image Using Deep Residual Aggregation Structure and Attention Mechanism. *Remote Sens.* **2023**, *15*, 2896. https://doi.org/10.3390/rs15112896

Academic Editors: Xiao-Hai Yan, Wenfang Lu and Hua Su

Received: 22 April 2023
Revised: 28 May 2023
Accepted: 30 May 2023
Published: 2 June 2023

1. Introduction

Multibeam echo sounding (MBES) is a high-precision underwater data measurement technique [1]. Compared with the traditional single-beam echo sounding technique, MBES uses multiple transmitters and receivers to collect echo signals in multiple directions simultaneously, obtaining more precise water depth and water body data. This technique has been widely applied in marine resource exploration [2], underwater pipeline laying [3], and underwater environmental monitoring [4]. The water column image (WCI) formed by water body data is an important means of underwater target detection, and the gas plumes in the water may be an indication of the presence of gas hydrates in the nearby seabed sediment layers [5]. Development and excavation of these resources will play a crucial role in alleviating current global issues such as energy scarcity and environmental degradation. Therefore, how to detect and locate gas plumes quickly and accurately has become an important research topic.

In traditional WCI target detection, the processing steps are image denoising, feature extraction, and target classification, in that order. Due to the influence of the working principle of MBES, some sidelobe beams are produced around the main lobe beam during transmission. When receiving the echo signal, the echo information of these sidelobe beams is mistaken for real signals, causing significant arc noise in the image [6], which is the main factor affecting image quality. In addition, the image also includes multisector noise and environmental noise. To make the collected raw data more accurate, [7] used weighted least squares to estimate the optimal beam incidence angle and corrected the difference in echo intensity under different water depths, thus obtaining normalized echo data. To effectively remove noise in the image, the image masking method was used to eliminate static noise interference in [8], and artificial thresholding was used to remove environmental noise. In [9], the data that fell within the minimum slant range were considered valid and used in the analysis, and the noise was removed by using the average echo intensity as the threshold. The second step in WCI target detection is to extract features from the denoised image. Feature extraction aims to obtain distinctive and representative image features, such as edges, morphology, and texture, to reduce data dimensions for subsequent classification. In [10], the authors used a clustering algorithm to extract information about regions containing gas plumes and then used feature histograms for feature matching to identify them. In [11], gas plume features were extracted using intensity and morphological characteristics to distinguish them from the surrounding environment. In [12], multiple features, such as color, gradient, and direction, were used for feature extraction, obtaining a set of feature vectors that can effectively distinguish between textures and nontextures. The final step is target classification. Based on the principle of feature invariance, the extracted features are input into a classifier for training. Commonly used classifiers include SVM (Support Vector Machine), Adaboost (Adaptive Boosting), and Random Forest. Then the training results are evaluated and optimized to achieve the high-precision detection of targets in the image. Traditional target detection methods for a WCI are complex, with human factors having a significant impact on image denoising and feature extraction algorithms being unable to extract representative features in complex images. The classifiers used in target classification are strongly influenced by lighting, angle, and noise, making it easy to miss or misidentify targets. Overall, target detection using traditional methods in a WCI is highly limited.

Convolutional neural networks (CNN) have proven effective for solving various visual tasks [13–15] in recent years, thus providing a new solution for multibeam WCI target detection. Compared to various machine learning classifiers, CNN not only automatically extracts image features, reducing human intervention, but can also learn more complex features, improving the model's robustness. In addition, the end-to-end advantage and the introduction of transfer learning [16] have increased the efficiency and accuracy of the model, and it has been applied to different scenarios and tasks. CNN detectors can be divided into one-stage and two-stage, which differ mainly in the order of detection and classification. One-stage detectors refer to the prediction of target category and position directly from feature maps, such as YOLO (You Only Look Once) [17–21], SSD (Single-Shot Multi-Box Detector) [22], RetinaNet [23], and EfficientDet [24]. Among them, YOLO adopts a multiscale feature map and anchor mechanism, which can detect multiple targets simultaneously. SSD adopts feature maps of different scales and multiple detection heads, which can detect targets of different sizes. The focal loss function is used in RetinaNet to reduce the effect of target class imbalance and has achieved excellent detection results. EfficientDet adopts a scalable convolutional neural network structure based on EfficientNet [25] and performs well in speed and accuracy. In [26], to adapt to the particularity of the underwater environment, the author introduced the CBAM (Convolutional Block Attention Module) based on YOLOv4 to help find attention regions in object-dense scenes. In [27], the author established direct connections between different levels of feature pyramid networks to better utilize low-level feature information, thereby increasing the capability and accuracy of feature representation. To make the model smaller, in [28], by pruning and fine-tuning the EfficientDet-d2 model, the author achieved a 50% reduction in model size without sacrificing detection accuracy. Two-stage detector initially generates candidate frames and then uses them for classification and position regression, such as Faster R-CNN (Region-based Convolutional Neural Network) [29], Mask R-CNN [30], and Cascade R-CNN [31]. Among them, Faster R-CNN uses the RPN (Region Proposal Network) to generate candidate boxes and then uses RoI (Region of Interest) pooling to extract features for classification and regression. Mask R-CNN extends Faster R-CNN by incorporating segmentation tasks, enabling it to perform target detection and instance segmentation simultaneously. Cascade R-CNN enhances the robustness and accuracy of the detector through a cascaded classifier. Wang [32] enhanced the Faster R-CNN algorithm by implementing automatic selection of difficult samples for training, thereby improving the model's ability to perform well and generalize on difficult samples. In [33], Song proposed Boosting R-CNN, a two-stage underwater detector featuring a new region proposal network for generating high-quality candidate boxes.

These methods have shown good improvements in their respective tasks but may not be applicable to gas plume object detection. This is because a gas plume generally consists of numerous bubbles, which are close together and interfere with each other. Compared to other objects, the reflection intensity of the gas plume is weaker, and the edges of the bubbles are not easily distinguishable. In addition, gas plumes in water often experience fracturing as they rise, which makes them difficult to accurately locate. This paper demonstrates through extensive experiments that the proposed method offers a viable solution to address the issues related to gas plume target detection. The main contributions are as follows:

1. ELAN-M and ELAN-F modules are designed to reduce model parameters, speed up model convergence, and alleviate the problem of insignificant accuracy gains in deep residual structures.
2. An ELAN-F1M3 backbone network structure is designed to fully utilize the efficiency of the ELAN-F and ELAN-M modules.
3. To reduce the effect of noise, the SimAM module is introduced to evaluate the weights of the neurons in each feature map of the neck network.
4. Extensive experiments show that the new model can accurately detect plume targets in complex water images, far outperforming other models in terms of accuracy.

2. Related Work

2.1. Data Augmentation

Data augmentation is a powerful technique used in neural networks to enhance the quantity and diversity of the training data by applying random transformations. By doing so, the model becomes more robust and better equipped to generalize to unseen data. This can alleviate the overfitting problem to some extent. Common augmentation methods include geometric transformations and color transformations. Geometric transformations include flipping, translation, rotation, scaling, etc., which have small changes in the original content. Color transformations include brightness, saturation, color inversion, histogram equalization, etc., which have significant changes in the original content and high diversity. In addition, there are a number of unique approaches being used to augment data. Some scholars have proposed data augmentation methods based on multisample interpolation, such as Sample Pairing [34], Mixup [35], Mosaic [36], etc. Sample Pairing randomly selects two images from the training set, averages the pixel values to synthesize a new sample, and uses one of the image labels as the correct label for the synthesized image; Mixup is an extension of Sample Pairing, which performs linear interpolation on both images and labels; Mosaic combines four training images in a way that is randomly scaled, cropped, and arranged to improve its ability to identify small targets. Some examples of WCI data augmentation are shown in Figure 1.

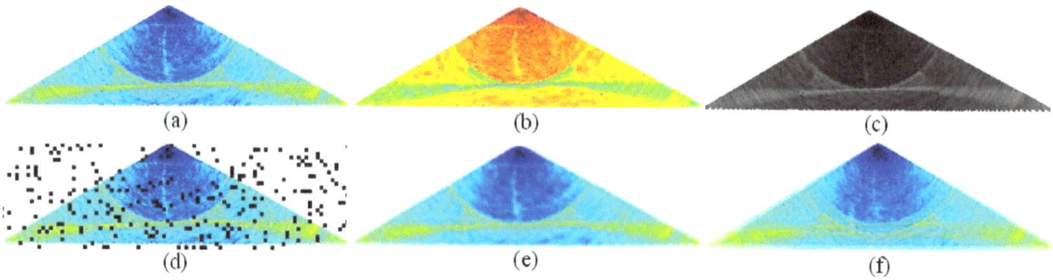

Figure 1. Partial results of data augmentation. (**a**) Original image. (**b**) Color transformation. (**c**) Grayscale transformation. (**d**) Random occlusion. (**e**) Blurring. (**f**) Mixup.

2.2. YOLOv7

The YOLOv7 model in the one-stage detector is another great achievement in the YOLO family, integrating E-ELAN (Extended Efficient Layer Aggregation Networks), structural reparameterization [37], positive and negative sample allocation strategies [17,18], and a training method with auxiliary heads, once again balancing the contradictions between the number of parameters, computational complexity, and performance. YOLOv7 has seven different versions, including YOLOv7-tiny, YOLOv7, YOLOv7x, YOLOv7-w6, YOLOv7-d6, YOLOv7-e6, and YOLOv7-e6e. Among them, YOLOv7-tiny, YOLOv7, and YOLOv7-w6 are the basic models of the network, and the other models are obtained by model scaling.

As shown in Figure 2, the YOLOv7 network structure consists of an input module, a backbone network, a neck network, and a head network.

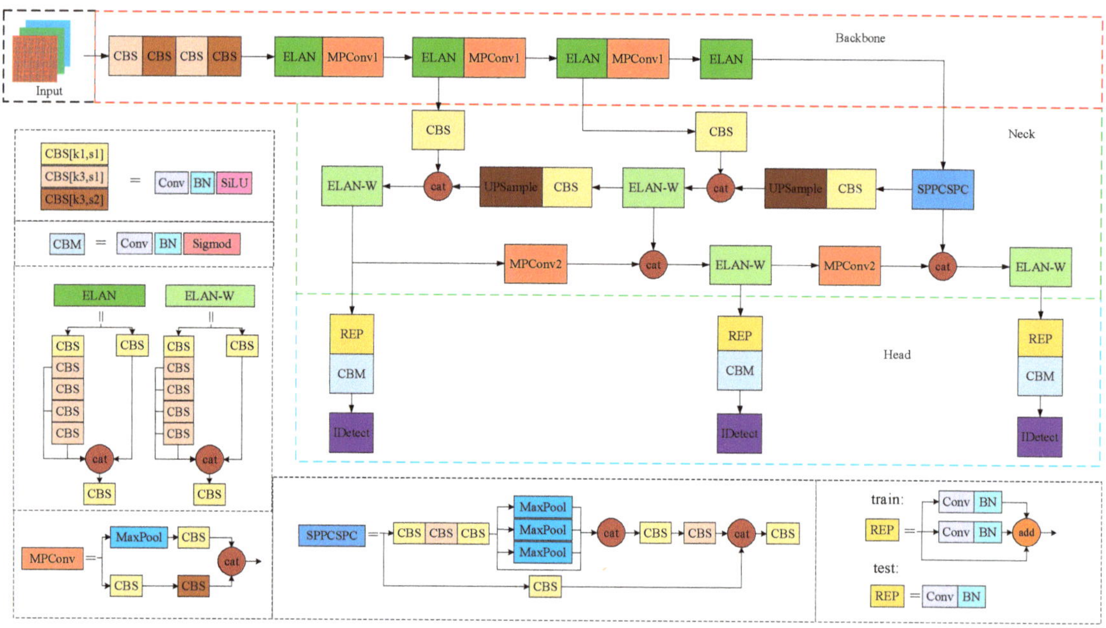

Figure 2. Structure of the YOLOv7.

2.2.1. Input Module

The input end of YOLOv7 continues to use the improvement points of YOLOv5, mainly utilizing the mosaic high-order data augmentation strategy to increase data diversity and reduce the computation cost of training. In addition, YOLOv7 uses an adaptive image adjustment strategy to calculate the input size of images. After calculating, the image is adaptively padded on all sides to obtain the final input image, thereby reducing the problem of increased inference time caused by excessive invalid information introduced by conventional image scaling and padding.

2.2.2. Backbone Network

The backbone network of YOLOv7 mainly consists of three types of modules: CBS, ELAN [38], and MPConv1. The CBS module includes Convolution (Conv), Batch Normalization (BN), and an SiLU activation function (k and s represent the size and stride of the convolution kernel). The ELAN module is a multibranch structure that effectively reduces the number of neurons in the network through multilayer aggregation, reducing computational and storage overheads, and by controlling the shortest and longest gradient paths to accelerate gradient propagation. Except for the first two 1×1 convolution kernel sizes of the CBS module, which achieve channel compression, the number of input and output channels of the remaining CBS modules is kept consistent, which has been proven to be an efficient network design principle in Shufflenet v2 [39]. The E-ELAN structure proposed in YOLOv7 is a grouped convolution of two ELAN structures with a group number of two, and the output results are concatenated in the channel direction. The MPConv1 structure is a two-branch structure composed of MaxPool (MP) and CBS, where one branch implements spatial downsampling through MP, and the other branch uses a 3×3 convolution with a stride of 2 to complete the downsampling. Finally, an enhanced version of the downsampling function is implemented through the connection operation. Both ELAN and MPConv1 are a sublimation of feature reuse, making the network better at capturing relationships between data.

2.2.3. Neck Network

The neck network of YOLOv7 introduces the SPPCSPC (Spatial Pyramid Pooling and Cross-Stage Partial Connection) module, which expands the receptive field and achieves multiscale feature fusion through MP operations with different pooling kernels. Immediately afterward, the enhanced feature extraction network structure of Feature Pyramid Network (FPN) + Pyramid Attention Network (PAN) is still adopted. The FPN + PAN architecture improves target discrimination at different scales by combining bottom-up feature extraction with top-down feature fusion. The ELAN-W and MPConv2 modules used here are similar to the ELAN and MPConv1 modules used in the backbone network. ELAN-W is simply an extension of ELAN, with the addition of two outputs in one of its branching structures for later concatenation; MPConv2 uses the same input and output channel numbers in the CBS module.

2.2.4. Head Network

The prediction part of YOLOv7 uses the reparameterization technique of RepVGG [37]. During model training, the REP structure consists of a 3×3 convolution branch for feature extraction, a 1×1 convolution branch for smoothing features, and an identity transformation branch. The three branches are combined through connection operation, improving network performance. In the inference stage, the REP structure is reparametrized into a 3×3 convolution operation to reduce model parameters and accelerate inference speed. In the prediction stage, the auxiliary head is used to supervise the output of the intermediate layer features so that the intermediate layer can more accurately represent the information in the input data, thus improving the expressiveness of the model.

3. Method

3.1. MBConv and Fused-MBConv Block

To achieve higher detection accuracy, we typically use deeper network models to capture more complex feature information. However, this often leads to longer training times, overfitting, gradient vanishing, and gradient exploding problems. Although residual connections [40] effectively solve these problems, the entire network model still requires significant resources. Therefore, a lightweight model has always been the goal of many researchers.

MobileNetV2 [41] introduces two modules: the linear bottleneck and the inverted residual. The linear bottleneck is a bottleneck block that removes the last activation function to avoid the loss of feature information; the inverted residual is used to form sparse features and reduce loss by first up-dimensioning and then down-dimensioning operations and then reducing model parameters and extracting high-dimensional features using depthwise convolution (DWConv). The combination of the two is called MBConv in other applications [42–44], and in later applications, channel attention [45] was also added. In Figure 3, the MBConv is composed of an expand convolution with BN and SiLU activation function, a DWConv with BN and SiLU activation function for parameter reduction, a channel attention block for calculating channel weights, a low-rank project convolution with BN, and a dropout layer. The channel attention block (Figure 4) consists of two operations: squeeze and excitation (SE). The squeeze operation uses Global Average Pooling (GAP) to compress the size of the feature map from H × W × C to 1 × 1 × C. This process compresses the height and width of each channel into a real number with global information, allowing the overall model to significantly reduce its number of parameters while preserving global features. The excitation operation first applies a fully connected (FC) layer to compress the channels (r denotes the compression ratio) to reduce the number of channels to further reduce the computational complexity; then after activation using the Relu activation function, the number of channels is restored to the original dimension by a second FC, followed by the Sigmoid activation function to obtain the final weight (different colors represent different weight values) to distinguish the importance of different channels. Finally, the total number of output features is obtained by multiplying the output weight coefficients on the branch with the original feature values of the model.

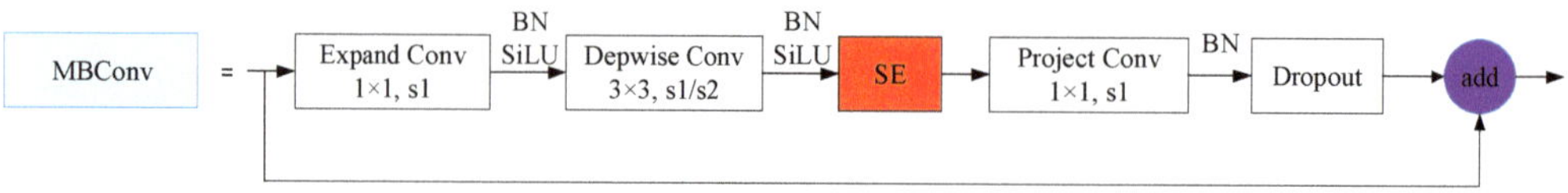

Figure 3. Structure of the MBConv.

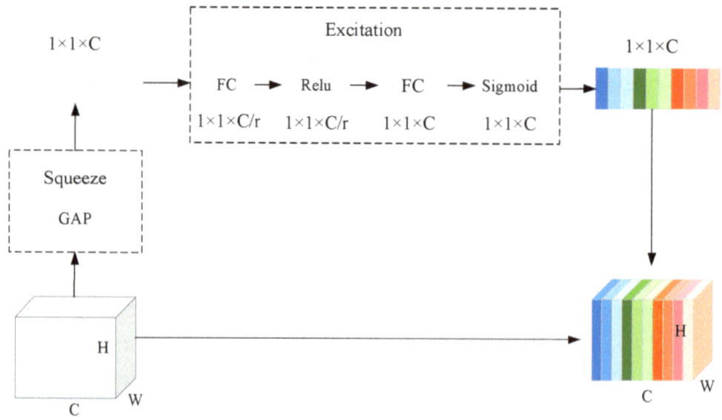

Figure 4. Structure of the squeeze and excitation operation.

Based on MBConv, Fused-MBConv is proposed. The authors of EfficientNet v2 [44] found that although DWConv can theoretically reduce the number of model training parameters, in practice it is slow to use on shallow networks, does not achieve the desired state, and does not fully utilize existing accelerators. Therefore, in the structure of MBConv, the DWConv block is removed. When channel expansion is not performed, a 3 × 3 ordinary convolution is used to replace the original expand convolution block, the SE block, and the low-rank project convolution block. When channel expansion is performed, only the original SE block is removed. The structural diagram of Fused-MBConv is shown in Figure 5.

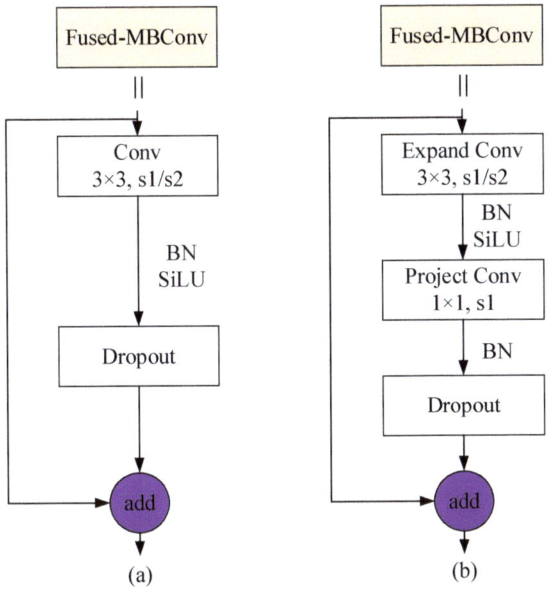

Figure 5. Structure of the Fused-MBConv: (**a**) the structure without channel expansion; (**b**) the structure with channel expansion.

3.2. ELAN-F and ELAN-M Module

To maximize the advantages of the MBConv block, Fused-MBConv block, and YOLOv7 itself, this paper embeds the MBConv and Fused-MBConv blocks into the ELAN structure of the backbone network and constructs the ELAN-F (ELAN based on the Fused-MBConv block) acceleration convergence module and the ELAN-M (ELAN based on the MBConv block) parameter reduction module (In Figure 6, c represents the number of channels). In the ELAN-F module, the original CBS block is replaced by the Fused-MBConv block, while the ELAN-M module is based on the ELAN-F, where the convolution kernel size of 3 × 3 is replaced by the MBConv block. In EfficientNet v2, the authors set the convolutional kernel sizes of both Fused-MBConv and MBConv to 3 × 3, while in the ELAN structure of this paper, the remaining 1 × 1 convolutional kernels of the CBS block are replaced by the Fused-MBConv block without channel expansion. The aim is to speed up model convergence and improve network accuracy without increasing the number of network parameters. Both the ELAN-F module and ELAN-M module are multibranch deep residual aggregation structures. After replacing the Fused-MBConv and MBConv blocks, the network depth is greatly increased, and deep semantic information is extracted through multiple residual connections. However, in general, as the depth of the network increases, the residual become less and less effective. In this paper, these two residual structures are added to the multilayer aggregation structure of ELAN, and the neurons of different layers are connected again through cross-layer connections to achieve information sharing, which alleviates the problem of deep residuals and enables the network to perform with better accuracy thanks to the combination of residual connections and multilayer aggregation.

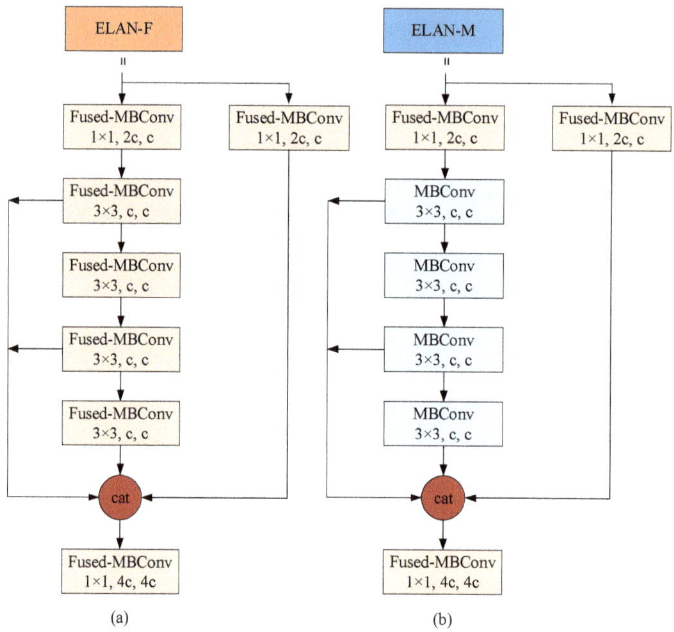

Figure 6. Structure of the ELAN-F and ELAN-M: (**a**) ELAN-F, (**b**) ELAN-M.

3.3. SimAM Attention Block

Many researchers [46–48] have demonstrated the effectiveness of attention mechanisms in helping models understand important features in images, reducing noise interference, and improving model robustness. As a plug-and-play block, it can be quickly applied at different positions in different networks, demonstrating its simplicity. SimAM is an attention block proposed by Yang [49] that has three-dimensional (3D) weights, as shown in Figure 7. By simultaneously considering the relationship between space and channels, the 3D weights of the neurons are generated and are assigned to the original feature map.

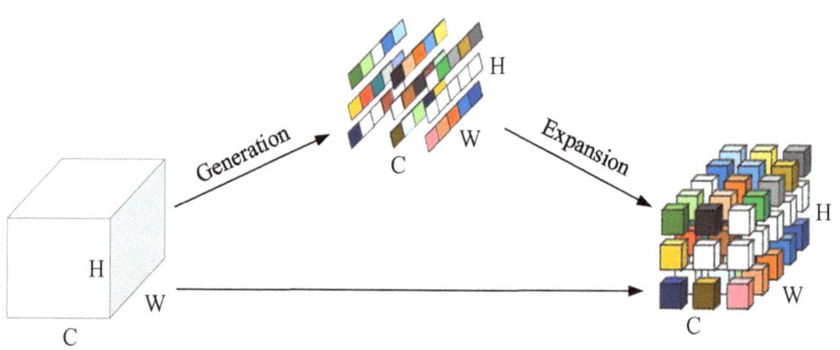

Figure 7. Structure of the SimAM attention block with 3D weights.

In neuroscience, neurons with important information often exhibit a different firing pattern than surrounding neurons and suppress the activity of surrounding neurons. Based on this, Yang defines an energy function for each neuron in the feature map to distinguish the target neuron from other neurons. The function (1) is shown below:

$$e_t(w_t, b_t, y, x_i) = (y_t - \hat{t})^2 + \frac{1}{M-1}\sum_{i=1}^{M-1}(y_0 - \hat{x}_i)^2 \tag{1}$$

where t and x_i are the target neurons and other neurons in a single channel in the input feature map X. $\hat{t}$ and $\hat{x}_i$ are obtained by a linear transformation of t and x_i with the transformation equations $\hat{t} = w_i t + b_t$ and $x_i = w_t x_i + b_t$, where i is the index on the spatial dimension and $M = H \times W$ is the number of neurons on that channel, and w_t and b_t are the weights and biases of the transformations. By solving for the minimum of (1), the linear separability of the target neuron t from all other neurons in the same channel can be obtained. For y_t and y_0, using binary labels and adding the regularizer $\lambda {w_t}^2$ to (1), the transformed energy expression is

$$\begin{aligned} e_t(w_t, b_t, y, x_i) = \tfrac{1}{M-1} \sum_{i=1}^{M-1} (-1 - (w_t x_i + b_t))^2 \\ + (1 - (w_t t + b_t))^2 + \lambda {w_t}^2 \end{aligned} \tag{2}$$

Equation (2) is a closed-form solution concerning w_t and b_t. The analytic equations for w_t and b_t are

$$w_t = -\frac{2(t - u_t)}{(t - u_t)^2 + 2{\sigma_t}^2 + 2\lambda} \tag{3}$$

$$b_t = -\frac{1}{2}(t + \mu_t) w_t \tag{4}$$

Assuming that all pixels in each channel follow the same distribution and that the mean $\mu_t = (1/(M-1))\sum_1^{M-1} x_i$ and variance ${\sigma_t}^2 = (1/(M-1))\sum_i^{M-1} (x_i - \mu_t)^2$ are known for all neurons except t, the minimizing neuron energy function is

$$e_t^* = \frac{4(\hat{\sigma}^2 + \lambda)}{(t - \hat{\mu})^2 + 2\hat{\sigma}^2 + 2\lambda} \tag{5}$$

where $\hat{\mu} = (1/M)\sum_{i=1}^{M} x_i$ and $\hat{\sigma}^2 = (1/M)\sum_{i=1}^{M} (x_i - \hat{\mu})^2$. When e^* is smaller, it means that neuron t is more distinct from peripheral neurons and that neuron t is more important and should be given a higher weight. Thus, the importance of each neuron can be obtained by $1/{e_t}^*$. Finally, feature refinement is carried out through the sigmoid function, and the entire refinement phase is

$$\widetilde{X} = \text{sigmoid}(\frac{1}{E}) \odot X \tag{6}$$

where E denotes the grouping of all ${e_t}^*$ in the spatial and channel dimensions, and $\odot$ indicates the multiplication operation.

The SimAM attention block obtains the 3D weights of each neuron by optimizing the energy function, avoiding the structural tuning work of other similar attention blocks. In addition, its parameter-free nature results in a minimal computational overhead.

3.4. YOLOv7 Neck Network with SimAM

In Figure 8, The CBS modules are replaced by SimAM in two downsampling blocks of the neck network. The SimAM module reduces feature loss during downsampling by re-evaluating each neuron in the feature map. Moreover, MPConv2 is located in the PAN structure of the neck network, which includes feature maps of multiple scales. Different scales of feature maps contain information about objects of different scales. The SimAM attention block can enhance the interaction and weight adjustment between different scales of feature maps, thus better capturing the features of target areas.

After replacing the ELAN-F, ELAN-M, and SimAM modules in YOLOv7, the proposed YOLOv7-FMS network structure is obtained. In the target detection task for gas plumes, which have scarce information and unclear contours, the YOLOv7-FMS network can fully use the extracted features and reduce feature loss during the feature extraction process, enabling the network to quickly and accurately locate useful regions in complex images and improve detection accuracy.

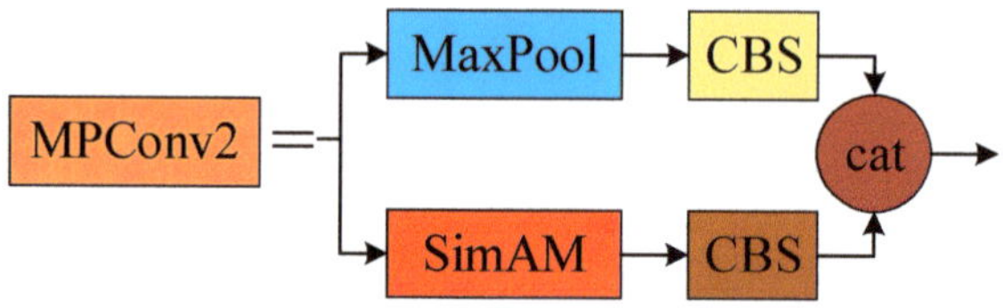

Figure 8. Structure of the SimAM module in MPConv2.

4. Experiments and Discussion

4.1. Preliminary Preparation

4.1.1. Dataset Preparation

This paper uses the Kongsberg EM710 multibeam bathymetry system to make in situ measurements in a marine area, obtaining a total of 320 images containing gas plume targets. During the data augmentation process, relevant studies [50,51] have shown that if the overall dataset is augmented first, the augmented data of the same image may be split into the training and validation sets during data splitting. This could cause the model to become overconfident and degrade its generalization ability. Therefore, this paper first divided the dataset in an 8:1:1 ratio, and then data augmentation was performed. The augmented dataset consisted of 1920 images, with 1536, 192, and 192 images used for training, validation, and testing. The dataset is labeled in YOLO format using LabelImg software.

4.1.2. Experimental Environment

This study was conducted on an Ubuntu 20.04 operating system with an Intel Xeon Platinum 8255C processor and an RTX 3090 (24 GB) graphics card. The GPU acceleration environment was created using CUDA 11.3, and the network framework was built using Python 3.8 and PyTorch 1.11.0. The development platform was Visual Studio Code 1.75.1.

4.1.3. Hyperparameter Setting

All experiments were performed with the same hyperparameters, which are listed in Table 1, to demonstrate the effectiveness of our method.

Table 1. Hyperparameter configuration.

Hyperparameter	Configuration
Initial learn rate	0.01
Optimizer	SGD
Weight decay	0.0005
Momentum	0.937
Image size	320×320
Batch size	16
Epochs	400

4.2. Model Evaluation Metrics

This study used parameters, computational complexity, FPS (Frames Per Second), F1, and mAP (mean Average Precision) to evaluate the performance of each model. Parameters and computational complexity measure the spatial and temporal complexity of a model, representing the number of learnable parameters and floating point operations, respectively; FPS refers to how many images the model can detect per second. F1 is a single score that evaluates the model and is a weighted average that considers precision and recall. Precision and recall indicate the proportion of false detections and missed detections in the dataset, respectively. Average precision (AP) indicates the performance of the model in a single category. mAP is the mean of AP for all categories, and we only studied the detection of gas plume, so AP is equal to mAP. mAP50 represents the mAP calculated using an

IoU (Intersection over Union) threshold of 0.50; mAP50:95 represents a set of mAP values calculated using multiple IoU thresholds from 0.50 to 0.95 and then averaged to full evaluation of model performance. The formulae are as follows, where TP (True Positive), FP (False Positive), and FN (False Negative) indicate the number of gas plume targets detected correctly, incorrectly, and not detected in the water column image.

$$\text{Precision} = \frac{\text{TP}}{\text{TP} + \text{FP}} \tag{7}$$

$$\text{Recall} = \frac{\text{TP}}{\text{TP} + \text{FN}} \tag{8}$$

$$\text{F1} = \frac{2 \times \text{Precision} \times \text{Recall}}{\text{Precision} + \text{Recall}} \tag{9}$$

$$\text{AP} = \int_0^1 \text{Precision}(\text{Recall})\text{dRecall} \tag{10}$$

4.3. Experimental Analysis

To verify the effectiveness of the YOLOv7-FMS model, a series of comparison and ablation experiments was conducted, and the results were compared in detail using the previously defined evaluation metrics.

4.3.1. The Selection of the Baseline Network

In the YOLOv7 model architecture, the three network structures selected for comparison were YOLOv7-tiny, YOLOv7, and YOLOv7x. From Table 2, it can be seen that although YOLOv7-tiny has a smaller number of parameters, a higher number of FLOPs, and a higher FPS, its accuracy is minimal and cannot meet the testing requirements. Although YOLOv7x has similar accuracy to YOLOv7, it consumes more resources and has the slowest inference speed. After comparing these models, YOLOv7 was chosen as our baseline network for further improvements.

Table 2. Performance Comparison of Different Models of YOLOv7 Network.

Method	Params. (M)	FLOPs (G)	F1 (%)	mAP50 (%)	mAP50:95 (%)	FPS Batch_Size = 1
YOLOv7-tiny	6.0	13.0	82.9	83.0	38.2	128.2
YOLOv7x	70.8	188.9	89.1	90.8	44.1	78.4
YOLOv7	36.5	103.2	90.8	91.0	46.8	84.7

4.3.2. Design of Backbone Network Based on ELAN-F and ELAN-M

The backbone of YOLOv7 contains four ELAN modules, which are replaced by the ELAN-F and ELAN-M modules. The new backbone network is designed in Figure 9, and the performance is shown in Table 3. The YOLOv7 backbone network with ELAN-F1M3 has the highest accuracy, with F1, mAP50, and mAP50:95 increasing by 3.6, 6.7, and 8.4%, respectively, compared to the baseline. Furthermore, the numbers of parameters and FLOPs are reduced by 16.7 and 14.2%. However, the inference speed of 57.8 is the slowest because the ELAN-M module uses DWConv, which wastes some time on reading and writing data from memory, and the GPU's computing power is not fully utilized. Moreover, even the worst YOLOv7-F2M2 has nearly the same accuracy as the YOLOv7 baseline network. To verify the accelerated convergence function of the model, Figure 10 shows the mAP50 change curve of these network structures throughout the training process. The curve results show that YOLOv7-F1M3 can achieve high accuracy in fewer batches and gradually become stable. This indicates that using the ELAN-F module only in the shallow layers of the network can achieve the best results, which is consistent with the conclusion of Efficient v2 obtained through NAS (Neural Architecture Search), where the Fused-MBConv module is used only in the shallow layers of the network.

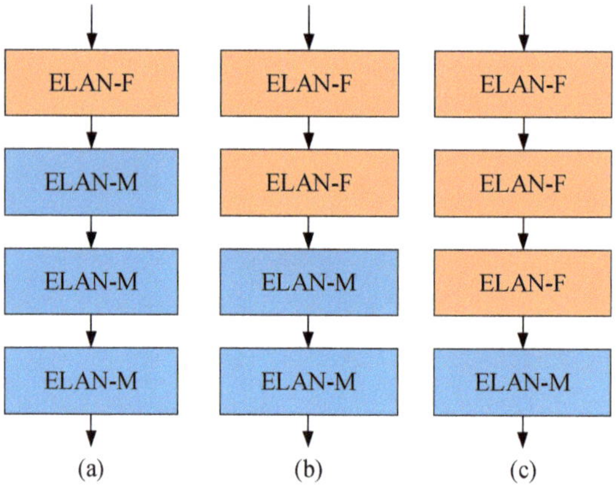

Figure 9. Comparison of Backbone Networks with Different Combinations. (**a**) One Fused-MBConv block and Three MBConv blocks (i.e., ELAN-F1M3). (**b**) Two Fused-MBConv blocks and Two MBConv blocks (i.e., ELAN-F2M2). (**c**) Three Fused-MBConv blocks and One MBConv block (i.e., ELAN-F3M1).

Table 3. Performance Comparison of Backbone Networks with Different Configurations.

Method	Params. (M)	FLOPs (G)	F1 (%)	mAP50 (%)	mAP50:95 (%)	FPS Batch_Size = 1
YOLOv7-F1M3	30.4	88.5	94.4	97.7	55.2	55.6
YOLOv7-F2M2	30.9	95.1	89.5	91.5	46.2	62.9
YOLOv7-F3M1	33.1	101.6	90.7	93.6	52.2	68.5

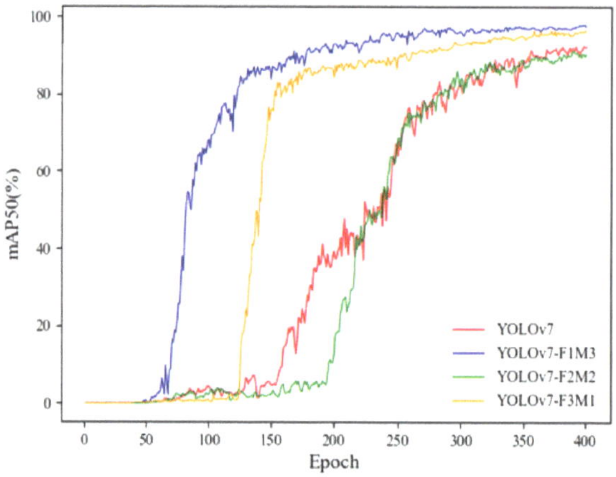

Figure 10. Comparison of the mAP curves of backbone networks under different configurations.

4.3.3. Experimental Analysis of the Proposed Method and Other Advanced Lightweight Networks

Table 4 shows the comparative results of the YOLOv7-F1M3 model and other lightweight networks such as MobileNet v3 [52], ShuffleNet v2 [39], GhostNet v2 [53], PP-LCNet (Positional Pyramid-based Lightweight Convolutional Network) [54], and MobileOne [55] by replacing the backbone network of the baseline. Although lightweight networks reduce the numbers of parameters and FLOPs, the corresponding feature representation capacity is reduced, resulting in a significant decrease in accuracy. Among them, MobileOne achieves the same lowest number of parameters and highest detection speed by implementing a branch-free network structure through reparameterization in the inference process. Although the YOLOv7-F1M3 has relatively high number of parameters and FLOPs, it has the highest detection accuracy, outperforming the second-placed GhostNet v2 by 3.1, 6.6, and 10.2% on F1, mAP50, and mAP50:95, respectively.

Table 4. Performance Comparison of Different Advanced Lightweight Networks in the Backbone Network.

Method	Params. (M)	FLOPs (G)	F1 (%)	mAP50 (%)	mAP50:95 (%)	FPS Batch_Size = 1
YOLOv7-MobileNetv3	24.8	36.9	80.1	82.5	35.7	22.6
YOLOv7-ShuffleNetv2	23.3	37.9	83.0	84.6	35.7	74.1
YOLOv7-GhostNetv2	29.6	75.2	91.3	91.1	45.0	38.5
YOLOv7-PPLCNet	28.9	63.9	83.9	86.1	43.5	63.3
YOLOv7-MobileOne	23.3	40.1	87.6	89.4	39.5	90.9
YOLOv7-F1M3	30.4	88.5	94.4	97.7	55.2	55.6

4.3.4. Experimental Analysis of the Proposed Method and Other Attention Blocks

In Table 5, the performance of SimAM in the baseline neck network is compared with other attention mechanisms, including SE [45], ECA (Efficient Channel Attention) [46], CBAM [47], and CA (Coordinate Attention) [48]. Compared to the baseline, the detection accuracy improves after the integration of ECA and SimAM. ECA, an improved version of SE, computes channel weights through a learnable 1D convolution kernel, avoiding the use of GPA, which does not capture long-range dependencies in the feature map well. The same GPA is used in CBAM, so it exhibits a relatively low detection accuracy. CA calculates attention weights based on the coordinate information of the target, whereas the gas plume target is randomly distributed in the sea and is often fractured and drifting in the water, making it difficult to detect accurately during the test. However, SimAM achieves the greatest improvement by directly considering the 3D weight relationship of the neurons through the energy function, with F1, mAP50, and mAP50:95 increasing by 2.0, 5.3, and 8.5%, respectively, compared to the baseline, with the same lowest number of FLOPs at an intermediate detection speed, In addition, this module does not require any parameters.

Table 5. Performance Comparison of Different Attention Mechanisms in the Neck Network.

Method	Block Params.	FLOPs (G)	F1 (%)	mAP50 (%)	mAP50:95 (%)	FPS Batch_Size = 1
YOLOv7-SE	163,840	102.9	77.0	78.7	33.2	79.4
YOLOv7-ECA	2	102.7	91.1	94.5	52.4	56.5
YOLOv7-CBAM	772	102.7	86.8	89.5	45.6	45.9
YOLOv7-CA	247,680	103.4	83.3	85.8	38.6	41.8
YOLOv7-SimAM	0	102.7	92.8	96.3	55.3	53.5

4.3.5. Ablation Study Based on YOLOv7

In previous sections, a series of horizontal comparisons among various improvements in the YOLOv7-FMS network have been conducted to demonstrate its superiority over other methods. In Table 6, a longitudinal comparison of the ablation experiments is made, where the model shows some improvement over the baseline in all metrics after adding the ELAN-F1M3 module, the SimAM module, or both. Compared to the SimAM module, the ELAN-F1M3 module, a multibranch deep residual aggregation structure, can learn more useful information and has a greater impact on model enhancement. Moreover, the

YOLOv7-FMS reduces its number of parameters and FLOPs by 17.0 and 14.5% and increases F1, mAP50, and mAP50:95 by 4.4, 7.4, and 10.7%, respectively. However, due to the increase in network depth, the side-connection span between the FPN and the three valid feature maps increases, and the dependency between the data is stronger, leading to a decrease in detection speed again. We also compared the performance of the improved module at different network locations. YOLOv7-F1M3 (Neck) refers to the replacement of four ELAN-W modules in the neck network with ELAN-M modules based on YOLOv7-F1M3. The test results show that the detection accuracy decreases significantly with the large reduction in parameters and computational complexity. In addition, the excessive use of DWConv modules slows down the detection speed of the network. YOLOv7-SimAM (Backbone) refers to the insertion of the SimAM attention module at the connection between the backbone network and the neck network. The detection results are slightly lower than YOLOv7, indicating that the SimAM module in the neck network can better capture the relationship between data and improve network performance.

Table 6. Ablation Study Based on YOLOv7.

Method	Params. (M)	FLOPs (G)	F1 (%)	mAP50 (%)	mAP50:95 (%)	FPS Batch_Size = 1
YOLOv7	36.5	103.2	90.8	91.0	46.8	84.7
YOLOv7-F1M3 (Neck)	27.0	81.8	63.4	62.3	21.2	37.5
YOLOv7-F1M3	30.4	88.5	94.4	97.7	55.2	55.6
YOLOv7-SimAM (Backbone)	35.4	102.3	87.7	89.2	44.5	51.0
YOLOv7-SimAM	36.3	102.7	92.8	96.3	55.3	53.5
YOLOv7-FMS	30.3	88.2	95.2	98.4	57.5	44.6

In Figure 11, the CAM (Class Activation Mapping) feature visualization technique is used to generate a weighted heat map for the various attention mechanisms to help us better understand and compare the detection performance and decision-making processes of different attention networks. In Figure 11b, the baseline network of YOLOv7 focuses more on the sidelobe noise and the seabed region and too little on the gas plume target. After adding attention to the network, the weights assigned to the gas plume target in the feature map are enhanced, and the coverage and attention of the region are significantly improved after adding SimAM (Figure 11g) compared with the other four mainstream attention mechanisms. Additionally, it effectively suppresses the sidelobe noise and irrelevant features in the image. This suggests that the model has better robustness with the addition of SimAM.

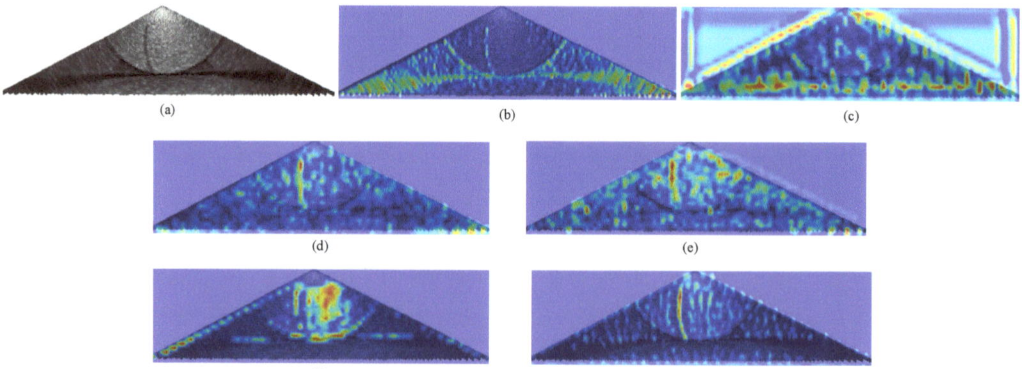

Figure 11. Comparison of heat map under different attention mechanisms. (**a**) Original image. (**b**) YOLOv7. (**c**) YOLOv7-SE. (**d**) YOLOv7-ECA. (**e**) YOLOv7-CBAM. (**f**) YOLOv7-CA. (**g**) YOLOv7-SimAM.

4.3.6. Experimental Analysis of the Proposed Method and Other CNN Methods

To further validate the performance of the proposed YOLOv7-FMS model, we selected YOLOv5 [17], YOLOX [18], YOLOv6 [19], SSD [22], RetinaNet [23], and EfficientDet [24], all of which are similarly sized detection models, for comparison experiments. Table 7 shows that as the number of model parameters increases, the accuracy of the model also increases. Although the lightweight YOLOv7-FMS model is still at a relatively high level in terms of parameters and computational complexity, in the accuracy metrics of F1, mAP50, and mAP50:95, our approach outperforms the second-ranked YOLOv6-m by 0.8, 0.8, and 7.5%, respectively, indicating that this method has good detection performance even at high detection confidence. The detection accuracy of SSD, RetinaNet, and EfficientDet is relatively low. This is because SSD requires separate prediction of the object's location and category at multiple scales during detection, which may result in some gas plume targets being missed or misclassified, whereas YOLOv7 uses the FPN + PAN structure, as in previous generations, to enable information sharing across multiple scales and pathways. RetinaNet uses focal loss to reduce the weight of easily classified samples in multiclass detection, but it cannot address the problem of size imbalance within the same class of objects. The D4 version of EfficientDet is similar in scale to other models, but its default image input size is 1024, resulting in larger feature map sizes generated during training. This requires more convolution and pooling operations during forward and backward propagation, resulting in increased FLOPs and inference time. In addition, we also trained Faster RCNN [29] and DETR (DEtection TRansformer) [56] on our dataset, but their accuracy was extremely low. The main reason for this is that Faster R-CNN uses too many anchors to generate candidate boxes with RPN, which can easily lead to redundancy. Moreover, for gas plume targets in multibeam water column images, their shapes and features differ greatly from those of other typical targets, making it difficult for RPN to generate sufficiently accurate candidate boxes, which may not be able to adapt to small targets in the gas plumes. Then, due to the relatively small size of the self-built dataset used in this paper and the limited computing resources available, it is difficult to fully leverage the performance of DETR, resulting in difficulty in optimizing the training results.

Table 7. Performance Comparison of the Proposed Method and other SOTA Models.

Method	Params. (M)	FLOPs (G)	F1 (%)	mAP50 (%)	mAP50:95 (%)	FPS Batch_Size = 1
YOLOv5-m	21.1	12.7	92.0	96.4	50.2	70.4
YOLOX-m	25.3	18.4	92.3	95.1	48.2	48.7
YOLOv6-m	34.8	21.4	94.4	97.6	50.0	67.4
SSD300	23.7	68.4	74.7	78.2	29.2	149.8
RetinaNet (resnet34)	29.9	38.3	19.8	22.5	6.1	53.6
EfficientDet-D4	20.5	104.9	69.3	71.5	23.1	12.6
Ours	30.3	88.2	95.2	98.4	57.5	44.6

4.3.7. Result of Detection and Recognition of Gas Plume Targets in WCI

Finally, we test the proposed YOLOv7-FMS model against the original YOLOv7 model on some representative images. Figure 12 shows a clear WCI of the target, where the YOLOv7 FMS bounding box can fit the gas plume target more closely and with better detection accuracy. Figure 13 shows an unclear WCI of the target, where YOLOv7 FMS still detects the target well. In Figure 14, there is an overlapping of targets. The two bounding boxes of YOLOv7-FMS can thus be closer and better represent the morphological of the gas plume target. In Figure 15, the WCI is greatly affected by sidelobe noise, resulting in YOLOv7 missing half of the gas plume targets. Figure 16 shows the phenomenon of gas plume targets fracturing during their upward motion due to the presence of internal waves on the seabed. After detection using YOLOv7, one gas plume target is missed, and two of the targets are mistakenly detected as a single target, whereas YOLOv7-FMS correctly detects all four gas plume targets. The detection results demonstrate that the improved method can accurately identify gas plume targets in WCI with high noise levels, blurred contours, and complex seabed environments, thereby enhancing the application of the WCI.

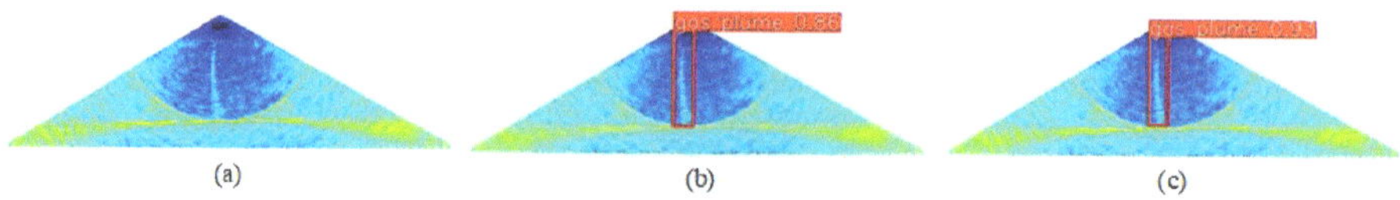

Figure 12. Detection and recognition results of the WCI with clear gas plume target: (**a**) original image, (**b**) YOLOv7, (**c**) YOLOv7-FMS.

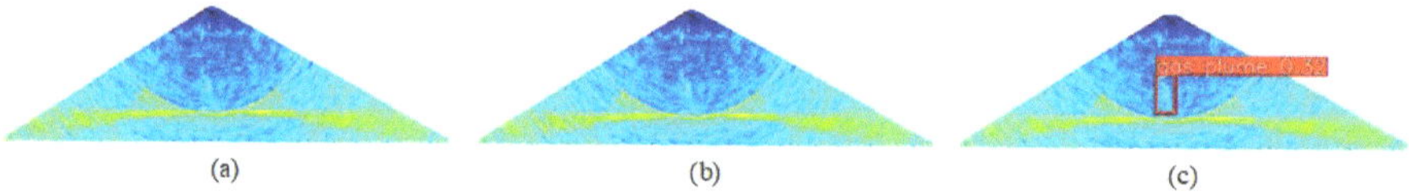

Figure 13. Detection and recognition results of the WCI with unclear gas plume target: (**a**) original image, (**b**) YOLOv7, (**c**) YOLOv7-FMS.

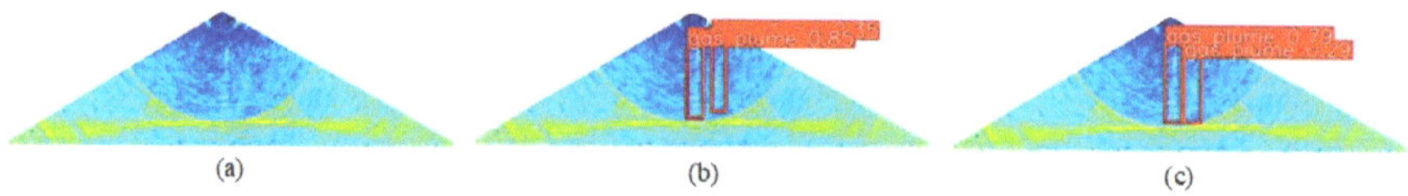

Figure 14. Detection and recognition results of the WCI with overlapping gas plume targets: (**a**) original image, (**b**) YOLOv7, (**c**) YOLOv7-FMS.

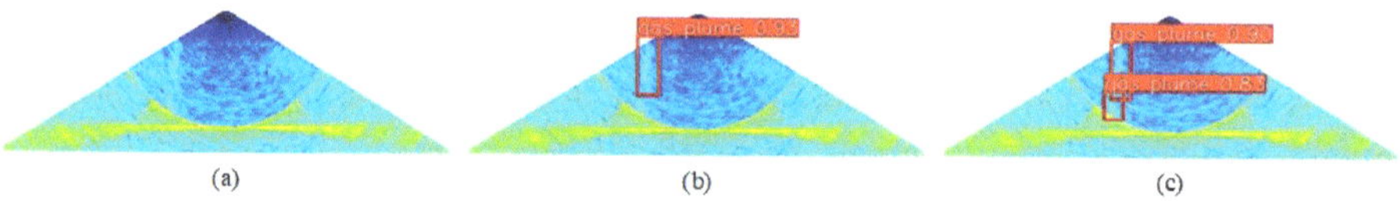

Figure 15. Detection and recognition results of the WCI affected by high noise levels: (**a**) original image, (**b**) YOLOv7, (**c**) YOLOv7-FMS.

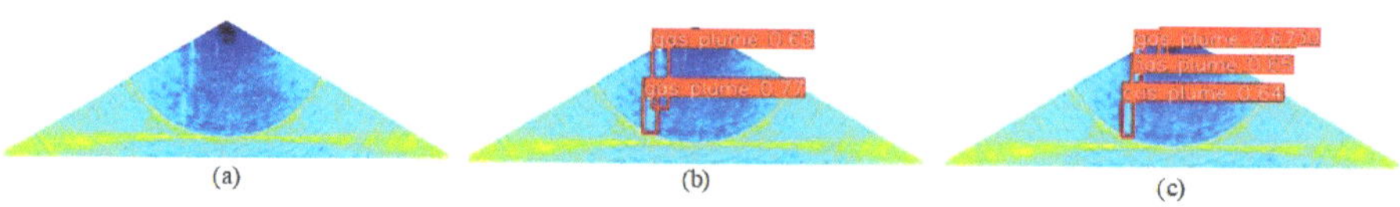

Figure 16. Detection and recognition results of the WCI with fractured gas plume targets: (**a**) original image, (**b**) YOLOv7, (**c**) YOLOv7-FMS.

5. Conclusions

In this paper, a YOLOv7-FMS model based on the YOLOv7 network structure was proposed. First, in the backbone network, we replaced the ELAN module with the ELAN-F and ELAN-M module and generated the ELAN-F1M3 backbone network, which reduces the numbers of parameters and FLOPs and accelerates the model convergence. The ELAN-F and ELAN-M modules are both internal residual and external aggregation structures. By repeatedly forming cross-layer connections, the model learns more complex mapping functions and reduces information loss. Then, by aggregating the outputs of the internal layers through a multilevel aggregation approach, the feature representation is enriched, and the expressiveness of the model is improved. In addition, the ELAN-M module uses the SE block to enable interaction between channels and enhance feature extraction. Next, we added the SimAM module to the neck network to evaluate the importance of each neuron in the multiscale feature maps, guided the network to focus on key features, and improved the robustness of the model. Experimental results show that the method outperforms other improvement points of the same type, it can adapt well to the morphological characteristics of the gas plume target, accurately locating the target's position, and that it has a strong anti-interference ability during the detection process. However, due to the significant increase in depth and complexity of the model in the improved network, the detection speed during the detection process has decreased. In future work, we will optimize the network structure through methods such as model pruning and distillation to improve detection speed and efficiency and achieve model deployment.

Author Contributions: Conceptualization, W.C. and X.W.; Data curation, W.C. and X.W.; Formal analysis, W.C. and X.W.; Funding acquisition, X.W., B.Y., T.J. and J.S.; Investigation, B.Y. and T.J.; Methodology, W.C. and B.Y.; Project administration, B.Y. and J.C.; Resources, X.W.; Software, W.C. and J.C.; Supervision, T.J. and J.S.; Validation, J.C., T.J. and J.S.; Visualization, W.C. and X.W.; Writing—original draft, W.C., X.W. and B.Y.; Writing—review and editing, J.C., T.J. and J.S. All authors have read and agreed to the published version of the manuscript.

Funding: This research was funded by the National Natural Science Young Foundation of China under Grant 41806117; the Marine Science and Technology Innovation Project of Jiangsu Province under Grant JSZRHYKJ202201; the National Natural Science Foundation of China under Grant 41004003; the Science and Technology Department Project of Jiangsu Province under Grant BE2016701; the Water Conservancy Science and Technology Project of Jiangsu Province under Grant 2020058 and 2021049; and the Lianyungang 521 Project Research Funding Project under Grant LYG06521202131.

Data Availability Statement: Not applicable.

Conflicts of Interest: The authors declare no conflict of interest.

References

1. Schimel, A.C.G.; Brown, C.J.; Ierodiaconou, D. Automated Filtering of Multibeam Water-Column Data to Detect Relative Abundance of Giant Kelp (*Macrocystis pyrifera*). *Remote Sens.* **2020**, *12*, 1371. [CrossRef]
2. Czechowska, K.; Feldens, P.; Tuya, F.; Cosme de Esteban, M.; Espino, F.; Haroun, R.; Schönke, M.; Otero-Ferrer, F. Testing Side-Scan Sonar and Multibeam Echosounder to Study Black Coral Gardens: A Case Study from Macaronesia. *Remote Sens.* **2020**, *12*, 3244. [CrossRef]
3. Guan, M.; Cheng, Y.; Li, Q.; Wang, C.; Fang, X.; Yu, J. An Effective Method for Submarine Buried Pipeline Detection via Multi-Sensor Data Fusion. *IEEE Access.* **2019**, *7*, 125300–125309. [CrossRef]
4. Zhu, G.; Shen, Z.; Liu, L.; Zhao, S.; Ji, F.; Ju, Z.; Sun, J. AUV Dynamic Obstacle Avoidance Method Based on Improved PPO Algorithm. *IEEE Access.* **2022**, *10*, 121340–121351. [CrossRef]
5. Logan, G.A.; Jones, A.T.; Kennard, J.M.; Ryan, G.J.; Rollet, N. Australian offshore natural hydrocarbon seepage studies, a review and re-evaluation. *Mar. Pet. Geol.* **2010**, *27*, 26–45. [CrossRef]
6. Liu, H.; Yang, F.; Zheng, S.; Li, Q.; Li, D.; Zhu, H. A method of sidelobe effect suppression for multibeam water column images based on an adaptive soft threshold. *Appl. Acoust.* **2019**, *148*, 467–475. [CrossRef]
7. Hou, T.; Huff, L.C. Seabed characterization using normalized backscatter data by best estimated grazing angles. In Proceedings of the International Symposium on Underwater Technology (UT04), Koto Ward, Tokyo, Japan, 7–9 April 2004; pp. 153–160. [CrossRef]
8. Urban, P.; Köser, K.; Greinert, J. Processing of multibeam water column image data for automated bubble/seep detection and repeated mapping. *Limnol. Oceanogr. Methods* **2017**, *15*, 1–21. [CrossRef]
9. Church, I. Multibeam sonar water column data processing tools to support coastal ecosystem science. *J. Acoust. Soc. Am.* **2017**, *141*, 3949. [CrossRef]
10. Ren, X.; Ding, D.; Qin, H.; Ma, L.; Li, G. Extraction of Submarine Gas Plume Based on Multibeam Water Column Point Cloud Model. *Remote Sens.* **2022**, *14*, 4387. [CrossRef]
11. Hughes, J.B.; Hightower, J.E. Combining split-beam and dual-frequency identification sonars to estimate abundance of anadromous fishes in the Roanoke River, North Carolina. *N. Am. J. Fish. Manag.* **2015**, *35*, 229–240. [CrossRef]
12. Fatan, M.; Daliri, M.R.; Shahri, A.M. Underwater cable detection in the images using edge classification based on texture information. *Measurement* **2016**, *91*, 309–317. [CrossRef]
13. Lu, S.; Liu, X.; He, Z.; Zhang, X.; Liu, W.; Karkee, M. Swin-Transformer-YOLOv5 for Real-Time Wine Grape Bunch Detection. *Remote Sens.* **2022**, *14*, 5853. [CrossRef]
14. Li, Z.; Zeng, Z.; Xiong, H.; Lu, Q.; An, B.; Yan, J.; Li, R.; Xia, L.; Wang, H.; Liu, K. Study on Rapid Inversion of Soil Water Content from Ground-Penetrating Radar Data Based on Deep Learning. *Remote Sens.* **2023**, *15*, 1906. [CrossRef]

15. Wu, J.; Xie, C.; Zhang, Z.; Zhu, Y. A Deeply Supervised Attentive High-Resolution Network for Change Detection in Remote Sensing Images. *Remote Sens.* **2023**, *15*, 45. [CrossRef]
16. Yosinski, J.; Clune, J.; Bengio, Y.; Lipson, H. How transferable are features in deep neural networks? *NIPS* **2014**, *27*, 3320–3328. [CrossRef]
17. YOLOv5 Models. Available online: https://Github.com/Ultralytics/Yolov5 (accessed on 13 January 2023).
18. Ge, Z.; Liu, S.; Wang, F.; Li, Z.; Sun, J. Yolox: Exceeding yolo series in 2021. *arXiv* **2021**, arXiv:2107.08430.
19. Li, C.; Li, L.; Jiang, H.; Weng, K.; Geng, Y.; Li, L.; Wei, X. YOLOv6: A single-stage object detection framework for industrial applications. *arXiv* **2022**, arXiv:2209.02976.
20. Wang, C.Y.; Bochkovskiy, A.; Liao, H.Y.M. YOLOv7: Trainable bag-of-freebies sets new state-of-the-art for real-time object detectors. *arXiv* **2022**, arXiv:2207.02696.
21. Redmon, J.; Divvala, S.; Girshick, R.; Farhadi, A. You only look once: Unified, real-time object detection. In Proceedings of the IEEE Conference on Computer Vision and Pattern Recognition, Las Vegas, NV, USA, 26 June–1 July 2016; pp. 779–788. [CrossRef]
22. Liu, W.; Anguelov, D.; Erhan, D.; Szegedy, C.; Reed, S.; Fu, C.Y.; Berg, A.C. Ssd: Single shot multibox detector. In Proceedings of the Computer Vision–ECCV 2016: 14th European Conference, Amsterdam, The Netherlands, 11–14 October 2016; pp. 21–37.
23. Lin, T.Y.; Goyal, P.; Girshick, R.; He, K.; Dollár, P. Focal loss for dense object detection. In Proceedings of the IEEE international Conference on Computer Vision, Venice, Italy, 22–29 October 2017; pp. 2980–2988. [CrossRef]
24. Tan, M.; Pang, R.; Le, Q.V. Efficientdet: Scalable and efficient object detection. In Proceedings of the IEEE/CVF Conference on Computer Vision and Pattern Recognition, Seattle, WA, USA, 14–19 June 2020; pp. 10781–10790. [CrossRef]
25. Tan, M.; Le, Q. Efficientnet: Rethinking model scaling for convolutional neural networks. In Proceedings of the International Conference on Machine Learning, Long Beach, CA, USA, 9–15 June 2019; pp. 6105–6114. [CrossRef]
26. Yu, K.; Cheng, Y.; Tian, Z.; Zhang, K. High Speed and Precision Underwater Biological Detection Based on the Improved YOLOV4-Tiny Algorithm. *J. Mar. Sci. Eng.* **2022**, *10*, 1821. [CrossRef]
27. Peng, F.; Miao, Z.; Li, F.; Li, Z. S-FPN: A shortcut feature pyramid network for sea cucumber detection in underwater images. *ESWA* **2021**, *182*, 115306. [CrossRef]
28. Zocco, F.; Huang, C.I.; Wang, H.C.; Khyam, M.O.; Van, M. Towards More Efficient EfficientDets and Low-Light Real-Time Marine Debris Detection. *arXiv* **2022**, arXiv:2203.07155.
29. Ren, S.; He, K.; Girshick, R.; Sun, J. Faster r-cnn: Towards real-time object detection with region proposal networks. *IEEE TPAMI* **2015**, *28*, 1137–1149. [CrossRef]
30. He, K.; Gkioxari, G.; Dollár, P.; Girshick, R. Mask r-cnn. In Proceedings of the IEEE International Conference on Computer Vision, Venice, Italy, 22–29 October 2017; pp. 2961–2969. [CrossRef]
31. Cai, Z.; Vasconcelos, N. Cascade r-cnn: Delving into high quality object detection. In Proceedings of the IEEE Conference on Computer Vision and Pattern Recognition, Salt Lake City, UT, USA, 18–22 June 2018; pp. 6154–6162.
32. Wang, H.; Xiao, N. Underwater Object Detection Method Based on Improved Faster RCNN. *Appl. Sci.* **2023**, *13*, 2746. [CrossRef]
33. Song, P.; Li, P.; Dai, L.; Wang, T.; Chen, Z. Boosting R-CNN: Reweighting R-CNN samples by RPN's error for underwater object detection. *NEUCOM* **2023**, *530*, 150–164. [CrossRef]
34. Inoue, H. Data augmentation by pairing samples for images classification. *arXiv* **2018**, arXiv:1801.02929.
35. Zhang, H.; Cisse, M.; Dauphin, Y.N.; Lopez-Paz, D. mixup: Beyond empirical risk minimization. *arXiv* **2017**, arXiv:1710.09412.
36. Bochkovskiy, A.; Wang, C.Y.; Liao, H.Y.M. Yolov4: Optimal speed and accuracy of object detection. *arXiv* **2020**, arXiv:2004.10934.
37. Ding, X.; Zhang, X.; Ma, N.; Han, J.; Ding, G.; Sun, J. Repvgg: Making vgg-style convnets great again. In Proceedings of the IEEE/CVF Conference on Computer Vision and Pattern Recognition, New York, NY, USA, 19–25 June 2021; pp. 13733–13742. [CrossRef]
38. Wang, C.Y.; Liao, H.Y.M.; Yeh, I.H. Designing Network Design Strategies Through Gradient Path Analysis. *arXiv* **2022**, arXiv:2211.04800.
39. Ma, N.; Zhang, X.; Zheng, H.T.; Sun, J. Shufflenet v2: Practical guidelines for efficient cnn architecture design. In Proceedings of the European Conference on Computer Vision, Munich, Germany, 8–14 September 2018; pp. 116–131. [CrossRef]
40. He, K.; Zhang, X.; Ren, S.; Sun, J. Deep residual learning for image recognition. In Proceedings of the IEEE Conference on Computer Vision and Pattern Recognition, Las Vegas, NV, USA, 26 June–1 July 2016; pp. 770–778. [CrossRef]
41. Sandler, M.; Howard, A.; Zhu, M.; Zhmoginov, A.; Chen, L.C. Mobilenetv2: Inverted residuals and linear bottlenecks. In Proceedings of the IEEE Conference on Computer Vision and Pattern Recognition, Salt Lake City, UT, USA, 18–22 June 2018; pp. 4510–4520. [CrossRef]
42. Tan, M.; Chen, B.; Pang, R.; Vasudevan, V.; Sandler, M.; Howard, A.; Le, Q.V. Mnasnet: Platform-aware neural architecture search for mobile. In Proceedings of the IEEE/CVF Conference on Computer Vision and Pattern Recognition, Long Beach, CA, USA, 16–20 June 2019; pp. 2820–2828. [CrossRef]
43. Wu, B.; Keutzer, K.; Dai, X.; Zhang, P.; Jia, Y. FBNet: Hardware-Aware Efficient ConvNet Design via Differentiable Neural Architecture Search. In Proceedings of the 2019 IEEE/CVF Conference on Computer Vision and Pattern Recognition, Long Beach, CA, USA, 16–20 June 2019; pp. 10726–10734. [CrossRef]
44. Tan, M.; Le, Q. Efficientnetv2: Smaller models and faster training. In Proceedings of the International Conference on Machine Learning, Graz, Austria, 18–24 July 2021; pp. 10096–10106. [CrossRef]

45. Hu, J.; Shen, L.; Sun, G. Squeeze-and-excitation networks. In Proceedings of the IEEE Conference on Computer Vision and Pattern Recognition, Salt Lake City, UT, USA, 18–22 June 2018; pp. 7132–7141. [CrossRef]
46. Wang, Q.; Wu, B.; Zhu, P.; Li, P.; Zuo, W.; Hu, Q. ECA-Net: Efficient channel attention for deep convolutional neural networks. In Proceedings of the IEEE/CVF Conference on Computer Vision and Pattern Recognition, Seattle, WA, USA, 14–19 June 2020; pp. 11534–11542. [CrossRef]
47. Woo, S.; Park, J.; Lee, J.Y.; Kweon, I.S. Cbam: Convolutional block attention module. In Proceedings of the European Conference on Computer Vision, Munich, Germany, 8–14 September 2018; pp. 3–19. [CrossRef]
48. Zhang, C.; Lin, G.; Liu, F.; Yao, R.; Shen, C. Canet: Class-agnostic segmentation networks with iterative refinement and attentive few-shot learning. In Proceedings of the IEEE/CVF Conference on Computer Vision and Pattern Recognition, Long Beach, CA, USA, 16–20 June 2019; pp. 5217–5226. [CrossRef]
49. Yang, L.; Zhang, R.Y.; Li, L.; Xie, X. Simam: A simple, parameter-free attention module for convolutional neural networks. In Proceedings of the International Conference on Machine Learning, Graz, Austria, 18–24 July 2021; pp. 11863–11874.
50. Hendrycks, D.; Mu, N.; Cubuk, E.D.; Zoph, B.; Gilmer, J.; Lakshminarayanan, B. Augmix: A simple data processing method to improve robustness and uncertainty. *arXiv* **2019**, arXiv:1912.02781.
51. Xie, Q.; Dai, Z.; Hovy, E.; Luong, T.; Le, Q. Unsupervised data augmentation for consistency training. *NeurIPS* **2020**, *33*, 6256–6268. [CrossRef]
52. Howard, A.; Sandler, M.; Chu, G.; Chen, L.C.; Chen, B.; Tan, M.; Wang, W.; Zhu, Y.; Pang, R.; Howard, V.; et al. Searching for mobilenetv3. In Proceedings of the IEEE/CVF International Conference on Computer Vision, Long Beach, CA, USA, 9–15 June 2019; pp. 1314–1324. [CrossRef]
53. Tang, Y.; Han, K.; Guo, J.; Xu, C.; Xu, C.; Wang, Y. GhostNetV2: Enhance Cheap Operation with Long-Range Attention. *arXiv* **2022**, arXiv:2211.12905.
54. Cui, C.; Gao, T.; Wei, S.; Du, Y.; Guo, R.; Dong, S.; Lu, B.; Zhou, Y.; Lv, X.; Liu, Q.; et al. PP-LCNet: A lightweight CPU convolutional neural network. *arXiv* **2021**, arXiv:2109.15099.
55. Vasu, P.K.A.; Gabriel, J.; Zhu, J.; Tuzel, O.; Ranjan, A. MobileOne: An improved one millisecond mobile backbone. *arXiv* **2022**, arXiv:2206.04040.
56. Carion, N.; Massa, F.; Synnaeve, G.; Usunier, N.; Kirillov, A.; Zagoruyko, S. End-to-end object detection with transformers. In Proceedings of the Computer Vision–ECCV 2020: 16th European Conference, Glasgow, UK, 23–28 August 2020; pp. 213–229. [CrossRef]

remote sensing

MDPI

Technical Note

Enhancing Water Depth Estimation from Satellite Images Using Online Machine Learning: A Case Study Using Baidu Easy-DL with Acoustic Bathymetry and Sentinel-2 Data

Zhongqiang Wu [1,2], Shulei Wu [1], Haixia Yang [3], Zhihua Mao [2,*] and Wei Shen [4,5]

[1] School of Information Science and Technology, Hainan Normal University, Haikou 571158, China; wuzhongqiang@hainnu.edu.cn (Z.W.); wsl@hainnu.edu.cn (S.W.)
[2] States Key Laboratory of Satellite Ocean Environment Dynamics, Second Institute of Oceanography, State Oceanic Administration, Hangzhou 310012, China
[3] China Siwei Surveying and Mapping Technology Co., Ltd., Beijing 100048, China; yanghaixia@chinasiwei.com
[4] School of Marine Science, Shanghai Ocean University, Shanghai 201306, China; wshen@shou.edu.cn
[5] Marine Surveying and Mapping Engineering and Technology Research Center, Shanghai 201306, China
* Correspondence: mao@sio.org.cn; Tel.: +86-13805744749

Abstract: Water depth estimation is paramount in various domains, including navigation, environmental monitoring, and resource management. Traditional depth measurement methods, such as bathymetry, can often be expensive and time-consuming, especially in remote or inaccessible areas. This study delves into the application of machine learning techniques, specifically focusing on the Baidu Easy DL model for water depth estimation leveraging satellite imagery. Utilizing Sentinel-2 satellite data over Rushikonda Beach in India and processing it into remote sensing reflectance using ACOLITE software, this research compares the performance of several machine learning algorithms, including the Stumpf model, Log-Linear model, and the Baidu Easy DL model, for accurate depth estimation. The results indicate that the Easy-DL model outperforms traditional methods, particularly excelling in the 0–11 m depth range. This study showcases the substantial potential of machine learning in remote sensing, offering robust water depth estimates, even in complex coastal environments. Furthermore, it underscores the critical role of comprehensive training datasets and ensemble learning techniques in enhancing accuracy. This research opens avenues for the further exploration of machine learning applications in remote sensing and highlights the promising prospects of online model APIs when streamlining remote sensing data processing.

Keywords: big model; machine learning; Baidu Easy-DL; water depth; satellite-based bathymetry

Citation: Wu, Z.; Wu, S.; Yang, H.; Mao, Z.; Shen, W. Enhancing Water Depth Estimation from Satellite Images Using Online Machine Learning: A Case Study Using Baidu Easy-DL with Acoustic Bathymetry and Sentinel-2 Data. *Remote Sens.* **2023**, *15*, 4955. https://doi.org/10.3390/rs15204955

Academic Editors: Xiao-Hai Yan and Wenfang Lu

Received: 20 August 2023
Revised: 28 September 2023
Accepted: 10 October 2023
Published: 13 October 2023

1. Introduction

Water depth is an important parameter for a wide range of applications, including navigation, resource management, and environmental monitoring [1]. Accurate and up-to-date information on water depth is essential for ensuring safe navigation, managing fisheries and other aquatic resources, and monitoring changes in the environment [2]. Remote sensing is a powerful tool that allows us to gather information about the Earth's surface without physically being there. It involves the use of satellites, aircraft, or drones to collect data on the environment using sensors that detect reflected or emitted electromagnetic radiation [3,4]. One of the many applications of remote sensing is water depth inversion, which is the process of estimating the depth of a body of water based on the characteristics of reflected light [5,6].

Traditionally, water depth inversion has been performed using methods such as bathymetry, which involves physically measuring the depth of a body of water using sonar or other instruments [7,8]. However, these methods can be time-consuming and expensive and may not always be feasible in remote or inaccessible areas. Remote sensing offers a more efficient and cost-effective alternative, allowing us to estimate water depth over large areas quickly and accurately [9].

In recent years, machine learning has emerged as a powerful tool for data analysis, with applications in a wide range of fields. Machine learning algorithms can learn from data to make predictions or decisions without being explicitly programmed to perform this [10]. This makes them well-suited for tasks such as water depth inversion, where traditional methods may be inadequate. Machine learning is a rapidly growing field that has seen many advances in recent years, particularly with the availability of online resources [11]. These resources, such as online courses, tutorials, and documentation, have made it easier for individuals and organizations to learn about and apply machine-learning techniques to various fields, including remote sensing bathymetry [12,13].

Large models, particularly large language models, have seen rapid development in recent years. These models are trained on vast amounts of data and have a large number of parameters, allowing them to generate human-like text and perform a wide range of tasks. Some examples of large language models include GPT-3, Megatron-Turing NLG, and Gopher [14]. Large models have also been applied to the field of remote sensing data processing. Remote sensing involves the collection and analysis of data regarding the Earth's surface using satellites, aircraft, or drones. The amount of data generated by remote sensing is increasing dramatically, creating challenges for storage, analysis, and visualization [15]. To address these challenges, researchers have developed frameworks and systems that process remote sensing big data using large models and parallel processing. These frameworks provide scalability, flexibility, and generalization without dependency on specific data or processing techniques [15,16]. They also provide reasonable results to quality criteria, such as the response time, efficiency, and performance development of large models, which have had a significant impact on many fields, including remote sensing data processing. These models offer new capabilities for analyzing and understanding large amounts of data, leading to new insights and discoveries [15,16].

Some people are currently using online remote sensing platforms to conduct some remote sensing research, such as using the Google Earth Engine [17]. However, the application of online AI remote sensing is limited by the confidentiality of field data and the legal review of countries where scientists are located. Baidu Easy-DL is a zero-threshold AI development platform that provides a simple and easy way to customize and deploy AI models. One of its features is table data prediction, which helps users discover potential patterns from tabular data through machine learning techniques, thereby creating machine learning models, processing new data based on machine learning models, and generating predictive results for business applications [18].

DL's easily structured data supports one-click customization, automatically processes data, generates machine learning models, and can achieve scenarios such as table data prediction. This feature can be used to solve binary classifications, multi-classification, regression, and other problems and is suitable for scenarios such as customer churn prediction, fraud detection, and price prediction. The Baidu Easy DL platform provides a simple and easy way to customize and deploy AI models, where the prediction feature of table data can help users quickly mine hidden patterns in data and generate predictive results for business applications.

This article employs Baidu Easy-DL to construct a water depth inversion model. Initially, satellite data were transformed into tabular data. Subsequently, the structured data processing platform developed by Baidu Easy DL was utilized for data prediction, culminating in the acquisition of predicted water depth values. While Baidu Easy DL's structured data prediction does not strictly qualify as a large model platform, it offers valuable insights for future remote sensing large model platforms when processing remote sensing data.

The objective of this paper is to explore the potential of online general artificial intelligence models in handling remote sensing tasks. The structure of this paper is as follows: an introduction, followed by "Section 2", which presents data "Section 3", which outlines the methodology "Section 4", which discusses the results "Section 5", which engages in a discourse about these findings, and finally Section 6. This structure allows readers to gain a clear understanding of the experimental setup, execution, and results and facilitates discussion on potential future research directions.

2. Experimental Data Collection

2.1. The Study Areas and Sentinel-2 Imagery

We acquired bathymetric data (https://github.com/wuzhenghan2022/ESAY-DL.git, accessed on 5 October 2023) using a modified jet ski with an acoustic survey system at Rushikonda Beach, a scenic c-shaped bay on the east coast of India, located approximately between Chennai and Kolkata (see Figure 1) [19]. The coastal area is mainly composed of fine sand with a median grain size of 0.45–0.5 mm [19]. Moreover, some parts of the beach have submerged and protruding rocky outcrops. Notably, Rushikonda Beach was selected as one of the 12 pilot beaches in India for the "Blue Flag Certification" by the Ministry of Environment, Forest and Climate Change (MoEF and CC). Therefore, the continuous monitoring of nearshore processes is important for tourism activities and safety.

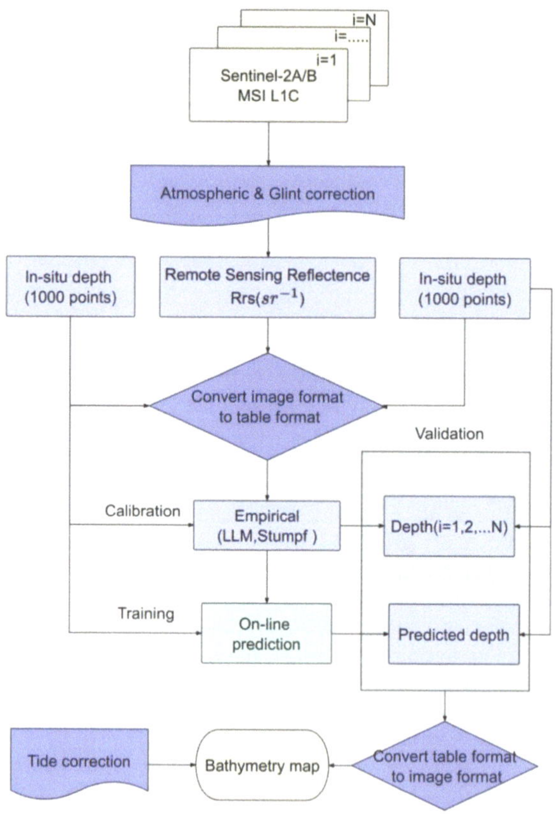

Figure 1. The general workflow of the proposed system for bathymetry from Sentienel-2 images.

In our study, we used data obtained on 24 November 2018. We processed the images into remote sensing reflectance (Rrs or Rw) [17,18] using the latest ACOLITE software (Python 20190326.0 version) provided by the Royal Belgian Institute of Natural Sciences (RBINS) [20,21]. ACOLITE provides Rrs data (sr-1) for all visible and near-infrared bands, which are resampled to a 10 m spatial resolution [19]. We predicted the bathymetry map based on the image data of multiple spectral reflectance bands (bands 1, 2, 3, 4, 5, 6, 8, and

10) from Sentinel-2 [19]. All spectral images were resampled to a resolution of 10 × 10 m. Finally, we mitigated the sun glint effect by resampling with an S2 view in ACOLITE software (Version 20210802.0) (Figure 2).

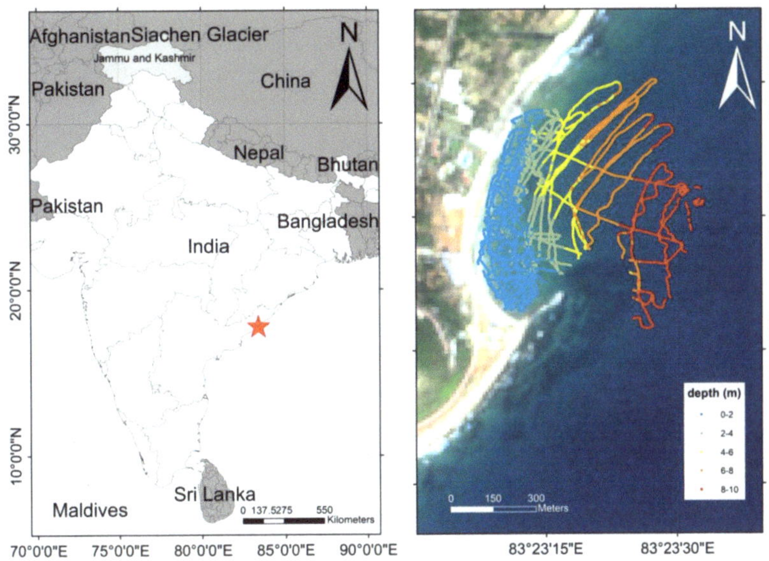

(a) (b)

Figure 2. The geographical location of the study area (**a**); Data collection area where different colors represent variations of in situ depth data (**b**).

2.2. In-Situ Data

We conducted two acoustic surveys on 23–24 October 2018 at Rushikonda using a modified jet ski to obtain bathymetric data. The jet ski was equipped with a 200 kHz CEESCOPETM echosounder and a 10 Hz Novatel OEMStar L1/L2 GNSS receiver, both provided by CEE Hydro systems. The echosounder had a high accuracy, with a vertical error of 1 cm ± 0.1% for the depth. The GNSS receiver had a horizontal error of about ±0.5 m. To improve the quality of data, the echosounder also had an inertial motion unit (IMU) sensor to record the three-dimensional motion. The IMU sensor had impressive accuracy, including roll and pitch angles ±0.1° (over 360°), heading angle ±1°, and heave distance ±5 cm. We applied wave correction to the echosounder depth data using heave data from the IMU sensor, following the method of Dugan et al. [22]. We applied a tidal correction to depth data using the tide gauge data located near Visakhapatnam Harbor (17°40′60″N, 83°16′60″E).

3. Proposed Machine Learning Algorithms for Bathymetry Mapping

3.1. Stumpf Model

In order to avoid the situation where the radiance received by the optical remote sensor and the radiance in the deep water was negative, Stumpf et al. [23]. proposed a model based on the log conversion ratio:

$$Z = m_1 \frac{ln[nR(\lambda_i)]}{ln[nR(\lambda_i)]} + m_0$$

where m_0 and m_1 are the regression coefficients; n is a fixed constant, usually taken as 1000; R (λ_i) and R (λ_j) are the remote sensing reflectance of the blue band i and the green band *j*.

3.2. Log-Linear Model

The dual-band log-linear model formula is as follows [5,6,24]:

$$Z = a_1 ln[L(\lambda_i) - L_\infty(\lambda_i)] + a_2 ln[L(\lambda_j) - L_\infty(\lambda_j)] + a_3$$

Here, a_1, a_2, and a_3 are the regression coefficients; L (λ_i) and L (λ_j) are the radiance of the blue band i and the green band j; $L\infty(\lambda_i)$ and $L\infty(\lambda_j)$ are the radiance of each band in deep water.

3.3. Baidu Easy DL Model

Easy-DL is an AI development platform for developers and data scientists, which was designed to help them quickly build high-quality AI models and implement their commercial applications. Its table prediction feature is an important part of the platform, used for predictions based on a given data table.

The table prediction feature consists of the following steps:

(a) Data preparation: Upload or import the data table that is to be used for prediction. Easy-DL supports multiple data formats, such as CSV, Excel, JSON, etc.
(b) Model selection: select a suitable pre-trained model for prediction based on the characteristics of the table data and the prediction requirements.
(c) Data preprocessing: preprocess the table data, including data cleaning, feature selection, data enhancement, and table format conversion, to improve the training effect of the model.
(d) Model training: Based on the uploaded table data and the selected model, Easy-DL automatically performs model training. During the training process, you can monitor the training progress and view performance indicators during training.
(e) Model evaluation: after the model training is complete, Easy-DL provides a series of evaluation indicators, such as accuracy, precision, and recall, to evaluate the performance of the model.
(f) Model deployment: After model evaluation is complete, the trained model can be deployed to the production environment. Easy-DL provides multiple deployment options, such as API, SDK, and a custom code, to meet different application needs.
(g) Prediction: The deployed model can be applied to actual scenarios for prediction. Through the API or SDK provided by Easy-DL, the prediction of the table can be easily performed.

The table prediction feature allows developers to build and apply AI models for prediction without in-depth knowledge of AI technology and algorithms. It is suitable for various scenarios, such as commercial prediction, disease prediction, recommendation systems, etc. By using Easy-DL's table prediction feature, developers can quickly and efficiently implement the development and deployment of AI applications.

3.4. Accuracy Evaluation Methods

The accuracy evaluation indexes of water depth accuracy are the mean absolute error (MAE), the mean relative error (MRE), and the root mean square error (RMSE), and the corresponding formula areas are as follows [7]:

$$\mathrm{MAE} = \frac{\sum_{i=1}^{n} \left| Z_i - Z_i' \right|}{n}$$

$$\mathrm{MRE} = \frac{\sum_{i=1}^{n} \left| \left(Z_i - Z_i' \right) / Z_i' \right|}{n}$$

$$\mathrm{RMSE} = \sqrt{\frac{\sum_{i=1}^{n} \left(Z_i - Z_i' \right)^2}{n}}$$

where z_i is the estimated water depth; $Z_i^{'}$ is the actual water depth; and n is the number of water depth points.

3.5. Bathymetry Mapping

Satellite-derived bathymetry (SDB) is a technique that uses remote sensing data to estimate water depth in shallow areas. One of the key factors that affect the accuracy of SDB is the selection of water depth control points, which are used to calibrate and validate the SDB models. Usually, about 1000 control points are selected from a single image of the study area, though this method may introduce variability in the SDB results depending on the number and locations of these control points.

In this study, we propose a comprehensive and robust process for water depth retrieval using SDB. The first step of our process is to select a high-quality remote sensing image that matches the timing of the scene and measured data. We then perform atmospheric correction and remove sunlight effects to obtain accurate reflectance data. Next, we applied various bathymetry algorithms to estimate water depth from the reflectance data and then corrected for tide effects to obtain consistent water depth values. Finally, we created a topographic map by integrating the estimated water depths. Our process consists of two stages for water depth estimation. In the first stage, we converted both remote sensing data and water depth data into a tabular format and used them for training purposes. In the second stage, we used Sentinel-2 data, which were also converted into the tabular format, for depth prediction. To evaluate the performance of our proposed SDB method, we compared the predicted depths with in situ depth values. We found that our method utilized water depth control point information effectively, reducing depth estimation errors and improving the accuracy of water depth inversion.

4. Experimental Setup and Results

In this comprehensive study, we meticulously explore the precision of water depth estimation through the application of machine learning algorithms and multiple training datasets derived from Sentinel-2 images. Our training dataset is extensive, comprising a total of 2000 data points.

The first phase involved the utilization of 1000 points for the initial estimation of water depth. This was followed by the application of an additional 1000-point training set for inversion, which facilitated the acquisition of preliminary values. Remote sensing reflectance data, corresponding to identical geographical coordinates as water depth data, were collected and systematically organized into a tabular format. The primary column represents water depth, while the subsequent columns correspond to data from various bands, specifically bands 1, 2, 3, 4, 5, 6, 8, and 10.

The process of water depth retrieval was initiated based on control points associated with one of the 1000-point training sets. This training set was subsequently employed for inversion to derive the final depth of results. The final results obtained through online prediction were utilized to compute the key evaluation metrics. These included the Mean Absolute Error (MAE), Mean Relative Error (MRE), Root Mean Square Error (RMSE), and the coefficient of determination (R^2).

Three models were selected for this study: the Stumpf model, the Log-Linear model, and the Baidu Easy-DL model. Each model was trained using remote sensing reflectance values from various bands as the input variables. The dataset was split using spatial random sampling to ensure a diverse range of data points for model training and verification.

Quality checks were conducted throughout the process to ensure the accuracy and reliability of our results. Possible branching was also considered in our study design. For instance, depth mapping was conducted within a certain range along the coast.

This rigorous approach ensured a comprehensive understanding of water depth estimation using machine learning algorithms and Sentinel-2 images. It provides valuable insights that could be used to further refine these techniques and improve their accuracy in future studies.

All data format conversions and experiments were carried out within the Matlab environment. For training points and verification points, we used a random sampling method to extract the samples. Our approach involved the random selection of calibration samples in accordance with the depth distribution. For each sample, we computed the MAE, MRE, RMSE, and R^2 values by comparing the estimated water depth values with ground-truth measurements. The ultimate results were presented as the mean values across all samples.

The results of water depth estimation exhibited significant variations among the different bathymetry algorithms employed. Notably, the accuracy of the Easy-DL model stood out as the highest among the three algorithms, followed by the Log-Linear model in second place, while the Stumpf model lagged behind in terms of accuracy (refer to Figure 3). These conclusions are substantiated by examining parameters such as the correlation coefficient, R^2, MAE, MRE, RMSE.

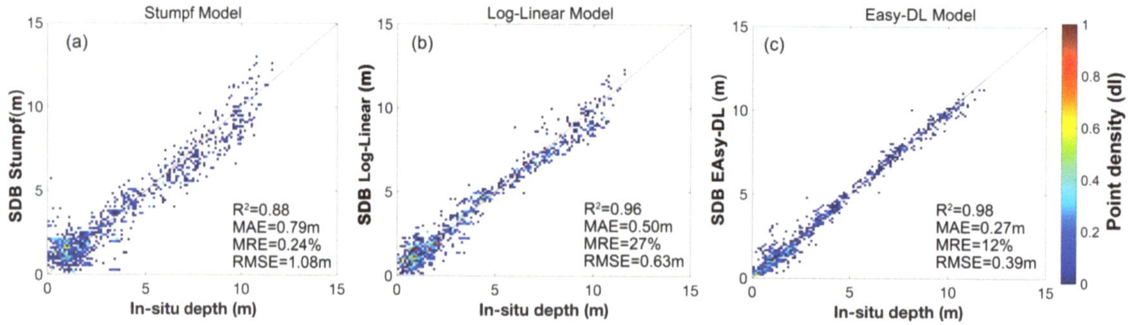

Figure 3. Correlation between the in situ depths and depth results based on different bathymetry methods: (**a**) Stumpf model; (**b**) Log-Linear model (**c**) Easy-DL model.

For the Easy-DL model, the water depth inversion results closely align with the 1:1 line, with fewer discrete data points. By contrast, the Stumpf model exhibited the lowest accuracy, characterized by a higher degree of data discreteness in the water depth inversion. Following this trend, the Log-Linear model fell in between the other two algorithms in terms of accuracy.

Notably, within the depth range of 0–3 m, the Easy-DL model demonstrated a high degree of alignment with actual measurements, resulting in a noticeably higher accuracy compared to the other two models. However, when the water depth exceeded 10 m in the Easy-DL model, the retrieved values started to decrease slightly, deviating marginally from the measured water depth values. Obviously, within the range of 10 m in the study area, the water depth inversion results given by the Baidu Easy-DL platform were the best, while the MAE, MRE, RMSE, and R^2 values performed better (Refer to Figure 3).

Figure 4 presents a comprehensive map of estimated water depths across the study area, spanning depths from 0 to 15 m. The results highlight variations in accuracy, notably showcasing lower accuracies in shallow depths (<0.5 m) and deeper depths (>15 m) within the study area, particularly in their proximity to the shoreline.

The topographic map derived from the Easy-DL model exhibited less noise, and the inversion results from shallow to deep waters closely mirrored the actual conditions. An examination of the scatter map reveals a noticeably superior accuracy in the inversion of water depth compared to the other two models. When compared to the measured data, the topographic map derived from satellite data effectively captures the general trend of water depth variability, albeit with some minor discrepancies in the finer details.

In the 0–5 m depth range, the inversion results of the three algorithms exhibit a similar trend, albeit with some localized variations. For instance, in specific areas, the Stumpf model manages to mitigate the influence of seabed geological heterogeneity on water depth inversion results, a feat not achieved by the Log-Linear algorithm. However, even though the Easy-DL model boasts high accuracy, it does exhibit errors in these regions.

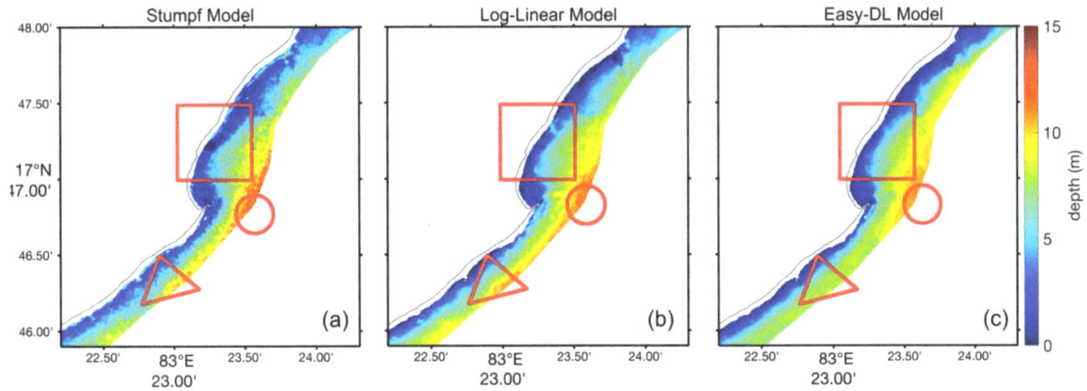

Figure 4. The water depth map estimated based on satellite image data and a different bathymetry method (**a**) Stumpf model; (**b**) Log-Linear model (**c**) Easy-DL model.

Yet, it is worth noting that the Stumpf model shows pronounced error bands in nearshore areas, which could potentially be attributed to wave-related factors. In the depth range of 6–10 m, the water depth changed trends, appearing consistent across models; however, the Stumpf model contained more noise points in a triangular region compared to the smoother results from the Easy-DL model.

In locations exceeding 10 m in depth, such as the circular section, data from the Easy-DL model tended to underestimate water depth inversion results, which is consistent with the scatter plot observations. Collectively, these results underscore the reliability of satellite-based water depth estimation. However, it is noteworthy that when the water depth in the Easy-DL model exceeded 10 m, the inverted water depth values began to exhibit a slight decline, deviating marginally from the measured water depth values.

5. Discussion

5.1. The Performance of Water Depth Inversion Model

As illustrated in the scatterplot presented in Figure 3, our proposed method exhibits a remarkably high level of accuracy in water depth inversion. This becomes particularly evident when compared to two classical algorithms: the Stumpf and Log-Linear algorithms. To comprehensively assess the bathymetry results across various water depth ranges, we calculated the root-mean-square errors (RMSE) for both the classical methods and our proposed online deep-learning method (see Table 1).

These methods enable precise water depth estimation under diverse conditions, encompassing factors such as human activities, pollution, and sediment accretion. Notably, our proposed online deep learning method consistently outperforms all other methods in terms of overall accuracy, boasting an RMSE that is 0.24 m less than the closest RMSE value among all other methods.

Moreover, the proposed online deep learning method excels in overall accuracy and demonstrates superior performance in the inversion accuracy of specific water depth ranges. Notably, within the 6–9 m range, our method achieved remarkable accuracy, with an RMSE as low as 0.23 m. In the case of the Stumpf algorithm, similar conclusions were obtained, where the RMSE was 0.94 m, albeit slightly lower than the overall accuracy. Conversely, the Log-Linear algorithm exhibited its smallest error in the 3–6 meter water depth range, with an RMSE of 1.01 m.

However, it is important to acknowledge that when the water depth exceeded 9 m, all algorithms tended to experience an increase in accuracy deviation and RMSE values, surpassing the overall results. This phenomenon can be attributed to the comprehensive approach employed in our method, where various machine learning algorithms were integrated to perform depth inversion. This enabled optimal depth estimation overall, yielding superior results in localized estimations as well.

Given that the study area encompasses an open coast, it is susceptible to significant influences from various environmental factors. Remote sensing images reveal valuable insights into the seabed quality of the nearshore sea, suggesting relatively high water transparency in this region.

As depicted in Figure 5, the results obtained through the approach proposed in this paper exhibit substantial improvements compared to those obtained through traditional methods. These improvements are noticeable across the entire depth range under consideration, which spanned from 0 to 15 m, with a particularly noteworthy enhancement at a depth of four meters.

Table 1. A comparison of the RMSE errors for different water depths and different bathymetry methods.

Training Method	RMSE				
	0–3 m (580 Points)	3–6 m (214 Points)	6–9 m (200 Points)	>9 m (95 Points)	Overall (1089 Points)
Stumpf	1.12	1.01	0.94	1.19	1.08
Log-Linear	0.59	0.58	0.66	0.89	0.63
Easy-DL	0.43	0.29	0.23	0.39	0.39

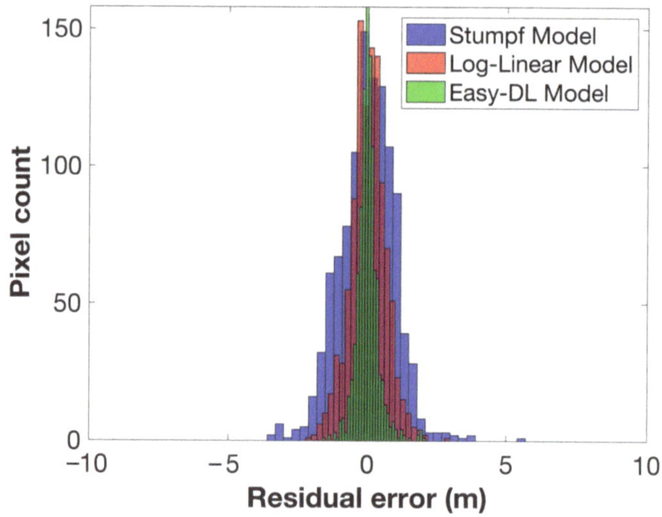

Figure 5. The histogram map of the residual error obtained from different methods.

Additionally, it is worth highlighting that the histogram depicting the residuals was limited to ±1 m in comparison to the previous method. In this context, the distributions observed for all three methods appeared to follow a normal pattern (see Figure 5). This reaffirms the feasibility and effectiveness of the method proposed in this paper.

5.2. The Uncertainty and Implications of Baidu Easy-DL Model

Obviously, within the range of 10 m in the study area, the water depth inversion results given by the Baidu Easy-DL platform were the best, while the MAE, MRE, RMSE, and R^2 values performed better. However, in the work of remote sensing for the inversion of water depth in turbid water bodies using the online artificial intelligence platform Baidu Easy-DL, there were certain uncertainties and impacts:

(a) **Model Selection:** The choice of the model may affect the accuracy of the inversion results. Although machine learning models generally have higher robustness than traditional semi-empirical, bio-optical, and semi-analytical models[2], different machine learning models may produce different results. For example, a study found that the Genetic Algorithm Optimized Extreme Learning Machine (GA-ELM) had a more compact network structure and better generalization ability than the Extreme Learning Machine (ELM).
(b) **Input Variables:** The choice of input variables may also affect the results. For example, using remote sensing reflectance values at different bands as input variables may lead to different inversion results.
(c) **Data Quality:** The quality of remote sensing data also affects the inversion results. For example, if remote sensing data contain noise or are affected by factors such as atmosphere and water turbidity, it may lead to inaccurate inversion results.

The methodology provided in this study holds significant implications for various research areas:

(a) **Depth Inversion**: This work is of paramount importance for depth inversion, offering valuable support for marine engineering, shipping, and maritime military security.
(b) **Environmental Monitoring**: This methodology can also be utilized for environmental monitoring, such as monitoring the water quality of inland bodies of water.
(c) **Scientific Advancement**: This work can propel scientific progress in related fields, such as enhancing the accuracy and robustness of remote sensing inversion models.

These implications underscore the practical application and scientific research value of this study, demonstrating its potential to contribute significantly to both practice and research in these areas.

In summary, despite certain uncertainties, the remote sensing inversion of turbid water depth using Baidu Easy DL still has important practical and scientific value. In future work, we can reduce uncertainty and improve the accuracy of inversion results by improving models, optimizing input variable selection, and improving the quality of data. At the same time, we need to pay attention to the various impacts that this work may bring in order to better utilize its application potential in practice and research.

6. Conclusions

This paper presents an online water depth estimation method that employs a comprehensive approach. This method uses a general large model, combining remote sensing data with measured training datasets, and incorporates multiple machine learning algorithms. The results achieved with this approach in water depth inversion have been promising. Moreover, using the online ensemble learning algorithm clearly shows different water depth estimations. In comparison, ensemble learning techniques can be further integrated with these algorithms to improve depth estimation accuracy, often resulting in a halving of the RMSE. Within the experimental area, our proposed method demonstrated superior precision, lower RMSE values, and higher R^2 values when compared to the classical Stumpf and Log-linear algorithms. The experimental results indicate that this method can effectively improve depth estimation within the range of 0 to 11 m, with an RMSE of 0.39 m. Remarkably, for water depths less than 9 m, the inversion accuracy is consistently high. The reduction in performance for depths exceeding 9 m may be attributed to similar water quality conditions and a substantial water depth, which might challenge the accurate reflection of depth changes through remote sensing reflectivity data.

It is worth noting that while the quantity of training samples significantly impacts the performance of depth estimation, this paper's algorithms are all based on a large volume of training data. Importantly, this method has the potential to be extended to estimate other physical parameters based on remote sensing image analysis, such as water turbidity and chlorophyll concentration. In summary, the method proposed in this paper effectively estimates the water depth from satellite images by leveraging the synergy between publicly available large-scale models and remote sensing depth retrieval. This method outperforms traditional remote sensing depth retrieval approaches. Due to the non-parametric nature of machine learning methods, it successfully achieves relatively high coherence and consistency from observed satellite images compared to depth estimation through acoustic methods.

Looking ahead, with the continuous advancement of large models, the method presented here, which involves invoking network online model APIs for remote sensing image processing, represents a promising direction in remote sensing applications. While many scholars have used remote sensing APIs for specific tasks in remote sensing image processing, the lack of an API for Satellite-based bathymetry is a challenge. Converting the formats of remote sensing data into the import and export formats of common online learning algorithms is crucial for future research and the widespread application of remote sensing online data processing.

Author Contributions: Z.W. and S.W. processed analysis and wrote manuscript. All authors reviewed and commented on the manuscript. Z.W. and H.Y. designed the algorithm with input from W.S. The BATHYMETRY project was conceived by Z.M., who also obtained funding and collected remote sensing data and Sonar data. All authors have read and agreed to the published version of the manuscript.

Funding: This research was funded by the Hainan Natural Science Foundation of China [No.620RC602], the National Natural Science Foundation of China [No.61966013], the 2023 Hainan Province "South China Sea New Star" Science and Technology Innovation Talent Platform Project [NHXXRCXM202316], and was funded by the Teaching Reform Research Project, Hainan Normal University [hsjg2023-07].

Data Availability Statement: The data and codes supporting the findings of this study can be found at the provided link: https://github.com/wuzhenghan2022/ESAY-DL.git.

Acknowledgments: Thanks to Zhao Yuchen, Zhang Xianyao, Luo Hui, Liu Pei-ran and Chen Huaze from the School of Information Science and Technology of Hainan Normal University for their valuable suggestions on paper revision.

Conflicts of Interest: The authors declare no conflict of interest.

References

1. Liu, Y.; Zhao, J.; Deng, R.; Liang, Y.; Gao, Y.; Chen, Q.; Xiong, L.; Liu, Y.; Tang, Y.; Tang, D. A downscaled bathymetric mapping approach combining multitemporal Landsat-8 and high spatial resolution imagery: Demonstrations from clear to turbid waters. *ISPRS J. Photogramm. Remote Sens.* **2021**, *180*, 65–81. [CrossRef]
2. Cao, B.; Deng, R.; Zhu, S. Universal algorithm for water depth refraction correction in through-water stereo remote sensing. *Int. J. Appl. Earth Obs. Geoinf.* **2020**, *91*, 102108. [CrossRef]
3. Niroumand-Jadidi, M.; Legleiter, C.J.; Bovolo, F. River Bathymetry Retrieval from Landsat-9 Images Based on Neural Networks and Comparison to SuperDove and Sentinel-2. *IEEE J. Sel. Top. Appl. Earth Obs. Remote Sens.* **2022**, *15*, 5250–5260. [CrossRef]
4. Liu, H.; Li, J.; Meng, X.; Zhou, B.; Fang, G.; Spencer, B.F. Discrimination Between Dry and Water Ices by Full Polarimetric Radar: Implications for China's First Martian Exploration. *IEEE Trans. Geosci. Remote Sens.* **2022**, *61*, 5100111. [CrossRef]
5. Lyzenga, D.R. Passive remote sensing techniques for mapping water depth and bottom features. *Appl. Opt.* **1978**, *17*, 379–383. [CrossRef] [PubMed]
6. Lyzenga, D.R.; Malinas, N.R.; Tanis, F.J. Multispectral bathymetry using a simple physically based algorithm. *IEEE Trans. Geosci. Remote Sens.* **2006**, *44*, 2251–2259. [CrossRef]
7. Brando, V.E.; Anstee, J.M.; Wettle, M.; Dekker, A.G.; Phinn, S.R.; Roelfsema, C. A physics based retrieval and quality assessment of bathymetry from suboptimal hyperspectral data. *Remote Sens. Environ.* **2009**, *113*, 755–770. [CrossRef]
8. Liu, Z.; Xu, J.; Liu, M.; Yin, Z.; Liu, X.; Yin, L.; Zheng, W. Remote sensing and geostatistics in urban water-resource monitoring: A review. *Mar. Freshw. Res.* **2023**, *74*, 747–765. [CrossRef]
9. Liu, Y.; Deng, R.; Qin, Y.; Cao, B.; Liang, Y.; Liu, Y.; Tian, J.; Wang, S. Rapid estimation of bathymetry from multispectral imagery without in situ bathymetry data. *Appl. Opt.* **2019**, *58*, 7538–7551. [CrossRef]
10. Wu, Z.; Mao, Z.; Shen, W. Integrating Multiple Datasets and Machine Learning Algorithms for Satellite-Based Bathymetry in Seaports. *Remote Sens.* **2021**, *13*, 4328. [CrossRef]

11. Traganos, D.; Poursanidis, D.; Aggarwal, B.; Chrysoulakis, N.; Reinartz, P. Estimating Satellite-Derived Bathymetry (SDB) with Google Earth Engine and Sentinel-2. *Remote Sens.* **2018**, *10*, 859. [CrossRef]
12. Zhou, W.; Tang, Y.; Jing, W.; Li, Y.; Yang, J.; Deng, Y.; Zhang, Y. A Comparison of Machine Learning and Empirical Approaches for Deriving Bathymetry from Multispectral Imagery. *Remote Sens.* **2023**, *15*, 393. [CrossRef]
13. Li, J.; Knapp, D.E.; Lyons, M.; Roelfsema, C.; Phinn, S.; Schill, S.R.; Asner, G.P. Automated Global Shallow Water Bathymetry Mapping Using Google Earth Engine. *Remote Sens.* **2021**, *13*, 1469. [CrossRef]
14. Wen, C.; Hu, Y.; Li, X.; Yuan, Z.; Zhu, X.X. Vision-Language Models in Remote Sensing: Current Progress and Future Trends. *arXiv* **2023**, arXiv:2305.05726.
15. Hu, Y.; Yuan, J.; Wen, C.; Lu, X.; Li, X. RSGPT: A Remote Sensing Vision Language Model and Benchmark. *arXiv* **2023**, arXiv:2307.15266.
16. Zhang, J.; Zhou, Z.; Mai, G.; Mu, L.; Hu, M.; Li, S. Text2Seg: Remote Sensing Image Semantic Segmentation via Text-Guided Visual Foundation Models. *arXiv* **2023**, arXiv:2304.10597.
17. Tian, H.; Huang, N.; Niu, Z.; Qin, Y.; Pei, J.; Wang, J. Mapping winter crops in China with multi-source satellite imagery and phenology-based algorithm. *Remote Sens.* **2019**, *11*, 820. [CrossRef]
18. Chen, M.; Zhang, B.; Topatana, W.; Cao, J.; Zhu, H.; Juengpanich, S.; Mao, Q.; Yu, H.; Cai, X. Classification and mutation prediction based on histopathology H&E images in liver cancer using deep learning. *NPJ Precis. Oncol.* **2020**, *4*, 14.
19. Arun Kumar, V.V. Numerical Modelling of Coastal and Nearshore Processes in the Vicinity of Shoreline Harbours with Special Reference to Visakhapatnam Coast India. Ph.D. Thesis, Andhra University, Visakhapatnam, India. 2012. Available online: http://hdl.handle.net/10603/407131 (accessed on 12 October 2023).
20. Ruddick, K.; Vanhellemont, Q.; Dogliotti, A.; Nechad, B.; Pringle, N.; Van der Zande, D. New opportunities and challenges for high resolution remote sensing of water colour. In Proceedings of the Ocean Optics XXIII, Victoria, BC, Canada, 23–28 October 2016; Volume 7.
21. Vanhellemont, Q.; Ruddick, K. Acolite for Sentinel-2: Aquatic applications of MSI imagery. In Proceedings of the 2016 ESA Living Planet Symposium, Prague, Czech Republic, 9–13 May 2016; pp. 9–13.
22. Dugan, J.; Morris, W.; Vierra, K.; Piotrowski, C.; Farruggia, G.; Campion, D. Jetski-based nearshore bathymetric and current survey system. *J. Coast. Res.* **2001**, *17*, 900–908.
23. Stumpf, R.P.; Holderied, K.; Sinclair, M. Determination of water depth with high-resolution satellite imagery over variable bottom types. *Limnol. Oceanogr.* **2003**, *48*, 547–556. [CrossRef]
24. Lyzenga, D.R. Shallow-water bathymetry using combined lidar and passive multispectral scanner data. *Int. J. Remote Sens.* **1985**, *6*, 115–125. [CrossRef]

remote sensing

MDPI

Article

Evaluating Ecosystem Service Value Changes in Mangrove Forests in Guangxi, China, from 2016 to 2020

Kedong Wang [1,2], Mingming Jia [2], Xiaohai Zhang [3,*], Chuanpeng Zhao [2], Rong Zhang [2] and Zongming Wang [2]

[1] School of Geomatics and Prospecting Engineering, Jilin Jianzhu University, Changchun 130118, China; wangkedong@student.jlju.edu.cn
[2] Key Laboratory of Wetland Ecology and Environment, Northeast Institute of Geography and Agroecology, Chinese Academy of Sciences, Changchun 130102, China; jiamingming@iga.ac.cn (M.J.); zhaochuanpeng@iga.ac.cn (C.Z.); zhangrong@iga.ac.cn (R.Z.); zongmingwang@iga.ac.cn (Z.W.)
[3] Haikou Marine Geological Survey Center, China Geological Survey, Haikou 571127, China
* Correspondence: zhangxiaohai@mail.cgs.gov.cn

Abstract: Mangrove forests play a vital role in maintaining ecological balance in coastal regions. Accurately assessing changes in the ecosystem service value (ESV) of these mangrove forests requires more precise distribution data and an appropriate set of evaluation methods. In this study, we accurately mapped the spatial distribution data and patterns of mangrove forests in Guangxi province in 2016 and 2020, using 10 m spatial resolution Sentinel-2 imagery, and conducted a comprehensive evaluation of ESV provided by mangrove forests. The results showed that (1) from 2016 to 2020, mangrove forests in Guangxi demonstrated a positive development trend and were undergoing a process of recovery. The area of mangrove forests in Guangxi increased from 6245.15 ha in 2016 to 6750.01 ha in 2020, with a net increase of 504.81 ha, which was mainly concentrated in Lianzhou Bay, Tieshan Harbour, and Dandou Bay; (2) the ESV of mangrove forests was USD 363.78 million in 2016 and USD 390.74 million in 2020; (3) the value of fishery, soil conservation, wave absorption, and pollution purification comprises the largest proportions of the ESV of mangrove forests. This study provides valuable insights and information to enhance our understanding of the relationship between the spatial pattern of mangrove forests and their ecosystem service value.

Keywords: mangrove forests; spatial distribution pattern; ecosystem service value; remote sensing; Guangxi

Citation: Wang, K.; Jia, M.; Zhang, X.; Zhao, C.; Zhang, R.; Wang, Z. Evaluating Ecosystem Service Value Changes in Mangrove Forests in Guangxi, China, from 2016 to 2020. *Remote Sens.* **2024**, *16*, 494. https://doi.org/10.3390/rs16030494

Academic Editors: Nikolay Strigul and Kenlo Nishida Nasahara

Received: 18 October 2023
Revised: 8 January 2024
Accepted: 25 January 2024
Published: 27 January 2024

1. Introduction

Mangrove forests, which are found in intertidal zones in tropical and subtropical regions, are among the most valuable and productive ecosystems on the earth [1]. They provide unique ecosystem services such as wave energy reduction, coastal erosion prevention, water purification, and biodiversity protection [2,3]. Mangrove forests also contribute to poverty alleviation and food security, including the provision of food and raw material provision, offering recreation and tourism opportunities, and moderating extreme events [4]. Thus, they are enormously relevant to sustainable development goals [5]. To better understand the services and benefits mangrove forests provide to people and how their services change under different scenarios, it is necessary to assess the economic value of mangrove forests as natural capital [6]. The valuation of the forests' ecosystem services is also a quantitative tool for decision-makers and conservation advocates in assessing the extent of recovery or degeneration [2].

Ecosystem valuation is an approach to assign monetary values to an ecosystem according to its key ecosystem goods and services, generally referred to as its Ecosystem Service Value (ESV) [7]. This approach can improve knowledge for informed decision-making to raise awareness of blue forest ecosystems and foster cooperation among blue forest stakeholders [8]. There are numerous studies on the ESV of coastal ecosystems [9–11].

In analyzing gains and losses in ecosystem services values (ESVs) in the coastal zones of Zhejiang Province during rapid urbanization, Cao et al. [9] found that changes in land use patterns, specifically disordered land-use changes from forestland and farmland to urban construction land, were a major cause of ESV loss. Ligate et al. [10] assessed temporal land cover and land-use changes, underlying socioeconomic drivers, and dynamics of ESV in the coastal zone of Tanzania, and identified population pressure and socioeconomic activities as key factors contributing to the degradation of coastal ecosystems. Yang et al. [11] proposed a detailed "donor-side" accounting approach based on the energy method, providing a "supply-side" evaluation of coastal and marine ecosystem services values (ESVs) that captures dynamic ecological processes and applies unified metrics.

However, three reasons make it challenging to accurately measure mangrove forests' ESV. The obstacles limiting mangrove measurements include deficiencies in global-scale assessment methods, previous studies focusing on case studies in specific regions, and a lack of attention to the spatial pattern of forests. Firstly, while global-scale assessment methodologies [12,13] can provide useful insights into the overall trends and patterns of ecosystem services, they may cause variability and inconsistencies in local-scale assessment due to differences in the ecological and socioeconomic contexts of each individual region. It is important to note that global-scale ESV assessment methods and results may not always be suitable for those at a local scale [14]. By considering the specific characteristics of the ecosystem, local-scale assessments can provide more accurate and comprehensive evaluations of ecosystem services. Secondly, previous studies on ESV in mangrove forests primarily focused on a nature reserve [15]. They provided a comprehensive valuation of ecosystem services of mangroves in a natural reserve. However, they tended to emphasize the linkages between land use/land cover and ESV change in a natural reserve [7,16], rather than focus specifically on the mangrove forests. Thirdly, the spatial pattern of these forests was often overlooked in previous studies. The ESV may be underestimated when their spatial structure and pattern are neglected [17]. According to Luke M. Brander [13], an increase in the abundance of mangroves within a region can lead to higher unit productivity. Furthermore, it is worth noting that most ecosystem services require certain minimum area thresholds to be achieved. Even if two habitats have similar total areas, the distribution and fragmentation of the patches can lead to significant differences in their ecological value [18]. As such, the importance of mangrove forests as ecosystem service providers is highly dependent on their spatial patterns.

Spatial distribution and landscape pattern are essential for accurately assessing the ESV of mangrove forests [13,18,19]. Conducting traditional surveys of mangroves can be a highly challenging and time-consuming task due to their muddy intertidal zone environment [20]. Remote sensing has been widely used to acquire spatial information about mangrove forests due to its unparalleled advantages in terms of multiscale capabilities [21,22]. To date, the Landsat series imagery is a widely used dataset for assessing the ESV in mangrove forests [7], since the images have a 30 m resolution and have been consistently available every 16 days since 1984 [23]. However, there are several shortcomings in the studies that have generated mangrove maps. First, it is difficult to obtain images during the low tide period due to the coarse temporal resolution (over 16 days). Second, landscape patterns of smaller mangrove forest patches might not be accurately discriminated with a 30 m spatial resolution. Thus, Sentinel-2 imagery, with its free access, 10 m resolution bands (Bands 2, 3, 4, and 8), and dense temporal resolution (2–5 days), is a better choice [24]. Particularly when combined with the computing capability provided by Google Earth Engine, a high-quality Sentinel image with more details can be obtained [25]. However, no mangrove forest ESV assessment has been conducted based on Sentinel-2 derived spatial data.

To address the above-mentioned issues, we assessed the ESV of mangrove forests based on the criterion of the MA and the precise spatial distribution of mangrove forests in Guangxi province derived from high spatial resolution Sentinel-2 imagery. The objectives of this study are to (1) obtain the spatial distribution and pattern of mangrove forests from 2016 to 2020 based on Sentinel-2 imagery; (2) construct a comprehensive evaluation system by drawing on the Millennium Ecosystem Assessment (MA) to estimate the ESV of mangrove forests based on their spatial patterns; and (3) analyze the ESV changes from 2016 to 2020 along the coasts of Guangxi, China.

2. Materials and Methods

2.1. Study Area

As illustrated in Figure 1, the study area is located in the southwest portion of mainland China and the northern region of the Beibu Gulf (21°24′–22°01′N and 107°56′–109°47′E). The mean annual temperature and precipitation vary from 22 °C to 23 °C and from 1500 mm to 2000 mm, respectively. It belongs to the tropical monsoon oceanic climate zone with high temperatures and rainy conditions. Tides across the study area are diurnal, with an average range of 2.24 m [26].

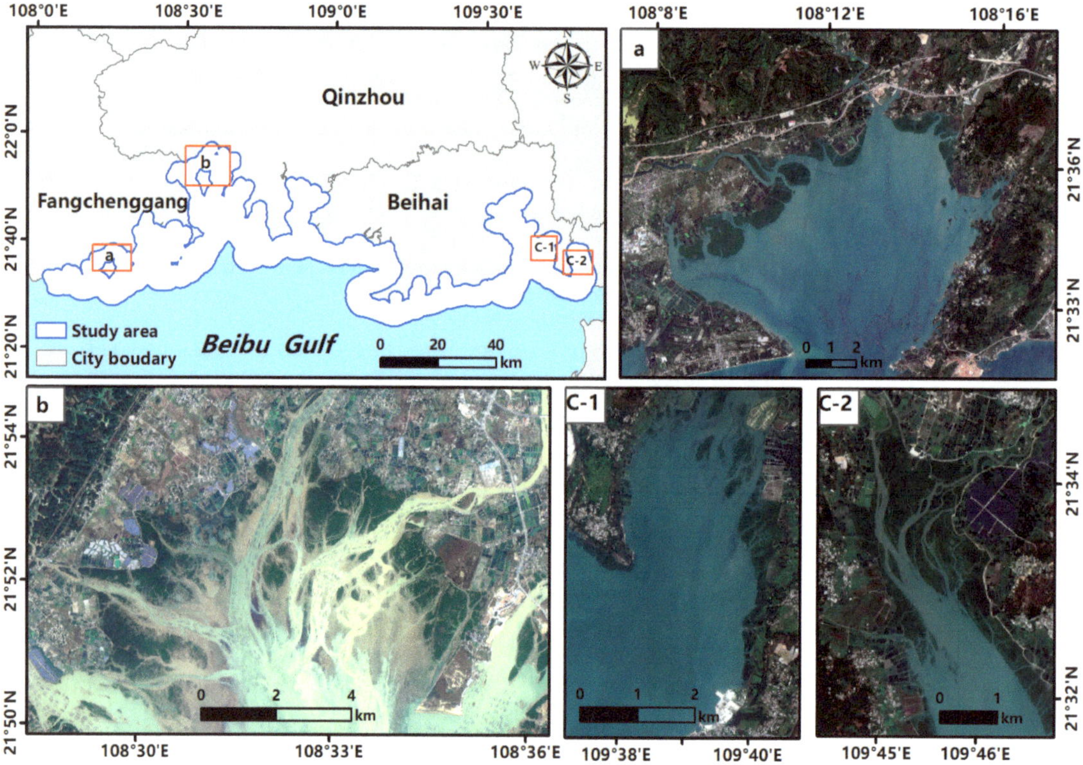

Figure 1. Location of the study area. (**a**) Beilun Eastuary National Mangrove Nature Reserve; (**b**) Maowei Sea Mangrove Reserve; (**c**) Shankou National Mangrove Nature Reserve (including **C-1** and **C-2**).

Along the coasts of Guangxi, there are two national mangrove reserves (Shankou National Mangrove Nature Reserve and Beilun Estuary National Mangrove Nature Reserve) and one provincial mangrove reserve (Maowei Sea Mangrove Reserve). Seven species of mangrove forests live along the coasts, among which Aegiceras comiculatum, Avicennia marina, Kandelia candel, and Aegiceras comiculatum occupy over 90% of the total area of mangrove forests [27]. Other species, such as Rhizophora stylosa and Bruguiear gymnorrhza, are sparsely distributed [28]. Mangrove forests distributed in Lianzhou Bay, Maowei Bay, and Zhenzhu Bay are typical estuary mangrove forests, which are found in the intertidal zone of estuaries where freshwater and seawater mix and create conditions with varying salinity levels. Qinzhou Bay has a unique island group of mangroves. The largest urban mangroves and sandy mangroves of China are distributing along the coasts of Beihai.

2.2. Sentinel-2 Data Acquisition and Pre-Processing

In this study, Sentinel-2 images were chosen to obtain information on mangrove forest distribution in Guangxi from 2016 to 2020. The Sentinel-2 mission has two polar-orbiting satellites (Sentinel-2A and Sentinel-2B) that provide high-resolution optical imagery. These satellites revisit the same place every 2–5 days. They both carry a MultiSpectral Instrument (MSI) sensor that offers 13 spectral bands. Only four bands (Bands 2, 3, 4, and 8) with a 10 m spatial resolution were employed, identifying, in particular, mangrove forest patches with small areas or narrow shapes [26].

In high-tide images, some low-lying mangroves may be submerged by water bodies, making mangrove forests difficult to identify and extract. To facilitate the extraction of accurate information on the regional extent and spatial pattern of mangroves, low-tide period and cloud-cover images were acquired in November and December of 2016 and 2020. The Level-2A product of the Sentine-2 Multispectral Instrument (MSI) images was downloaded from the Copernicus Open Access Hub (https://scihub.copernicus.eu/dhus (accessed on 13 June 2022)). The Level-2A product underwent radiometric, geometric, orthorectified, and atmospheric corrections. It can provide per-pixel radiometric measurements of surface reflectance [29]. In order to ensure consistency throughout the study and obtain accurate mangrove extraction ranges, we manually drew coastline data from 2016 to 2020 using Google Earth Pro software (version 7.3.6), using artificial embankments as reference points. Lastly, each image was clipped using a 5 km buffer zone along the coastline.

2.3. Field Investigation and Other Data

Three field investigations were conducted during the periods of 1–15 November 2016, 13–25 September 2019, and 17–27 December 2020. The ground survey work was conducted along designated walkways, and each field point's location was established by Real-Time Kinematic (IRTK5) with a global positioning system accurate to within 1 m, which can be affected by the number of available satellites and prevailing weather conditions. Aerial photographs were also taken with unmanned aerial vehicles during low tide. Given that much of the mudflat areas where the mangrove forests were located were inaccessible, some sample points were selected using Google Earth and unmanned aerial vehicles.

We collected 961 sample points in each of the years 2016 and 2020. Out of these, 200 mangrove and 200 non-mangrove points, respectively, were collected as training samples during the classification process. The remaining 224 mangrove points and 337 non-mangrove points were used for image validation in 2016 and 2020.

2.4. Classification Methods and Accuracy Assessment

In this study, object-based image analysis and the Random Forest classification method were applied, in conjunction with visual modification, to classify the mangrove and non-mangrove in 2016 and 2020, respectively.

Object-based image analysis involves setting certain homogeneous standard parameters according to the spectral information and shape information of the image [30]. It also segments the remote sensing image to form an image object. Image segmentation can directly influence the efficiency and accuracy of classification results [31]. The classification results avoid salt-and-pepper noise, have good integrity, and have a high classification accuracy [32].

In this study, multi-scale segmentation, which is one of the most useful segmentation algorithms, was selected, and the eCognition software (version 9.0) was used as the operating platform [33]. Through visual judgement and by systematically adjusting different segmentation scales and segmentation parameters until the mangrove forests regions were separated from water [24,30], the segmentation scale, segmentation shape, and tightness parameters were established as 20, 0.2, and 0.8, respectively.

Random Forest is an ensemble learning algorithm based on decision trees, that has demonstrated its usefulness and robustness in image classification [34]. It includes two critical parameters: the number of decision trees (ntree), which is established by randomly selecting samples from the training dataset, and the number of predictive variables (mtry), which defines the best partition in each node of decision trees and is determined as the square root of the number of input features [35].

When using a Random Forest classifier model, a wide range of features can be used as input variables. Compared to pixel-based methods, object-based image analysis can provide more spatial features. In this study, 15 spectral, spatial, and vegetation index features were used as input variables. A detailed list of these features is presented in Table 1.

Table 1. Features used in Random Forest classification.

Feature Type	Classification Feature
Spectral feature	Mean value of band 2 3 4 8, Standard deviation of band 2 3 4 8
Spatial feature	Shape index, Compactness index, Border index, Homogeneity, Contrast
Vegetation index	Normalized Difference Vegetation Index, Normalized Difference Water Index

In this study, Random Forest was also run in eCognition (version 9.0). After segmenting the image into multi-scale segmentation, we set the parameter ntree to 150 and the parameter mtry to 4. After obtaining the initial interpretation results, we inspected the results and adjusted the omitted or incorrect mangrove forest objects via visual modification.

To validate mapping accuracies, the accuracy of the classification results of 2016 and 2020 was assessed by the sample points (described in Section 2.3). The overall accuracy represents the proportion of correctly mapped points compared to ground points. The Kappa coefficient, a harmonic mean of user's accuracy and producer's accuracy, represents the classification performance of a single class.

2.5. Spatial Pattern of Mangrove Forests

In this study, combined with the spatial pattern of mangrove forest distribution in Guangxi [36], the indices shown in Table 2 were used to describe the spatial pattern of mangrove forests. On a landscape scale, the spatial pattern of mangrove forests refers to their spatial distribution pattern within regions (such as bays, etc.), including the spatial distribution and combination of mangrove forest patches with different sizes, shapes, and attributes. Cultivating a good spatial pattern and realizing its maximum comprehensive value is the goal of mangrove forest protection, management, and development. Abundance of mangrove refers to the area of mangroves per unit length of coastline in a bay or region. The number of patches is positively correlated with landscape fragmentation. Mangrove shoreline refers to the coastline effectively protected by mangroves. Finally, the relatively ideal distribution of mangroves highlights the contribution of mangroves to ecosystem services.

Table 2. Spatial pattern of mangrove.

Indices	Description
Abundance of mangrove	The area of mangroves per unit length of coastline (ha/km).
Number of patches	The number of mangrove patches.
Average patch area	The average area of all mangrove patches (ha).
Mangrove shoreline	Shoreline with mangroves (km)
Coastwards mangrove	Mangroves with a minimum distance between the landward boundary and the coastline less than 30 m.
Ideally distributed mangrove	Shoreline mangrove with a patch width $\geq$ 100 m and coverage $\geq$ 0.4

2.6. Assessment of Ecosystem Service Value

In this study, the ecosystem services of Guangxi's mangrove forests were organized into four categories and 10 types based on the criterion of the MA, as shown in Table 3. To ensure the accuracy of the ESV assessments for 2016 and 2020, the reference values of the evaluation indices and results were standardized to a common metric of 2016 USD per ha per year. Given that the reference values for the selected indicators came from different years, we used the GDP deflators to adjust them to 2016 [37], and then converted them to 2016 USD. This approach ensured that the reference values were comparable and consistent, which was necessary for accurate and meaningful ESV assessments.

Table 3. Indicators, calculation criterion, and data source for evaluating ESV of mangrove.

Category	Type	Evaluation Index	Equation
Provisioning service	Material production value	Wood production	$V_{wood} = G \times P \times (A_1 \times d_1 + A_2 \times d_2)$
		Fishery	$V_{Fishery} = P_f \times (A_1 \times d_1 + A_2 \times d_2)$
Regulating service	Soil conservation value	Soil conservation	$V_{Soil} = (A_1 \times d_1 + A_2 \times d_2) \times (X_1 - X_2) \times P_1/P_b$
		Fertilizer conservation	$V_F = (A_1 \times d_1 + A_2 \times d_2) \times S_{NPK} \times d \times P_b \times P$
	wave absorbing revetment	Mangrove shoreline	$V_{wave} = (L_1 \times d_1 + L_2 \times d_2) \times (C_1 + C_2)$
	Climate regulation	CO_2	$V_{CO_2} = (A_1 \times d_1 + A_2 \times d_2) \times T \times C$
		O_2	$V_{O_2} = (A_1 \times d_1 + A_2 \times d_2) \times M \times P_0$
		CH_4	$V_{CH_4} = (A_1 \times d_1 + A_2 \times d_2) \times Q \times 21 \times T$
	Pollution purification	Degrade pollutants	$V_{Purification} = (A_1 \times d_1 + A_2 \times d_2) \times S$
	Water conservation	Water	$V_{Water} = A \times R \times P_w$
Supporting service	Biodiversity Conservation	Habitat	$V_{Habitat} = (A_1 \times d_1 + A_2 \times d_2) \times P_h$
	Nutrient accumulation	Nutrient	$V_{Nutrient} = (A_1 \times d_1 + A_2 \times d_2) \times S_t \times P$
Cultural service	Cultural	Scientific Research and education	$V_{Science} = A \times P_s$
	Recreation	Recreation	$V_{Recreation} = A \times P_r$

To analyze and compare the ecological values and services provided by mangrove forests, we divided them into two categories: ideally distributed mangroves and remaining mangroves. We assigned weight values of 0.7 and 1 to the remaining mangroves and ideally distributed mangrove categories, respectively [38,39], which enabled us to conduct a more comprehensive analysis of their respective ecosystem service values. The following is a description of the 10 types of ESV that were selected for evaluation. Note that all the reference values provided in the description below have been adjusted to reflect the 2016 values using GDP deflators.

(1) Material production value

The material production function refers to the various products that can obtained from the ecosystem, including fresh water, food fuel, medical supplies, and so on. The material production function is closely related to human activity, and the shortage of these products can have direct or indirect adverse effects on human well-being. This study mainly considers the wood production value and natural aquatic product output value of mangrove forests.

① Wood production

In Guangxi, logging of mangroves is not allowed in mangrove reserves, and it is subject to strict supervision and restrictions in other areas. Therefore, the value of wood production is calculated based on the growth of living standing trees, and the market value method is used to calculate the value of wood production. The value of the growth of mangrove forests' living trees can be expressed as follows [40,41]:

$$V_{wood} = G \times P \times (A_1 \times d_1 + A_2 \times d_2) \tag{1}$$

where V_{wood} is the value of the wood production service, A_1 is the area of ideally distributed mangroves, A_2 is the area of the remaining mangroves; d_i is the weighting factor (0.7–1.0) (here, we define the values of d_1 and d_2 as 1.0 and 0.7, respectively), G is the annual volume growth of standing trees (4.98 m^3/(ha*a)), and P is the market price (USD 110.52/(ha*a) in 2016 and USD 92.22/(ha*a) in 2020).

② Fishery

Mangrove forests can provide a wealth of aquatic products, mainly including Sipunculus, Phascolosma esculenta, Ostrea rivularis, Meretrix meretrix, and other fishes. Aquaculture is generally widely distributed on tidal flats. Considering the availability of data, we used the fishery output value per unit area to calculate the fishery value provided by mangrove forests. The equation for calculating the fishery value is as follows [42]:

$$V_{Fishery} = P_f \times (A_1 \times d_1 + A_2 \times d_2) \tag{2}$$

where $V_{Fishery}$ is the fishery value, and P_f is the value of mangrove fishery per unit area (USD 19,945.15/(ha*a)).

(2) Soil conservation value

Soil conservation has the most directly positive effect on the growth and development of trees and the control of soil erosion. It mainly refers to reducing soil erosion and maintaining soil. The value of soil consolidation can be calculated based on the alternative engineering method. Fertilizer conservation mainly refers to protecting the soil from the fertility loss caused by soil erosion. It can be measured by multiplying the sum of the total amount of N, P, and K in the topsoil (0–31 cm). The conservation value of the soil can be expressed as follows [38,39]:

$$V_{Soil} = (A_1 \times d_1 + A_2 \times d_2) \times (X_1 - X_2) \times P_1 / P_b \tag{3}$$

$$V_{Fertilization} = (A_1 \times d_1 + A_2 \times d_2) \times S_{NPK} \times d \times P_b \times P \tag{4}$$

where V_{Soil} and $V_{Fertilization}$ are the values of the soil consolidation and fertilizer conservation, X_1 is the erosion index of bare soil (74.06 t/ha), X_2 is the erosion index of woodland (47.69 t/ha), P_1 is the cost of excavating earthwork (USD 0.57/m^2), P_b is the density of the topsoil (0.77 t/m^3), S_{NPK} is the contents of N, P, and K (1.39%), d is the topsoil thickness (0.31 m), and P is the price of the fertilizer (USD 391.43/t).

(3) Wave absorbing revetment

Mangrove forests can absorb a large amount of tidal energy and significantly slow down water flow. They have unique morphological characteristics and develop root systems that form a stable network system, which enables mangrove forests to grow more firmly on the tidal flat and form a tight fence on the beach. The value of wave-absorbing revetment can be estimated by applying the shadow engineering method. The equation for calculating the value of wave-absorbing revetment is as follows [28,39]:

$$V_{wave} = (L_1 \times d_1 + L_2 \times d_2) \times (C_1 + C_2) \tag{5}$$

where V_{wave} is the total value of the wave-absorbing revetment, L_1 is the length of the ideally distributed mangrove shoreline, L_2 is the length of the two remaining mangrove shorelines, C_1 is the ecological benefits provided by mangrove forests per unit distance per year (USD 13,300/km), and C_2 is the cost of repairing the dam.

(4) Climate regulation

The climate regulation of mangrove forests has both positive and negative effects. The positive effect mainly refers to their carbon fixation and oxygen release function, that is, the function of absorbing CO_2 in the atmosphere through photosynthesis and releasing O_2. Additionally, the negative effect mainly refers to their emission of greenhouse gas CH_4. In this study, the afforestation cost and carbon tax method were used to evaluate the value of climate regulation. The equation is as follows [42,43]:

$$V_{CO_2} = (A_1 \times d_1 + A_2 \times d_2) \times T \times C \tag{6}$$

$$V_{O_2} = (A_1 \times d_1 + A_2 \times d_2) \times M \times P_0 \tag{7}$$

$$V_{CH_4} = (A_1 \times d_1 + A_2 \times d_2) \times Q \times 21 \times T \tag{8}$$

$$V_{Climate} = V_{CO_2} + V_{O_2} - V_{CH_4} \tag{9}$$

where T is the carbon tax (USD 182.82/t in 2016 and USD 195.58/t in 2020), C is the average annual carbon sequestration in mangrove forests (14.139 t/(ha*a)), M is the average annual oxygen release from mangrove forests (30.31 t/(ha*a)), P_o is the industrial oxygen price (USD 63.27/t in 2016 and USD 91.23/t in 2020), Q is the annual emission flux of mangrove methane per unit area (USD 0.0077/t), and 21 is the warming potential value of methane.

(5) Pollution purification

The pollution purification value service refers to the value generated by the decomposition of and reduction in various invasive harmful substances in mangrove forests. Mangrove forests and understory soil have the ability to absorb and purify various pollutants, purify water quality, and reduce red tides [34]. The pollution prevention cost method was used to evaluate the value of pollution purification. The equation is as follows [41,42]:

$$V_{Purification} = (A_1 \times d_1 + A_2 \times d_2) \times S \tag{10}$$

where $V_{Purification}$ is the value of the pollution purification, and S is the purification value of mangrove forest pollution per unit area (USD 6151.66/ha).

(6) Water conservation

Mangrove forests can accumulate excess precipitation and release it slowly, so that precipitation can be redistributed in time and space. The water conservation of mangrove forests provides water for residents in the form of shallow groundwater, so its value can be calculated by storing the same amount of water in the reservoir. The shadow price method was chosen to calculate the value of surface water resources. The equation is as follows [44,45]:

$$V_{Water} = A \times R \times P_w \tag{11}$$

where V_{Water} is the water conservation value, A is the area of mangrove forests, R is the water storage capacity of mangrove forests per unit area (8100 m^3/ha), and P_w is the cost of unit water storage capacity (USD 0.39/t).

(7) Habitat

Mangrove forests provide ideal living environments for various marine organisms, benthos, and seabirds. They are rich in biological species, playing an important role in ecosystem succession and biological evolution. Therefore, the protection value of biodiversity is crucial and cannot be ignored. The outcome reference method was used in this paper to calculate the value of the habitat. The equation is as follows [45,46]:

$$V_{Habitat} = (A_1 \times d_1 + A_2 \times d_2) \times P_h \tag{12}$$

where $V_{Habitat}$ is the value of the habitat, and P_h is the value of biodiversity per unit area (USD 1791.44/ha).

(8) Nutrient accumulation

Mangrove forests are characterized by their strong ability to cycle and recycle nutrients within the ecosystem. This high productivity is an essential feature of mangrove forests that supports their ecological functions and ESV. The accumulation of nutrients is mainly the accumulation of N, P, and K, so their value can be calculated with the same amount of fertilizer. The value of nutrient accumulation can be expressed as follows [40]:

$$V_{Nutrient} = (A_1 \times d_1 + A_2 \times d_2) \times S_t \times P \quad (13)$$

where $V_{Nutrient}$ is the value of the nutrient accumulation, S_t is the total nutrient retention in mangrove forests (0.291 t/ha), and P is the price of the fertilizer (USD 91.43/t).

(9) Scientific research and education

Mangrove forests have attracted experts and scholars from different fields to conduct research due to their viviparous phenomena, rich species diversity, high biomass, and productivity. However, the necessary research funds and time investments are difficult to obtain, and their values are difficult to quantify. Therefore, the outcome reference method was used in this paper to calculate the scientific research and education value. The equation is as follows [47]:

$$V_{Science} = A \times P_s \quad (14)$$

where $V_{Science}$ is the scientific research and education value, A is the area of the mangroves, and P_s is the scientific and educational value of mangrove forests per unit area (USD 474.90/ha).

(10) Recreation

The rich animal and plant resources of the mangrove forests provide good conditions for the development of tourism activities. Calculating the tourism value of mangrove forests is challenging due to various factors. In Guangxi, most of the scenic spots are located within nature reserves, and access is free to the public. Therefore, we took research results from previous studies as a reference to calculate the value generated by recreation. The calculation equation is as follows [46]:

$$V_{Recreation} = A \times P_r \quad (15)$$

where $V_{Recreation}$ is the recreation value, A is the area of the mangrove forests, and P_r is the recreation value per unit of wetland area in Guangxi (USD 1076.68/ha in 2016 and USD 1118.08/ha in 2020).

3. Results

3.1. Accuracy Assessment of Mangrove Forests Map

Based on the verification points, two confusion matrices were generated to assess the accuracy of the 2016 and 2020 mangrove forest classification results (Table 4). The overall accuracies all exceeded 90%, and the Kappa coefficients all exceeded 0.8. In 2016, the mangrove forests map had a user accuracy and producer accuracy of 94% and 89%, respectively. In 2020, the mangrove forests map had a user accuracy and producer accuracy of 96% and 93%, respectively. The accuracy assessment results indicated that the classification results and the verification data have good consistency.

Table 4. Confusion matrix of mangrove classification results.

Year	Actual Type	Mangrove	Non-Mangrove	Total	User's Accuracy	Producer's Accuracy	Overall Accuracy	Kappa Coefficient
2016	mangrove	210	14	224	93.75%	89.36%	93.05%	0.86
	non-mangrove	25	312	337	92.58%	95.71%		
2020	mangrove	215	9	224	95.98%	92.67%	95.37%	0.90
	non-mangrove	17	320	337	94.96%	97.26%		

3.2. Spatial Distribution and Pattern of Guangxi's Mangrove Forests

We obtained the spatial distribution and pattern of mangrove forests from 2016 to 2020 based on Sentinel-2 imagery. The spatial distribution of mangrove forests in Guangxi is shown in Figure 2. The mangrove forests are mainly concentrated in Zhenzhu Harbour, Fangcheng Bay, Maowei Sea, the Dafeng River, Lianzhou Bay, Tieshan Harbour, and Dandou Bay. Additionally, the area of mangrove forests has increased by 8% from 6245.15 ha in 2016 to 6750.01 ha in 2020. This increase was mainly concentrated in Lianzhou Bay, Tieshan Harbour, and Dandou Bay. In addition, we compared our mangrove forests map with the Guangxi mangrove forests maps created by Zhang et al. [24] and Hu et al. [48]. Our result was close to the result of Hu et al. (7089 ha) and much lower than the area of Zhang et al. (7528 ha). Hu used 30 m spatial resolution Landsat images, which led to mixed pixels in the mangrove forests and reduced the precision of the analysis. Zhang used one-meter spatial resolution Gaofen-2 imagery, which allowed for the identification of numerous small mangrove forests.

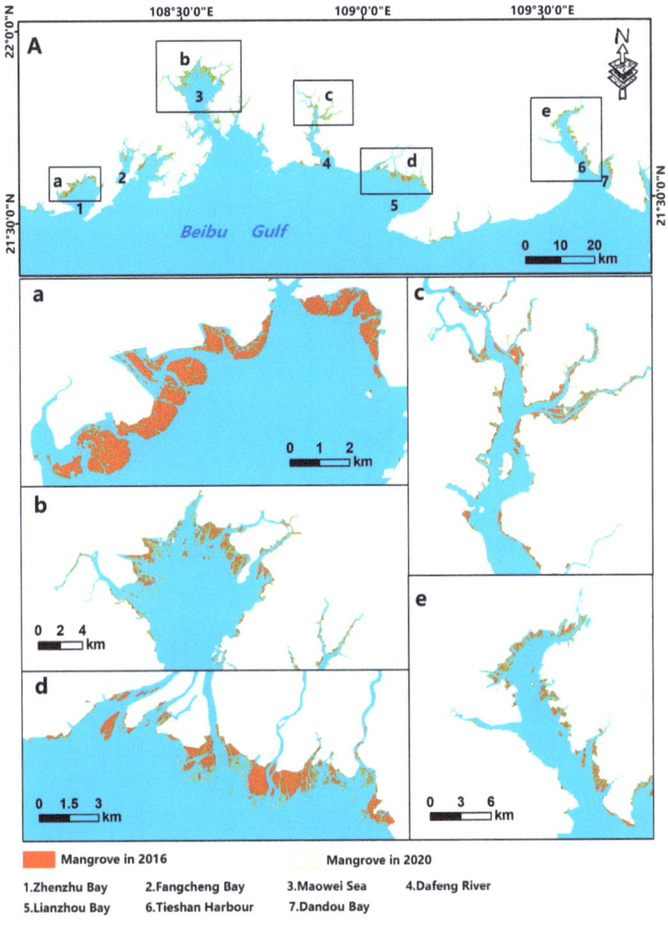

Figure 2. (**A**) Spatial dynamics of mangrove forests along the coasts of Guangxi in 2016 and 2020. (The data from 2020 for mangrove forests were superimposed on the data from 2016 and subfigures (**a**–**e**) show the areas of concentrated or highly dynamic changes).

The spatial patterns of mangrove forests in Guangxi from 2016 to 2020 are presented in Table 5. From 2016 to 2020, based on the changes in coastlines and mangrove forests, the corresponding mangrove shorelines and coastward mangroves increased by 3.19% and 4.69%, respectively. The abundance of mangroves increased by 5.71%, and the number of patches increased by 4.13%. The average area of mangrove patches increased by 3.8%, and ideally distributed mangroves increased by 4.20%. The Guangxi coastline increased slightly from 1686.66 km to 1724.48 km from 2016 to 2020.

Table 5. Changes in spatial pattern during 2016–2020.

Spatial Indices	Year of 2016	Year of 2020	Proportion of Changes
Abundance of mangrove (ha/km)	3.70	3.91	5.71%
Number of patches (pcs)	1018	1060	4.13%
Average patch area (ha)	6.14	6.37	3.80%
Mangrove shoreline (km)	578.90	597.37	3.19%
Coastwards mangrove (ha)	5436.97	5692.19	4.69%
Ideally distributed mangrove (ha)	5114.972	5201.398	4.20%

3.3. Variations in ESV

Table 6 shows the value changes for different ecosystem services. The total service value of mangrove forests changed from USD 363.78 million in 2016 to USD 390.74 million in 2020. The proportion of each service is obtained by dividing its own value by the total service value. This allows us to determine the relative contribution of each ecosystem service to the overall value of mangrove forests. As illustrated in Figure 3, the provisioning service value accounted for more than 33% of the total value, which proportionately constituted a decrease. The value of fishery remained at about 32.1%, but the value of wood decreased. The main reason is that the decline in the market price of logs in Guangxi has exceeded the increase in the areas of mangrove forests. Provisioning services accounted for 33.30% of the total value in 2016, and their proportion in 2020 decreased by 0.45% in comparison.

Table 6. Changes in ESV during 2016–2020 (unit: million USD).

Service	2016		2020	
	Value	Proportion	Value	Proportion
Wood	3.25	0.89%	2.89	0.74%
Fishery	117.89	32.41%	125.47	32.11%
Soil consolidation	0.11	0.03%	0.12	0.03%
Fertilizer conservation	76.71	21.09%	81.63	20.89%
Wave absorbing revetment	54.20	14.90%	55.60	14.23%
Carbon fixation	24.45	6.72%	25.75	6.59%
Oxygen release	11.33	3.11%	17.38	4.45%
Methane release	−1.24	−0.34%	−1.32	−0.34%
Pollution purification	36.33	9.99%	38.66	9.89%
Water conservation	19.80	5.44%	21.40	5.48%
Habitat	10.58	2.91%	11.70	2.99%
Nutrient accumulation	0.67	0.18%	0.72	0.18%
Scientific research	2.97	0.82%	3.21	0.82%
Recreation	6.72	1.85%	7.55	1.93%
Total	363.78		390.74	

The proportion of regulating services remained at 60%, only slightly increasing from 2016 to 2020. Among the regulating services, fertilizer conservation and wave-absorbing revetment remained the main service functions, which indicates that mangrove forests have unique ecosystem services. The reason for the increase in oxygen release is that the average price of the Chinese oxygen market in 2020 (USD 91.23) increased significantly, compared with 2016 (USD 63.27), reaching 44.2%. In 2016, the value of regulating services accounted for 60.94% of the total services value; in 2020, it increased to 61.22%.

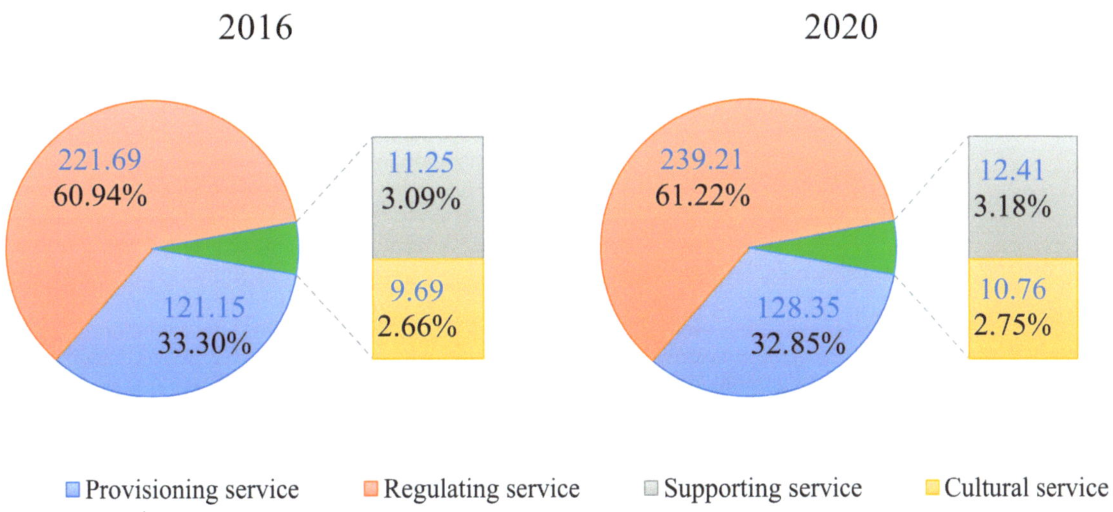

Figure 3. Spatial dynamics of ecosystem ESV in Guangxi, 2016–2020 (in blue color, USD million).

The cultural service and supporting service values accounted for only 2.7% and 3.1% of the total value, respectively. The reason for the increase in habitat and recreation is that the value of ecosystem services per unit area in China in 2020 (USD 236.38) increased, compared with 2016 (USD 227.63), by up to 3.9%. Due to the significant increase in the area of mangrove forests, the net change in the ESV was found to be positive. However, the annual ESV changed slightly, decreasing from USD 58,250 to 57,886.

4. Discussion

4.1. Factors Driving Changes in Spatial Pattern

To improve the protection and management of mangroves, optimize their spatial layout, and realize their ecological and environmental value, in-depth research on their spatial pattern on a landscape scale is essential [49]. In addition to the three basic landscape pattern indices—mangrove area, patch number, and patch area—this study also analyzes shoreline mangroves, ideally distributed mangroves, and mangrove abundance.

Table 5 illustrates that, over the five-year period from 2016 to 2020, the mangrove area was widely used as the most basic spatial indicator in spatial structure analysis. Due to the joint efforts of local governments and the Chinese government, a series of laws and regulations have been formulated and implemented. The mangroves have shown a steady increasing trend, indicating a positive condition between development and recovery. Additionally, the average patch area of mangroves has increased, which further supports our observations of positive growth and recovery in mangrove forests.

The change in the abundance of mangroves can be attributed to two main reasons. Firstly, the continued increase in mangrove area, and, secondly, the construction of various infrastructures, such as reclamation projects, salt pans and breeding ponds, seawalls, urban development, and port and terminal construction, has extended the length of the coastline. The increase in abundance provides a more intuitive indicator of mangrove growth in comparison to measuring their number by area, which can lead to vague and incomparable results at the scale of bays or protected areas. The relative abundance not only facilitates the comparison of mangroves in different regions during the same period but also of mangroves in the same region with significant differences in different periods.

Since mangroves are often distributed along the coastline, the length of the coastline can serve as an indicator of mangrove distribution. The reasons for the increase in mangrove shorelines are multifaceted and can be attributed to several factors. One of the primary reasons is the construction of many breeding ponds on the tidal flats between the original natural shorelines and mangroves. This has brought the mangroves closer to the shoreline. Additionally, other factors, such as the artificial afforestation of new areas and the destruction of existing mangroves, can lead to changes in the extent of mangrove shorelines.

In addition to an increase in mangrove area, the number of shoreline mangroves has also increased. The proportion of shoreline mangroves to total mangrove area has remained at 85%, indicating that the vast majority of mangroves are located close to the shore and have good wave dissipation and shoreline protection characteristics. Furthermore, the ratio of ideally distributed mangrove areas to total mangrove area remains at 80%. This suggests that the mangroves are primarily clustered rather than evenly distributed across the area. The causes of changes in mangrove patch patterns are similar to those of mangrove shorelines. These include a large shift in the spatial location of the coastline, the expansion of mangrove patches, and damage to mangroves caused by natural evolution.

The measurement of coastwards mangroves and ideally distributed mangroves provides a more intuitive depiction of the spatial scale and ecological value of mangroves. For instance, the efficacy of mangroves in wave-absorbing revetment is related to characteristics such as the stand structure, the distance from the embankment, and the patch width [36]. From this perspective, it is easy to understand why certain indicators were selected to indicate a more ideal spatial distribution of mangroves and how the spatial structure impacts the ecological value of mangroves.

4.2. The Rationality and Existing Problems of Selecting Evaluation Index

In our study, we built an evaluation system for mangrove forest ESV by incorporating spatial pattern analysis and the Millennium Ecosystem Assessment framework. Our approach involved categorizing ecosystem services into four main types: provisioning services, regulating services, supporting services, and cultural services. Following the principles of scientific, representative, comprehensive, concise, and operational criteria, we selected 10 indicators for the quantitative evaluation of ecosystem services. Each of these 10 indicators was chosen to fulfill the evaluation objectives while also being appropriate and relevant to the evaluation of mangrove ecosystems [50]. Moreover, each indicator is independent from the others to prevent any double-counting of data caused by information overlap.

Due to significant differences in regional and local contexts, environmental factors, and social dynamics that can affect the provision and valuation of ecosystem services in different locations, local-scale reference values can provide a more accurate and suitable basis for estimating mangrove ESV and informing management and policy decisions. To account for the challenges related to data collection and time constraints in the evaluation process, the result–reference method has been used for some parameters (fisheries, pollution purification, habitat, recreation, scientific research, and education) in this study. This method considers the similarity between the evaluated object and the reference object. The higher the similarity, the better the result. However, according to Lautenbach et al. [51], errors in the valuation of ecosystems can arise due to their diversity and spatial heterogeneity. For instance, the coastal area of Guangxi has a significant number of aquaculture ponds, making it challenging to assess the value of mangroves in terms of their contribution to fisheries in the corresponding area. Despite the inclusion of fishery as an indicator of mangrove ecosystem service value, the direct impact of mangroves on aquaculture cannot be fully measured [52]. Research has shown that the presence of mangroves in coastal areas may increase the survival rate of coastal shrimp farming by 15–35% compared to areas without mangroves [53]. Thus, the calculated results in this regard are most likely lower than the actual value.

The relationship between the size of mangroves and their value per unit area is complex. On the one hand, increasing the area of mangroves may lead to reduced marginal returns, while, on the other hand, most ecosystem services require a threshold area for good functioning, implying that value increases with size [13]. These factors must be considered in more detailed research in the future. Furthermore, there is a general trend that larger mangrove patches can provide a greater ecosystem service supply compared to smaller fragmented patches. Future research using appropriate methods and parameters will be necessary in order to further assess the practical value of mangrove ecosystems.

4.3. Threatened Situations

The protection, management, and restoration of wetlands have become important global issues to be addressed. The issues surrounding wetlands' protection, management, and restoration are still evident [54]. These include disease and insect pest risks from single-community structures, as well as biological hazards such as barnacles. Additionally, invasive alien species like non-native plants can pose threats to wetlands. Furthermore, human factors such as coastal development, excessive pollution, overuse, and seawall construction also contribute to the challenges currently facing wetlands.

Over the time scale of this study, the impact of human socioeconomic activities on the land use types and landscape structures of mangrove wetland ecosystems in Guangxi was evident. In China, during the early 1990s, the ecological and economic values of mangrove ecosystems began to gain widespread recognition and public acknowledgement. As a result, a series of relevant laws and regulations were formulated during this period to protect mangrove resources. In 1982, the "Marine Environmental Protection Law of the People's Republic of China" was adopted [55], which clearly stated that "destroying coastal protection forests, mangroves, and coral reefs is prohibited". Since 2002, the State Forestry Administration has launched a series of mangrove protection and restoration projects. Most recently, in 2020, the Chinese government launched the "Special Action Plan for Red Forest Protection and Restoration (2020–2025)".

The protective measures have played a positive role in the conservation of mangroves in China, and according to Jia et al. [30], the area of mangroves in China has increased from 22,674.22 ha in 2016 to 23,420.34 ha in 2020. With the findings of this study, we have reason to believe that the ESV of mangroves in China is continuously rising. However, on a global scale, the situation is still not optimistic. According to the Global Mangrove Watch, the area of global mangroves decreased from 802,419 ha in 2016 to 775,337 ha in 2020. This once again reminds us of the need to strengthen the protection of mangroves globally to ensure the sustainable growth of the area and the effective protection of the ecosystem.

5. Conclusions

During the period from 2016 to 2020, the mangrove area in Guangxi increased from 6245.15 ha to 6750.01 ha, with a net increase of 504.81 ha, which was mainly concentrated in Lianzhou Bay, Tieshan Harbour, and Dandou Bay. This study aims to explore the spatial distribution and structural changes in mangroves in Guangxi from the perspectives of mangrove abundance, mangrove coastline, ideally distributed mangroves, and other related factors. The results indicate that the average area of mangroves, ideally distributed mangroves, mangrove coastline, and mangrove abundance in Guangxi all increased, suggesting that the mangrove ecosystem in Guangxi is developing well and undergoing a process of recovery. Moreover, the fragmentation degree of the mangrove ecosystem has reduced.

In this study, the ESV of Guangxi mangrove forests were evaluated for the period from 2016 to 2020. The total ESV of mangroves increased from USD 363.78 million to USD 390.74 million. The fishery value, soil conservation value, wave-absorbing revetment, and pollution purification occupy the largest proportion; in addition to the increase in the area of mangrove forests, people's awareness of its ecological value is also an important reason for these changes and trends. The proposed approach and present results of this study could contribute significantly to a better understanding of the relationship between the spatial pattern and distribution of mangroves in Guangxi and their ecological value.

Author Contributions: Conceptualization, X.Z.; data curation, R.Z.; formal analysis, R.Z.; investigation, K.W., C.Z. and R.Z.; methodology, K.W.; project administration, X.Z. and C.Z.; resources, X.Z., Z.W. and M.J.; software, M.J.; supervision, X.Z., Z.W. and M.J.; validation, C.Z. and Z.W.; visualization, C.Z. and M.J.; writing—original draft, K.W.; writing—review and editing, K.W. and X.Z. All authors have read and agreed to the published version of the manuscript.

Funding: This research was funded by the National Natural Science Foundation of China (42171379, 42201422), the China Postdoctoral Science Foundation (No. 2022M713132), and the National Earth System Science Data Center (www.geodata.cn (accessed on 12 June 2023)).

Data Availability Statement: Sentinel-2 satellite data are sourced from European Space Agency (https://scihub.copernicus.eu/dhus (accessed on 13 June 2022)). The data presented in this study are available upon request from the corresponding author.

Acknowledgments: The authors would like to acknowledge the comments of the reviewers and the Editorial Board. They also would like to thank Mingming Jia from the Northeast Institute of Geography and Agroecology for generously providing the field data used in this study.

Conflicts of Interest: The authors declare no conflicts of interest.

References

1. Myint, S.W.; Giri, C.P.; Wang, L.; Zhu, Z.; Gillette, S.C. Identifying Mangrove Species and Their Surrounding Land Use and Land Cover Classes Using an Object-Oriented Approach with a Lacunarity Spatial Measure. *GIScience Remote Sens.* **2008**, *45*, 188–208. [CrossRef]
2. Mukherjee, N.; Sutherland, W.J.; Dicks, L.; Hugé, J.; Koedam, N.; Dahdouh-Guebas, F. Ecosystem Service Valuations of Mangrove Ecosystems to Inform Decision Making and Future Valuation Exercises. *PLoS ONE* **2014**, *9*, e107706. [CrossRef] [PubMed]
3. Zeng, H.; Jia, M.; Zhang, R.; Wang, Z.; Mao, D.; Ren, C.; Zhao, C. Monitoring the Light Pollution Changes of China's Mangrove Forests from 1992–2020 Using Nighttime Light Data. *Front. Mar. Sci.* **2023**, *10*, 1187702. [CrossRef]
4. Cherrington, E.A.; Griffin, R.E.; Anderson, E.R.; Hernandez Sandoval, B.E.; Flores-Anderson, A.I.; Muench, R.E.; Markert, K.N.; Adams, E.C.; Limaye, A.S.; Irwin, D.E. Use of Public Earth Observation Data for Tracking Progress in Sustainable Management of Coastal Forest Ecosystems in Belize, Central America. *Remote Sens. Environ.* **2020**, *245*, 111798. [CrossRef]
5. Lovelock, C.E.; Barbier, E.; Duarte, C.M. Tackling the Mangrove Restoration Challenge. *PLoS Biol.* **2022**, *20*, e3001836. [CrossRef] [PubMed]
6. Pendleton, L.; Mongruel, R.; Beaumont, N.; Hooper, T.; Charles, M. A Triage Approach to Improve the Relevance of Marine Ecosystem Services Assessments. *Mar. Ecol. Prog. Ser.* **2015**, *530*, 183–193. [CrossRef]
7. Sannigrahi, S.; Chakraborti, S.; Joshi, P.K.; Keesstra, S.; Sen, S.; Paul, S.K.; Kreuter, U.; Sutton, P.C.; Jha, S.; Dang, K.B. Ecosystem Service Value Assessment of a Natural Reserve Region for Strengthening Protection and Conservation. *J. Environ. Manag.* **2019**, *244*, 208–227. [CrossRef] [PubMed]
8. Himes-Cornell, A.; Grose, S.O.; Pendleton, L. Mangrove Ecosystem Service Values and Methodological Approaches to Valuation: Where Do We Stand? *Front. Mar. Sci.* **2018**, *5*, 376. [CrossRef]
9. Cao, L.; Li, J.; Ye, M.; Pu, R.; Liu, Y.; Guo, Q.; Feng, B.; Song, X. Changes of Ecosystem Service Value in a Coastal Zone of Zhejiang Province, China, during Rapid Urbanization. *Int. J. Environ. Res. Public Health* **2018**, *15*, 1301. [CrossRef]
10. Ligate, E.J.; Chen, C.; Wu, C. Evaluation of Tropical Coastal Land Cover and Land Use Changes and Their Impacts on Ecosystem Service Values. *Ecosyst. Health Sustain.* **2018**, *4*, 188–204. [CrossRef]
11. Yang, Q.; Liu, G.; Hao, Y.; Zhang, L.; Giannetti, B.F.; Wang, J.; Casazza, M. Donor-Side Evaluation of Coastal and Marine Ecosystem Services. *Water Res.* **2019**, *166*, 115028. [CrossRef]
12. De Groot, R.; Brander, L.; Van Der Ploeg, S.; Costanza, R.; Bernard, F.; Braat, L.; Christie, M.; Crossman, N.; Ghermandi, A.; Hein, L.; et al. Global Estimates of the Value of Ecosystems and Their Services in Monetary Units. *Ecosyst. Serv.* **2012**, *1*, 50–61. [CrossRef]
13. Brander, L.M.; Wagtendonk, A.J.; Hussain, S.S.; McVittie, A.; Verburg, P.H.; De Groot, R.S.; Van Der Ploeg, S. Ecosystem Service Values for Mangroves in Southeast Asia: A Meta-Analysis and Value Transfer Application. *Ecosyst. Serv.* **2012**, *1*, 62–69. [CrossRef]
14. Gaodi, X.; Lin, Z.; Chunxia, L.; Yu, X.; Wenhua, L. Applying Value Transfer Method for Eco-Service Valuation in China. *J. Resour. Ecol.* **2010**, *1*, 51–59. [CrossRef]
15. Camacho-Valdez, V.; Ruiz-Luna, A.; Ghermandi, A.; Berlanga-Robles, C.A.; Nunes, P.A.L.D. Effects of Land Use Changes on the Ecosystem Service Values of Coastal Wetlands. *Environ. Manag.* **2014**, *54*, 852–864. [CrossRef] [PubMed]
16. Yushanjiang, A.; Zhang, F.; Kung, H.; Li, Z. Spatial–Temporal Variation of Ecosystem Service Values in Ebinur Lake Wetland National Natural Reserve from 1972 to 2016, Xinjiang, Arid Region of China. *Environ. Earth Sci.* **2018**, *77*, 586. [CrossRef]

17. Zhang, H.; Chen, C.; Zhang, H.; Zhang, L.; Jia, G. Evaluation of Value of Wetland Ecosystem Services of Zhangjiang Estuary Mangrove National Nature Reserve. *Wetl. Sci.* **2013**, *11*, 108–113. [CrossRef]
18. Harborne, A.R.; Mumby, P.J.; Micheli, F.; Perry, C.T.; Dahlgren, C.P.; Holmes, K.E.; Brumbaugh, D.R. The Functional Value of Caribbean Coral Reef, Seagrass and Mangrove Habitats to Ecosystem Processes. In *Advances in Marine Biology*; Elsevier: Amsterdam, The Netherlands, 2006; Volume 50, pp. 57–189. ISBN 978-0-12-026151-2.
19. Li, Y.; Wen, H.; Wang, F. Analysis of the Evolution of Mangrove Landscape Patterns and Their Drivers in Hainan Island from 2000 to 2020. *Sustainability* **2022**, *15*, 759. [CrossRef]
20. Pham, T.; Yokoya, N.; Bui, D.; Yoshino, K.; Friess, D. Remote Sensing Approaches for Monitoring Mangrove Species, Structure, and Biomass: Opportunities and Challenges. *Remote Sens.* **2019**, *11*, 230. [CrossRef]
21. Tian, J.; Wang, L.; Li, X.; Gong, H.; Shi, C.; Zhong, R.; Liu, X. Comparison of UAV and WorldView-2 Imagery for Mapping Leaf Area Index of Mangrove Forest. *Int. J. Appl. Earth Obs. Geoinf.* **2017**, *61*, 22–31. [CrossRef]
22. Jia, M.; Wang, Z.; Zhang, Y.; Mao, D.; Wang, C. Monitoring Loss and Recovery of Mangrove Forests during 42 Years: The Achievements of Mangrove Conservation in China. *Int. J. Appl. Earth Obs. Geoinf.* **2018**, *73*, 535–545. [CrossRef]
23. Matejicek, L.; Kopackova, V. Changes in Croplands as a Result of Large Scale Mining and the Associated Impact on Food Security Studied Using Time-Series Landsat Images. *Remote Sens.* **2010**, *2*, 1463–1480. [CrossRef]
24. Zhang, R.; Jia, M.; Wang, Z.; Zhou, Y.; Wen, X.; Tan, Y.; Cheng, L. A Comparison of Gaofen-2 and Sentinel-2 Imagery for Mapping Mangrove Forests Using Object-Oriented Analysis and Random Forest. *IEEE J. Sel. Top. Appl. Earth Obs. Remote Sens.* **2021**, *14*, 4185–4193. [CrossRef]
25. Zhao, C.; Jia, M.; Wang, Z.; Mao, D.; Wang, Y. Toward a Better Understanding of Coastal Salt Marsh Mapping: A Case from China Using Dual-Temporal Images. *Remote Sens. Environ.* **2023**, *295*, 113664. [CrossRef]
26. Ghorbanian, A.; Zaghian, S.; Asiyabi, R.M.; Amani, M.; Mohammadzadeh, A.; Jamali, S. Mangrove Ecosystem Mapping Using Sentinel-1 and Sentinel-2 Satellite Images and Random Forest Algorithm in Google Earth Engine. *Remote Sens.* **2021**, *13*, 2565. [CrossRef]
27. Zhang, L.; Xu, S. Discussion about the Eco-function of Mangrove Wetlands of Beibu gulf of Guangxi Zhuang Nationality Autonomous Region. *Anhui Agri. Sci. Bull.* **2010**, *16*, 134–136. [CrossRef]
28. Li, C. Quantitative distribution of mangroves in Guangxi Zhuang Autonmous Region. *J. Beijing For. Univ.* **2004**, *26*, 47–52. [CrossRef]
29. Jia, M.; Wang, Z.; Mao, D.; Ren, C.; Wang, C.; Wang, Y. Rapid, Robust, and Automated Mapping of Tidal Flats in China Using Time Series Sentinel-2 Images and Google Earth Engine. *Remote Sens. Environ.* **2021**, *255*, 112285. [CrossRef]
30. Jia, M.; Wang, Z.; Mao, D.; Ren, C.; Song, K.; Zhao, C.; Wang, C.; Xiao, X.; Wang, Y. Mapping Global Distribution of Mangrove Forests at 10-m Resolution. *Sci. Bull.* **2023**, *68*, 1306–1316. [CrossRef]
31. Powers, R.P.; Hay, G.J.; Chen, G. How Wetland Type and Area Differ through Scale: A GEOBIA Case Study in Alberta's Boreal Plains. *Remote Sens. Environ.* **2012**, *117*, 135–145. [CrossRef]
32. Guo, X.; Zhang, C.; Luo, W.; Yang, J.; Yang, M. Urban Impervious Surface Extraction Based on Multi-Features and Random Forest. *IEEE Access* **2020**, *8*, 226609–226623. [CrossRef]
33. Liu, M.; Mao, D.; Wang, Z.; Li, L.; Man, W.; Jia, M.; Ren, C.; Zhang, Y. Rapid Invasion of Spartina Alterniflora in the Coastal Zone of Mainland China: New Observations from Landsat OLI Images. *Remote Sens.* **2018**, *10*, 1933. [CrossRef]
34. Belgiu, M.; Drăguţ, L. Random Forest in Remote Sensing: A Review of Applications and Future Directions. *ISPRS J. Photogramm. Remote Sens.* **2016**, *114*, 24–31. [CrossRef]
35. Vuolo, F.; Neuwirth, M.; Immitzer, M.; Atzberger, C.; Ng, W.-T. How Much Does Multi-Temporal Sentinel-2 Data Improve Crop Type Classification? *Int. J. Appl. Earth Obs. Geoinf.* **2018**, *72*, 122–130. [CrossRef]
36. Li, C.; Xia, Y.; Dai, H. Temporal Analysis on Spatial Structure of Mangrove Distribution in Guangxi, China from 1960 to 2010. *Wetl. Sci.* **2015**, *13*, 265–275. [CrossRef]
37. Kang, N.; Hou, L.; Huang, J.; Liu, H. Ecosystem Services Valuation in China: A Meta-Analysis. *Sci. Total Environ.* **2022**, *809*, 151122. [CrossRef] [PubMed]
38. Zhang, Q. *Research on the Biologic Seashore of South China Sea*; Guangdong Economic Press: Guangzhou, China, 2008.
39. Yanagisawa, H.; Koshimura, S.; Goto, K.; Miyagi, T.; Imamura, F.; Ruangrassamee, A.; Tanavud, C. The Reduction Effects of Mangrove Forest on a Tsunami Based on Field Surveys at Pakarang Cape, Thailand and Numerical Analysis. *Estuar. Coast. Shelf Sci.* **2009**, *81*, 27–37. [CrossRef]
40. Han, W.; Gao, X.; Lu, C.; Lin, P. The Ecological Values of Mangrove Ecosystems in China. *Ecol. Sci.* **2000**, *19*, 40–45.
41. *China Forestry and Grassland Statistical Yearbook*; China Forestry Publishing House: Beijing, China, 2020.
42. Fan, H.; Zhang, Y.; Zou, L.; Pan, L. A study on the baseline value of the Chinese mangrove services and allocation of the value to individual tree. *Acta Ecol. Sin.* **2022**, *42*, 1262–1275. [CrossRef]
43. Chen, S.; Wen, Z. Estimating Forest Ecosystem Service Function of Carbon Sequestration and Oxygen Release in Guangxi Province. *J. Agro-For. Econ. Manag.* **2016**, *15*, 557–563. [CrossRef]
44. Lu, X. *Wetland Protection and Management*; Chemical Industry Press: Beijing, China, 2004.
45. *Guangxi Statistical Yearbook*; China Statistics Press: Beijing, China, 2020.
46. Xie, G.; Zhang, C.; Lei, M.; Chen, W.; Li, S. Improvement of the Evaluation Method for Ecosystem Service Value Based on Per Unit Area. *J. Nat. Resour.* **2015**, *30*, 1243–1252. [CrossRef]

47. Chen, Z.; Zhang, X. Value of Ecosystem Services in China. *Chin. Sci. Bull.* **2000**, *45*, 870–876. [CrossRef]
48. Hu, L.; Li, W.; Xu, B. Monitoring Mangrove Forest Change in China from 1990 to 2015 Using Landsat-Derived Spectral-Temporal Variability Metrics. *Int. J. Appl. Earth Obs. Geoinf.* **2018**, *73*, 88–98. [CrossRef]
49. Ellison, A.M.; Felson, A.J.; Friess, D.A. Mangrove Rehabilitation and Restoration as Experimental Adaptive Management. *Front. Mar. Sci.* **2020**, *7*, 327. [CrossRef]
50. Costanza, R.; d'Arge, R.; de Groot, R.; Stephen, F.; Monica, G. The Value of the World Ecosystem Services and Natural Capital. *Nature* **1997**, *387*, 253–260. [CrossRef]
51. Lautenbach, S.; Kugel, C.; Lausch, A.; Seppelt, R. Analysis of Historic Changes in Regional Ecosystem Service Provisioning Using Land Use Data. *Ecol. Indic.* **2011**, *11*, 676–687. [CrossRef]
52. Kelleway, J.J.; Cavanaugh, K.; Rogers, K.; Feller, I.C.; Ens, E.; Doughty, C.; Saintilan, N. Review of the Ecosystem Service Implications of Mangrove Encroachment into Salt Marshes. *Glob. Chang. Biol.* **2017**, *23*, 3967–3983. [CrossRef] [PubMed]
53. Anneboina, L.R.; Kavi Kumar, K.S. Economic Analysis of Mangrove and Marine Fishery Linkages in India. *Ecosyst. Serv.* **2017**, *24*, 114–123. [CrossRef]
54. Wang, X.; Xiao, X.; Xu, X.; Zou, Z.; Chen, B.; Qin, Y.; Zhang, X.; Dong, J.; Liu, D.; Pan, L.; et al. Rebound in China's Coastal Wetlands Following Conservation and Restoration. *Nat. Sustain.* **2021**, *4*, 1076–1083. [CrossRef]
55. Jia, M.; Wang, Z.; Mao, D.; Huang, C.; Lu, C. Spatial-temporal changes of China's mangrove forests over the past 50 years: An analysis towards the Sustainable Development Goals (SDGs). *Chin. Sci. Bull.* **2021**, *66*, 3886–3901. [CrossRef]

remote sensing

Article

DAENet: Deformable Attention Edge Network for Automatic Coastline Extraction from Satellite Imagery

Buyun Kang [1], Jian Wu [2], Jinyong Xu [3,*] and Changshang Wu [4]

[1] School of Geomatics and Urban Spatial Informatic, Beijing University of Civil Engineering and Architecture, Beijing 100044, China; 2108570022069@stu.bucea.edu.cn
[2] Department of Cloud Computing Technology and Applications, School of Industrial Internet, Beijing Information Technology College, Beijing 100018, China
[3] Aerospace Information Research Institute, Chinese Academy of Sciences, Beijing 10094, China
[4] Department of Geography, University of Wisconsin-Milwaukee, Milwaukee, WI 53211, USA
* Correspondence: xujy@aircas.ac.cn

Abstract: Sea–land segmentation (SLS) is a crucial step in coastline extraction. In CNN-based approaches for coastline feature extraction, downsampling is commonly used to reduce computational demands. However, this method may unintentionally discard small-scale features, hindering the capture of essential global contextual information and clear edge information necessary for SLS. To solve this problem, we propose a novel U-Net structure called Deformable Attention Edge Network (DAENet), which integrates edge enhancement algorithms and a deformable self-attention mechanism. First of all, we designed a multi-scale transformation (MST) to enhance edge feature extraction and model convergence through multi-scale transformation and edge detection, enabling the network to capture spatial–spectral changes more effectively. This is crucial because the deformability of the Deformable Attention Transformer (DAT) modules increases training costs for model convergence. Moreover, we introduced DAT, which leverages its powerful global modeling capabilities and deformability to enhance the model's recognition of irregular coastlines. Finally, we integrated the Local Adaptive Multi-Head Attention-based Edge Detection (LAMBA) module to enhance the spatial differentiation of edge features. We designed each module to address the complexity of SLS. Experiments on benchmark datasets demonstrate the superiority of the proposed DAENet over state-of-the-art methods. Additionally, we conducted ablation experiments to evaluate the effectiveness of each module.

Keywords: coastline; deep learning; adaptive edge detection; global modeling; deformable features

Citation: Kang, B.; Wu, J.; Xu, J.; Wu, C. DAENet: Deformable Attention Edge Network for Automatic Coastline Extraction from Satellite Imagery. *Remote Sens.* **2024**, *16*, 2076. https://doi.org/10.3390/rs16122076

Academic Editors: Xiao-Hai Yan, Hua Su and Wenfang Lu

Received: 6 April 2024
Revised: 22 May 2024
Accepted: 30 May 2024
Published: 7 June 2024

1. Introduction

The coastline holds significant importance in topographic maps and maritime charts and has been officially recognized as one of the 27 terrestrial features by the International Geographic Data Committee [1–3]. Since the 20th century, economic hubs in coastal nations globally have progressively migrated toward coastal regions [4,5]. The natural attributes of coastlines have rapidly diminished due to the proliferation of coastal development projects, substantial population growth, rapid economic expansion, and escalating geographical significance. The original productive capacities and ecological functions of coastlines have undergone substantial changes [6,7], leading to severe challenges for the natural ecological environment in coastal areas [8,9]. Coastal regions have become some of the areas with the most frequent and intense human activities [10–12]. In light of this, the objective analysis of the temporal and spatial evolution characteristics of coastlines, as well as the quantitative assessment of their impacts and dynamic responses to human interventions, has emerged as a core concern in the academic community.

Coastal-related information is crucial across multiple applications, including coastal management [13], ship detection [14], and water resource management [15,16]. With the introduction of satellites equipped with visible light and infrared sensors [17], remote sensing imagery has become a viable alternative for coastline extraction, replacing laborious techniques such as photogrammetry and GPS field surveys [1,18,19]. Nevertheless, the extraction process faces obstacles primarily due to challenges in effective SLS. These challenges encompass irregular band intensity, complex land textures, and a restricted contrast between sea and land [20]. From an image classification perspective, these methods mainly rely on object-oriented and pixel-oriented classification [21]. Pixel-based segmentation methods [3,22–24] exacerbate these issues by dealing with mixed pixels, leading to noisy classification results and undermining the accuracy of coastline extraction. Moreover, the complexity of rules within the object-oriented methodology poses additional challenges. Currently, developing models tailored explicitly for intricate coastal structures remains a significant challenge [25–27]. Consequently, there is an urgent need to explore innovative approaches for intelligent analysis and extraction of coastlines in complex environments.

Given that deep learning aims to assign each pixel in the input image to fixed categories in semantic segmentation tasks, its objectives align closely with those of sea–land segmentation. In recent years, researchers have endeavored to integrate deep learning techniques into coastline extraction. Liu et al. [28] utilized convolutional neural networks (CNNs) for SLS, leading to enhanced image segmentation accuracy. Following this advancement, several effective SLS networks based on CNNs have emerged. These include a deep convolutional neural network (DeepUNet) [29], squeeze and excitation rank faster R-CNN [30], fully convolutional DenseNet (FC-DenseNet) [31], a multi-scale sea–land segmentation network (MSRNet) [32], a more comprehensive range of batch sizes network (WRBSNet) [33], and a deep learning model based on the U_2-Net deep learning model [34]. For example, WRBSNet [33] integrates a broader range of batch sizes to enhance performance. MSRNet [32] integrates squeeze and attention modules to bolster features across different scales, thereby reinforcing weak sea–land boundary information. Furthermore, many researchers also take into account the semantic and edge characteristics of sea–land segmentation and have devised networks for coastline extraction, such as dual-branch structures [35], multi-scale transformation [36], and edge-semantic fusion [37]. Recent studies primarily focus on optimizing and innovating based on CNNs. During the feature extraction process, CNN-based models frequently down-sample features to reduce computational requirements, potentially resulting in the loss of small-scale features [38–40]. Land objects from various semantic categories may share similar sizes, materials, and spectral characteristics, thereby posing challenges in their differentiation. Furthermore, occlusions within the network model frequently result in semantic ambiguities. Consequently, there is an urgent requirement for more comprehensive global contextual information and refined spatial features to serve as cues for semantic reasoning [41].

Recently, transformers have demonstrated significant benefits in natural language processing and computer vision, exhibiting outstanding performance across numerous tasks. The first introduction of transformers in this domain was by Yang et al. [42]. They first introduced transformers into this domain by experimenting with the pure Transformer architecture SETR [43], achieving performance comparable to existing CNN methods in land-sea segmentation. Another hybrid approach, SegFormer [44], which combines CNNs and transformers, surpassed state-of-the-art CNN methods and exhibited strong robustness. This illustrates the capability of Transformer architectures in the field of SLS. Subsequently, Yang et al. [45] employed a Transformer model to forecast alterations in the coastline of Weitou Bay, China. Zhu et al. [46] perform parallel feature extraction using both the Swin Transformer and ResNet branches concurrently. However, research in the field of SLS has been relatively limited. With the continuous evolution of Transformer structures, numerous attention-based variant structures [47–54] have emerged. Among these approaches, Swin Transformer [50] utilizes window-based local attention to confine attention within local windows, whereas Pyra-mid Vision Transformer (PVT) [51] decreases the resolution of key and value feature maps to reduce computation. Although effective, hand-crafted attention patterns are agnostic to data and may not be optimal. Relevant keys/values may be dropped, while less important ones are retained. Ideally, the candidate key/value set for a given query should be flexible and capable of adapting to each individual input, thereby alleviating issues with hand-crafted sparse attention patterns. In fact, in CNN literature, learning a deformable receptive field for convolution filters has been demonstrated to effectively focus on more informative regions based on data [52]. The coastline frequently displays irregular shapes. Additionally, due to the typically indistinct boundary between the ocean and land, its demarcations often appear blurred in remote sensing images. Consequently, coastline extraction requires both more global contextual information and clear edge information. Currently, there is a scarcity of results from such studies. To address this issue, our research investigates integrating a self-attention mechanism into sea–land segmentation to emphasize more global contextual information. Meanwhile, considering the linear characteristics of coastlines, our study aims to utilize edge enhancement algorithms and integrate the Deformable Attention Transformer mechanism [55], leading to the development of a Deformable Attention Edge Network (DAENet). The core structure of this network comprises an encoder, decoder, bottleneck region, and skip connections. Experimental results demonstrate that our network structure can accurately extract coastline features, closely matching the actual coastline. The main contributions of this study can be summarized as follows:

(1) Introduction of the multi-scale edge detection module: Upon image input, we introduced the multi-scale transformation (MST) module, applying canny edge detection across multiple spatial scales and stacking them as input channels. This not only enhances the model's robustness but also facilitates faster convergence.
(2) Construction of an adaptive edge detection module: During training, we developed a Local Adaptive Multi-Head Attention-based Edge Detection (LAMBA) module to enhance the disparities in edge features in the spatial dimension, thus reducing semantic ambiguities that may arise from similar features among objects across different semantic categories.
(3) Exploration of the Deformable Attention (DAT) Application: In order to enhance DAENet's receptive field, we incorporated deformability into the U-shaped structure. This integration serves to alleviate constraints imposed by the fixed convolutional kernel in CNNs and the conventional patch generation in Transformers. Additionally, edge maps are utilized to compute an edge-aware loss, optimizing a novel edge loss function and accelerating model convergence.

2. Methods

2.1. Overall Framework

To tackle the complexities of SLS environments, our proposed DAENet incorporates a series of specialized modules, each designed to address specific features of SLS. The MST in DAENet is strategically designed for use in SLS. MST enhances edge feature extraction and model convergence through multi-scale transformation and edge detection, enabling the network to more effectively capture spatial–spectral changes and ensuring accurate depiction of SLS with varying spatial–spectral characteristics. Specifically, the core principle of multi-scale edge detection involves step-by-step downsampling of a two-dimensional image and applying edge detection to capture edge information at various scales. This is crucial because the deformability of DAT modules leads to significant training costs for model convergence. Our MST module filters the information for validity before it enters the DAT module. After the filtered information enters DAT, the relationships between tags are modeled, sampled, and projected, guided by important regions in feature mapping, obtaining the keys and values after deformation. The exchange of information between edge details at different scales, captured by MST, and the global context, captured by the converter, is enhanced using standard multi-head attention that focuses on sampling keys and aggregating features. This is achieved by calculating attention weights that reflect the importance of each feature in the context of the entire image, enabling the model to make informed predictions based on both local and global features.

Finally, in the output generation phase, we integrated the LAMBA module to improve the spatial differentiation of edge features, consequently decreasing semantic ambiguities resulting from similarities among features from disparate semantic categories. Specifically, we use gradient information and a multi-head self-attention mechanism to generate a projection function. This function calculates the relationship between the current pixel and its surrounding pixels. Weight parameters are introduced to compute comprehensive features, ensuring consideration of local characteristics and rotation invariance. Furthermore, our study utilizes skip connections to combine multi-scale features from the encoder with upsampled features, concatenating shallow and deep features to reduce spatial information loss caused by downsampling. During network training, the edge-aware loss is calculated using the binary Dice Loss function, thereby improving the model's performance. These design choices are based on our understanding of the spectral, spatial, and edge characteristics of SLS. Together, they enable DAENet to excel in the challenging task of semantic segmentation in SLS.

As shown in Figure 1, the proposed deep learning network model, D DAENet, a hybrid of DAT and UNet, inherits the robust structure of UNet. DAENet employs skip connections between encoders and decoders, forming an encoder–decoder structure with MST, LAMBA, and DAT.

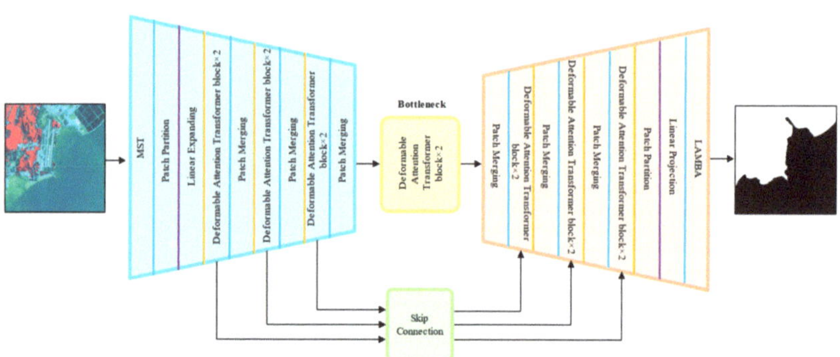

Figure 1. DAENet network structure.

2.2. Multi-Scale Transformation Module

To address the complexity and variability of SLS, capturing both edge and semantic information is crucial for generating informative features. To achieve this goal, we designed the multi-scale transformation (MST). The fundamental principle of multi-scale edge detection involves analyzing the image at different scales to capture edge information of various granularities. Specifically, this approach utilizes the concept of scale-space and is implemented through a Gaussian pyramid. We use a linear Gaussian kernel to construct the image scale to add another dimension to the 2D image without introducing noise. The input image is convolved with a Gaussian filter with an increasing standard deviation. After generating a series of smoothed images, edge detection is performed on each image using the Canny algorithm to capture edge information at different scales. The detailed structure of this module is illustrated in Figure 2.

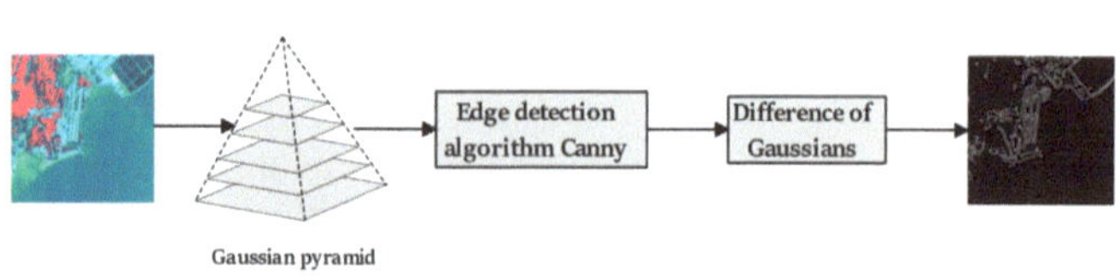

Figure 2. Multi-scale transformation module.

In the first instance, with regard to a specific image $I(x,y)$, its spatial representation within the Gaussian scale-space is:

$$L(x,y,\sigma) = G(x,y,\sigma) * I(x,y) \quad (1)$$

Here, L denotes the scale-space representation, G is the Gaussian function with a variance of σ^2, and $*$ signifies the convolution operation.

When transitioning across scales, the Difference of Gaussians (DoG) can be expressed as:

$$D(x,y,\sigma) = L(x,y,k\sigma) - L(x,y,\sigma) \quad (2)$$

In this equation, k is a constant multiplier, typically greater than 1.

This module builds upon the principles of differential Gaussians, which encompass a range of scales. It utilizes the summation of cross-scale disparities to integrate information across these scales. The governing expression is elaborated as follows:

$$A(x,y) = \sum_{\sigma} D(x,y,\sigma) \quad (3)$$

In this equation, $A(x,y)$ denotes the cumulative response.

2.3. Deformable Attention Transformer

Xia et al. [55] used the Vision Transformer with a Deformable Attention module to create a robust pyramid skeleton network called DAT, suitable for image classification and various dense prediction tasks. Unlike Deformable Convolutional Networks (DCN), which learn different offsets for different pixels across the entire feature map, this module learns a set of offsets independent of the query. These offsets direct the keys and values towards significant areas, as shown in Figure 3a. This design retains linear spatial complexity and introduces a Deformable Attention pattern to the Transformer backbone. Each attention module initially generates reference points as a uniform grid, which remains consistent across the input data. Subsequently, an offset network uses the query features as input to generate corresponding offsets for all the reference points. As a result, the candidate keys and values are shifted towards essential regions, enhancing the flexibility and efficiency of the original self-attention module to capture richer information features. The detailed structure of this module is described as follows:

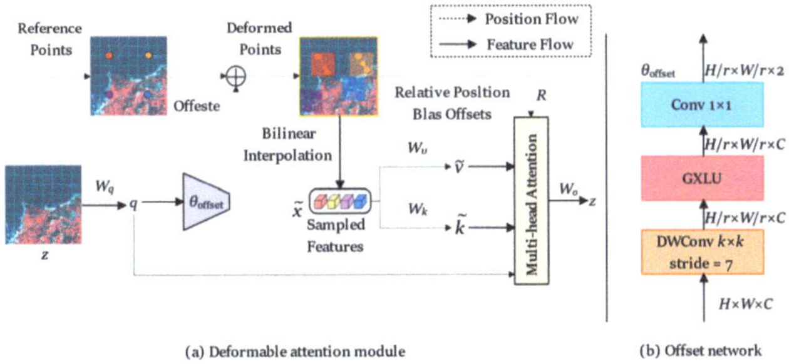

Figure 3. An illustration of our Deformable Attention mechanism. (**a**) presents the information flow of Deformable Attention. In the left part, a group of reference points is placed uniformly on the feature map, whose offsets are learned from the queries by the offset network. Then, the deformed keys and values are projected from the sampled features according to the deformed points, as shown in the right part. Relative position bias is also computed by the deformed points, enhancing the multi-head attention, which outputs the transformed features. We show only four reference points for a clear presentation; there are many more points in real implementation de facto. (**b**) Reveals the detailed structure of the offset generation network, marked with sizes of feature maps.

Firstly, as illustrated in Figure 3a, given an input feature map $x \in \mathbb{R}^{H\times W\times C}$, a uniform grid composed of points $p \in \mathbb{R}^{H_G\times W_G\times 2}$ is generated as a reference. Specifically, the grid size is downsampled by a factor of r from the input feature map, so that the grid dimensions are r, $H_G = H/r, W_G = W/r$. The reference points are linearly spaced two-dimensional coordinates ranging from $\{(0,0),\ldots,(H_G-1,W_G-1)\}$. These coordinates are then normalized to lie within the range $[-1,+1]$, based on the grid shape $H_G \times W_G$, where $(-1,-1)$ represents the top–left corner and $(+1,+1)$ signifies the bottom–right corner.

Subsequently, to obtain the offset for each reference point, the feature map is linearly projected to the query tokens as $q = xW_q$. Then, the feature map is fed into a lightweight sub-network $\theta_{offset}(\cdot)$ generate the offsets $\triangle p = \theta_{offset}(q)$. To stabilize the training process, we scale the amplitude of $\triangle p$ by some predefined factor s to prevent the offset from becoming too large, i.e., $\triangle p \leftarrow stanh(\triangle p)$. Next, the features are sampled at the positions corresponding to the deformed points, serving as both keys and values, and subsequently processed using projection matrices:

$$q = xw_q, \tilde{k} = \tilde{x}W_k, \tilde{v} = \tilde{x}W_v \tag{4}$$

$$\text{with } \triangle p = \theta_{\text{offset}}(q), \tilde{x} = \varphi(x; p + \triangle p) \tag{5}$$

The symbols $\tilde{k}$ and $\tilde{v}$ represent the deformed keys and values embeddings, respectively. Specifically, the sampling function $\phi(\cdot;\cdot)$ made differentiable by translating it into a bilinear interpolation.

$$\phi\left(z; \left(p_x, p_y\right)\right) = \sum_{(r_x, r_y)} g(p_x, r_x) g\left(p_y, r_y\right) z\left[r_y, r_x, :\right] \tag{6}$$

The function $g(a,b) = max(0, 1-|a-b|)$, and (r_x, r_y) index all the positions on $z \in R^{H_G\times W_G\times 2}$. Given that g is non-zero only at the four integral points closest to $\left(p_x, p_y\right)$, it simplifies Equation (5) to the weighted average of these four positions. In line with established methodologies, multi-head attention is employed, along with the adoption of relative position offsets. The resulting attention-head output can be formulated as follows:

$$z^{(m)} = \sigma\left(\frac{q^{(m)}\tilde{k}^{(m)^T}}{\sqrt{d}} + \varphi(\hat{B}; R)\tilde{v}^{(m)}\right. \tag{7}$$

Here $\varphi(\hat{B}; R) \epsilon R^{HW\times H_G W_G}$.

In conclusion, the Deformable Multi-Head Attention (DMHA) has a computational cost that is analogous to that of the Pyramid Vision Transformer (PVT) or the Swin Transformer. The sole additional overhead is attributed to the subnetwork designated for offset generation. The computational complexity of the entire module can be encapsulated by the following equation:

$$\Omega(\text{DMHA}) = 2HWN_SC + 2HWC^2 + 2N_SC^2 + \left(k^2 + 2\right)N_SC \tag{8}$$

Here, $N_S = H_G W_G = \frac{HW}{r^2}$ denotes the number of sampling points.

2.4. Local Adaptive Multi-Head Attention-Based Edge Detection Module

The fundamental principle of LAMBA is to adapt to different image features by comprehensively considering gradients, textures, and intensity. Unlike most self-attention mechanisms, in the initial stages of the task, we employ local binary pattern (LBP) features and adaptive edge detection to capture local texture and edge information, guiding the multi-head self-attention mechanism to focus on various directional information, thereby enhancing the model's ability to extract coastlines. In LAMBA, the spatial information of pixels is leveraged to integrate multi-directional feature modules. The following section outlines the detailed steps Figure 4:

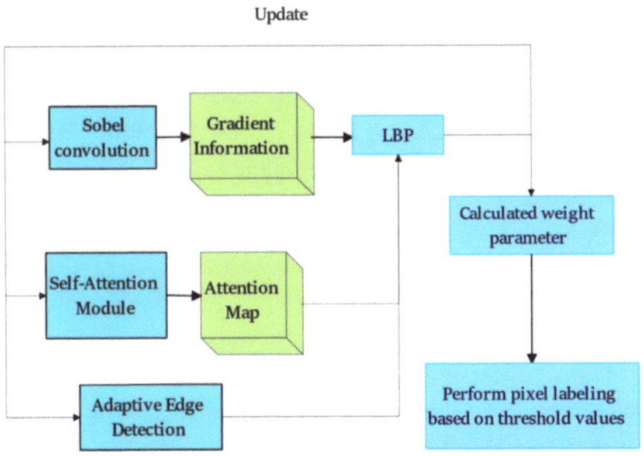

Figure 4. Local Adaptive Multi-Head Attention-based Edge Detection Module.

Initially, the input image undergoes prediction to generate the model's output image. Let the prediction result be denoted as P (x, y), where (x, y) represents pixel coordinates. The Sobel convolution kernel is utilized to compute horizontal and vertical gradients, and the gradient information, along with the multi-head self-attention mechanism, is employed to adjust the projection direction.

$$\begin{aligned} \nabla_x I(x,y) &= \mathrm{Sobel}_x * P(x,y) \\ \nabla_y I(x,y) &= \mathrm{Sobel}_y * P(x,y) \end{aligned} \tag{9}$$

$$A_{ij} = \mathrm{Concat}(\mathrm{Attention1}(\nabla I), \mathrm{Attention2}(\nabla I), \ldots, \mathrm{AttentionK}(\nabla I)) \tag{10}$$

where Sobel_x and Sobel_y represent the Sobel convolution kernels, and A_{ij} denotes the projection adjustment function.

Next, for each pixel, compute its local binary pattern (LBP):

$$\mathrm{LBP}(x,y) = \sum_{p=0}^{P-1} s(I_p - I(x,y)) \cdot 2^P \tag{11}$$

where Ip represents the grayscale value of the neighborhood points with the current pixel as the center, P is the number of neighborhood points, and s(x) is the step function.

Consider dynamically adjusting high and low thresholds based on local image properties and calculating the average intensity value within the local window.

$$W(x,y) = \frac{1}{N} \sum_{i=x-W_x}^{x+W_x} \sum_{j=y-W_y}^{y+W_y} P(i,j) \tag{12}$$

$$T_{high}(x,y) = k_{high} \cdot \mathrm{avg}(W(x,y)) \tag{13}$$

$$T_{low}(x,y) = k_{low} \cdot \mathrm{avg}(W(x,y)) \tag{14}$$

Finally, we introduce weight parameters and combine LBP, adaptive edge detection, and self-attention mechanisms to compute comprehensive features.

$$F(x,y) = \delta \cdot \mathrm{LBP}(x,y) + \gamma \cdot \mathrm{avg}(W(x,y)) + (1 - \gamma - \delta) \cdot \mathrm{DA} \tag{15}$$

$$DA(\nabla I, A_{ij}) = \alpha \cdot \nabla_x I + (1-\alpha) \cdot A_{ij}, \beta \times \nabla_y I + (1-\beta) \times A_{ij} \tag{16}$$

where $LBP(x, y)$ represents the LBP feature for each pixel, $avg(W(x, y))$ is the average intensity value within the local window, $DA(\nabla I, A_{ij})$ is the function obtained after adjusting the projection direction using gradient information and the multi-head self-attention mechanism, where α, δ, γ and β are weight parameters.

Perform pixel labeling based on the threshold values.

$$\begin{cases} \text{StrongEdge} & F(x,y) \geq T_{high}(x,y) \\ \text{PotentialEdge} & T_{low}(x,y) \leq F(x,y) < T_{high}(x,y) \\ \text{NonEdge} & F(x,y) < T_{low}(x,y) \end{cases} \tag{17}$$

2.5. Auxiliary Loss Function

For adaptive thresholds, the threshold is not static but is determined based on the local attributes of the image. Consequently, we have redesigned the auxiliary function. This involves the incorporation of edge information to further modify the Dice Loss. The underlying principle is that the edges or boundaries of objects within an image are particularly crucial in many segmentation tasks. By incorporating these boundaries into the loss function, the network can be directed toward generating enhanced segmentations, particularly along the edges of objects. The formulation is as follows:

$$L = L_D + k \times l_E \tag{18}$$

L_D is computed between the prediction and the ground truth.

l_E represents the Dice Loss calculated between the predicted edge map (obtained through canny edge detection) and the ground truth.

k is a hyperparameter employed to determine the degree of emphasis on the edge Dice Loss within the final loss value.

In this paper, numerical stability is maintained by introducing the 'smooth' parameter. In instances where both the prediction and the ground truth lack discernible features, the Dice Loss may assume an indeterminate form of 0/0. By adding a small "smooth" value to both the numerator and the denominator, such scenarios are circumvented, ensuring the stability of the loss value.

In summary, this enhanced Dice Loss leverages both global segmentation information and local edge details to yield superior segmentation outcomes, especially around object boundaries. Incorporating an edge-weighted term in the loss function is anticipated to guide the model to better handle boundaries and provide more accurate segmentation.

3. Experiment

3.1. Study Area

China, located in the southeastern part of the Asian continent, borders the Northwest Pacific Ocean. With nearly 3 million square kilometers of maritime territory and a coastline stretching over 32,000 km, China ranks sixth in the world. This coastline is bifurcated into mainland coastlines and island coastlines, with the former accounting for 18,000 km [56–58]. From 1980 to 2015, China's coastline expanded by 3000 km. However, its natural coastline decreased by approximately 50% during the same period [57,59,60]. This transformation can be attributed to various factors, including monsoonal waves, tectonic uplift and subsidence, and human activities. Notably, the coastal regions, known for their rich resources and transportation convenience, accommodate only 40% of the country's population but contribute 70% to China's GDP, showcasing one of the most robust economic activities globally.

This study primarily examines the mainland coastline of China, as illustrated in Figure 5. The study area extends from the Yalu River Estuary in the north, situated on the China–North Korea border, to the Beilun Estuary in the south, which marks the China–Vietnam border. It encompasses four seas: the Bohai, Yellow, East, and South China Seas. It spans across 14 primary provinces, municipalities, and autonomous regions, cutting through three climatic zones: temperate, subtropical, and tropical. Based on geological and geomorphic features, coastlines can primarily be categorized into four types: sandy coasts, bedrock coasts, muddy coasts, and artificial coasts. From this, it is evident that China's coastline is elongated and intricate, making extraction quite challenging.

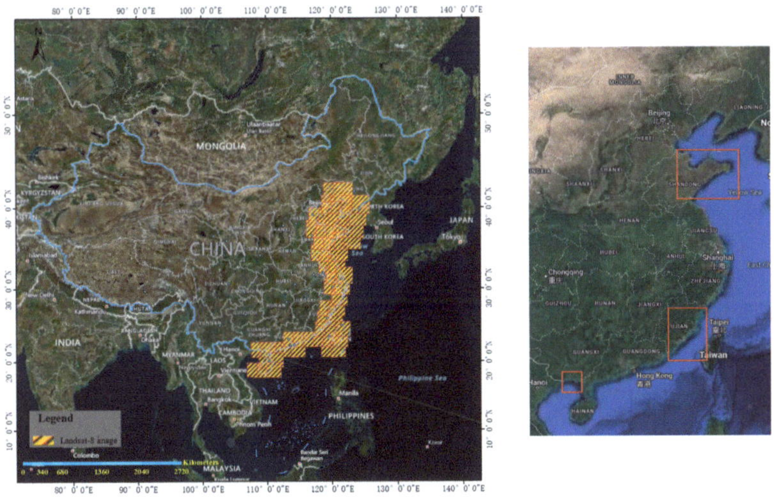

Figure 5. Study area of China and its adjacent maritime regions.

Due to substantial topographical differences between northern and southern China, we chose datasets from Fujian Province, Guangxi Province, and Shandong Province to be the test set for evaluating our model's generalizability comprehensively. The datasets from the other provinces were randomly divided into training and validation sets at a 4:1 ratio. The validation set adjusts model hyperparameters and monitors model performance during training. After each training epoch, the model is evaluated on the validation set to assess its ability to generalize to unseen data. The test set, consisting of data not encountered during training and validation, is used for the model's final performance evaluation. It assesses the model's ability to generalize to real-world data and evaluates the overall performance of the model. The experimental data in this study come from 103 multispectral images of the mainland China coastal zone captured by the "Landsat 8" satellite in 2020 [61,62].

After acquiring the images, various preprocessing steps were undertaken. These steps encompass orthorectification, image fusion, mosaicking, and cropping. A standard false-color composite image composed of Band 5, Band 4, and Band 3 was chosen for a more precise capture of coastline features. It was ensured that each selected image exhibited diverse coastline characteristics. Ground truth maps for land–sea segmentation of these images were created through expert visual interpretation. Following cropping of the preprocessed remote sensing images to create samples measuring 256 × 256 pixels, a total of 65,389 sample images were obtained. However, given the extensive scale of the data and the high costs associated with training, this study additionally employed a threshold method and expert visual recognition to filter out images containing solely a single landform feature. Within each remote sensing image, pixel values in the binary label map are set to 0 and 255. The threshold segmentation algorithm removes label maps with uniform gray values of 0 or 255, along with their original remote sensing images. Moreover, black borders in downloaded remote sensing images may misinterpret label images containing 0 and 255 gray values during screening, requiring manual scrutiny of the original image. Only images that concurrently showcased both land and sea features were retained. After this selection process, the final training set consisted of 4637 samples, the validation set contained 989 samples, and the test set had 1922 samples.

3.2. Coastline Dataset

Extracting coastlines through deep learning necessitates the creation of precise remote-sensing image labels, constituting a pivotal step in the process. Ensuring accuracy in label creation is imperative for effectively enabling deep networks to discern terrain features during the training phase. In the past, when extracting coastlines from satellite remote sensing images, the instantaneous water boundary observed during satellite passes has frequently been regarded as the accurate coastline [63,64]. This differs significantly from the definition of the actual coastline as the sea–land boundary at the moment of the average high tide over several years. In order to objectively and accurately extract coastlines that closely resemble the actual coastline, this study utilizes the five coastal-type (Table 1) remote sensing interpretation markers proposed by Sun Weifu to construct a deep learning dataset [65].

Table 1. Five types of coastline.

Coastline	Sample	Interpretation Signs	Location
Bedrock Coastline	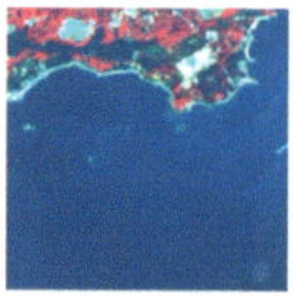	Distinct concave-convex and mountainous texture features	The evident land–water boundary
Silty Coastline		Appears grey or white with a smooth texture	A boundary that is located in an estuary, delta, or low-lying area and has a marked contrast in vegetation density
Sandy Coastline	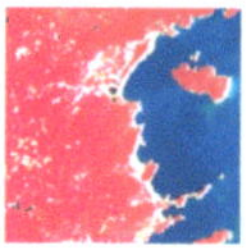	Clear dividing line between white and other colors	Beach ridges and lack of beach ridges, the beach is directly adjacent to the cliffs of the bedrock shoreline
Biogenic Coastline		Red tones, darker textures, and irregular shapes	It is mainly distributed in Guangdong, Guangxi, Fujian, Taiwan, and parts of Hainan Island
Artificial Coastline	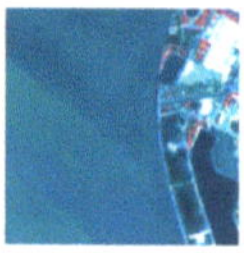	Bright and white structures, smooth textures and narrow stretches, regular layouts, and colors ranging from light beige to tan or even white	A man-made coastline exhibits a multitude of features, which are often intricate and require thorough analysis and consideration

3.3. Implementation Details

The experiments were conducted on a computer equipped with an NVIDIA RTX 3060 GPU with 12 GB of VRAM and running the Ubuntu 18 operating system. All models were trained and tested using the PyTorch framework. During the training process, the Adam optimizer was employed to minimize the loss. The initial learning rate was set to 1×10^{-3}, with the number of iterations set at 100 epochs and a batch size of 8.

3.4. Evaluation Metrics

In this study, we utilize Intersection over Union (IoU), mean Intersection over Union (MioU), Frequency Weighted Intersection-over-Union (FWIoU), and Overall Accuracy (OA) to evaluate the performance of the model. These five evaluation metrics are derived from the confusion matrix, which comprises true positives (TP), false positives (FP), true negatives (TN), and false negatives (FN). Below are the computational formulas for each metric.

For each class, IoU is defined as the ratio of the intersection to the union of the predicted and true values. The formula is as follows:

$$\mathrm{IoU} = \frac{\mathrm{TP}}{\mathrm{TP} + \mathrm{FP} + \mathrm{FN}} \tag{19}$$

The MIoU represents the average result of the IoU for each class. Its calculation is given by:

$$\mathrm{MIoU} = \frac{\sum_{\mathrm{i}}^{\mathrm{n}} \mathrm{IoU}}{\mathrm{n}} \tag{20}$$

FWIoU assigns weights based on the frequency of occurrences for each class. The weight is multiplied by the IoU for each class and then summed up. The formula for this is as follows:

$$\mathrm{FWIoU} = \frac{\mathrm{TP} + \mathrm{FN}}{\mathrm{TP} + \mathrm{FP} + \mathrm{TN} + \mathrm{FN}} \times \frac{\mathrm{TP}}{\mathrm{TP} + \mathrm{FP} + \mathrm{FN}} \tag{21}$$

The $\mathrm{F_1}$-Score ($\mathrm{F_1}$) takes into account both precision and recall, aiming for a balance that maximizes both. The formula for the $\mathrm{F_1}$ or each class is as follows:

$$\mathrm{F_1} = 2 \times \frac{\mathrm{precision} \times \mathrm{recall}}{\mathrm{precision} + \mathrm{recall}} \tag{22}$$

$$\mathrm{Precision} = \frac{\mathrm{TP}}{\mathrm{TP} + \mathrm{FP}} \tag{23}$$

$$\mathrm{Recall} = \frac{\mathrm{TP}}{\mathrm{TP} + \mathrm{FN}} \tag{24}$$

Accuracy (ACC) indicates the proportion of correctly predicted samples among all samples. This includes both true positive and true negative predictions. The calculation for accuracy is as follows:

$$\mathrm{Accuracy} = \frac{\mathrm{TP} + \mathrm{FN}}{\mathrm{TP} + \mathrm{FP} + \mathrm{TN} + \mathrm{FN}} \tag{25}$$

4. Results

4.1. Performance of Daenet

This paper contrasts the performance of DAENet with several existing methods, including Segmenter [66], Mask2Former [67], and Swin-UNet. All three methods are built upon the Transformer architecture. Specifically, Mask2Former is a hybrid structure based on Mask R-CNN and Transformer, while Swin-UNet constitutes a UNet structure formed purely from Swin Transformer modules. The three methods, PIDNet [68], DDRNet [69], and SegNeXt [70], are state-of-the-art models established based on CNN. This study will analyze the accuracy and adaptability of the DAENet model from two perspectives:

(1) Conducting an analysis using evaluation metrics and results to ascertain and validate its accuracy.

The results for the Shandong Province dataset are presented in Table 2, which showcases the numerical outcomes for each semantic segmentation method. The findings indicate that DAENet outperforms other techniques in metrics such as IoU, MIoU, FWIoU, and overall OA.

Table 2. Comparison of segmentation accuracy on the Shandong Province dataset.

Method	IoU(%)		F1 (%)		Evaluation Index		
	Ocean	Land	Ocean	Land	MioU (%)	OA (%)	FWIoU (%)
Segmenter	90.49	94.09	92.28	92.21	85.61	92.25	85.61
SegNeXt	95.67	93.84	94.71	94.77	90.01	94.74	90.01
PIDNet	91.74	92.06	95.69	95.86	91.90	95.78	91.90
DDRNet	92.18	92.48	95.93	96.09	92.33	96.66	92.33
Swin-UNet	88.59	88.33	93.95	93.80	88.46	93.88	88.46
Mask2Former	88.53	88.44	93.91	93.86	88.48	93.89	88.48
DAENet	**93.85**	**94.04**	**96.82**	**96.93**	**93.94**	**96.88**	**93.94**

For the Mask2Former model, which has achieved SOTA results in semantic, instance, and panoptic segmentation as a unified segmentation structure, Tables 2–4 reveal its performance on coastline data extraction is not especially commendable. As a comprehensive, large-scale model designed to incorporate a wide array of features, it unavoidably sacrifices precision when isolating specific ones. Consequently, it loses its SOTA edge when extracting the requisite edge information for coastline detection.

Table 3. Comparison of segmentation accuracy on the Fujian Province dataset.

Method	IoU (%)		F1 (%)		Evaluation Index		
	Ocean	Land	Ocean	Land	MioU (%)	OA (%)	FWIoU (%)
Segmenter	82.03	84.20	89.90	91.16	83.12	90.82	83.16
SegNeXt	83.71	85.64	91.13	92.26	84.68	91.13	84.72
PIDNet	84.99	86.89	91.88	92.98	85.94	92.47	85.97
DDRNet	85.43	87.30	92.14	93.22	86.37	93.53	86.40
Swin-UNet	83.22	85.29	90.84	92.06	84.25	91.49	83.30
Mask2Former	82.08	84.68	90.16	91.70	83.38	91.00	83.44
DAENet	**87.28**	**88.84**	**93.20**	**94.09**	**88.06**	**93.68**	**88.09**

Table 4. Comparison of segmentation accuracy on the Guangxi Province dataset.

Method	IoU (%)		F1 (%)		Evaluation Index		
	Ocean	Land	Ocean	Land	MioU (%)	OA (%)	FWIoU (%)
Segmenter	69.28	77.90	81.85	87.58	73.59	85.25	74.28
SegNeXt	69.70	77.88	82.14	87.56	73.79	85.34	74.45
PIDNet	72.03	81.63	83.74	89.88	77.68	87.53	77.68
DDRNet	72.72	82.02	84.20	90.12	77.37	87.84	78.16
Swin-UNet	74.54	82.88	85.41	90.63	78.71	88.59	79.38
Mask2Former	71.20	81.21	83.18	89.63	76.21	87.17	77.01
DAENet	**76.39**	**84.91**	**86.61**	**91.84**	**80.65**	**89.86**	**81.33**

In reference to the Swin-UNet, predominantly employing the Swin Transformer module for semantic segmentation, the model, despite its superior global modeling capability, necessitates revision when configured according to the Swin-UNet design for remote sensing imagery [58]. This deficiency becomes apparent in Figures 6–8, illustrating fragmented semantic segments in boundary extraction, a common issue stemming from ambiguity in edge information. However, upon integrating the LAMBA and MST modules into our DAENet, Figures 6–8 demonstrate more accurate edge extraction and the absence of fragmented semantic segments, thereby reinforcing the model's robustness in edge detection.

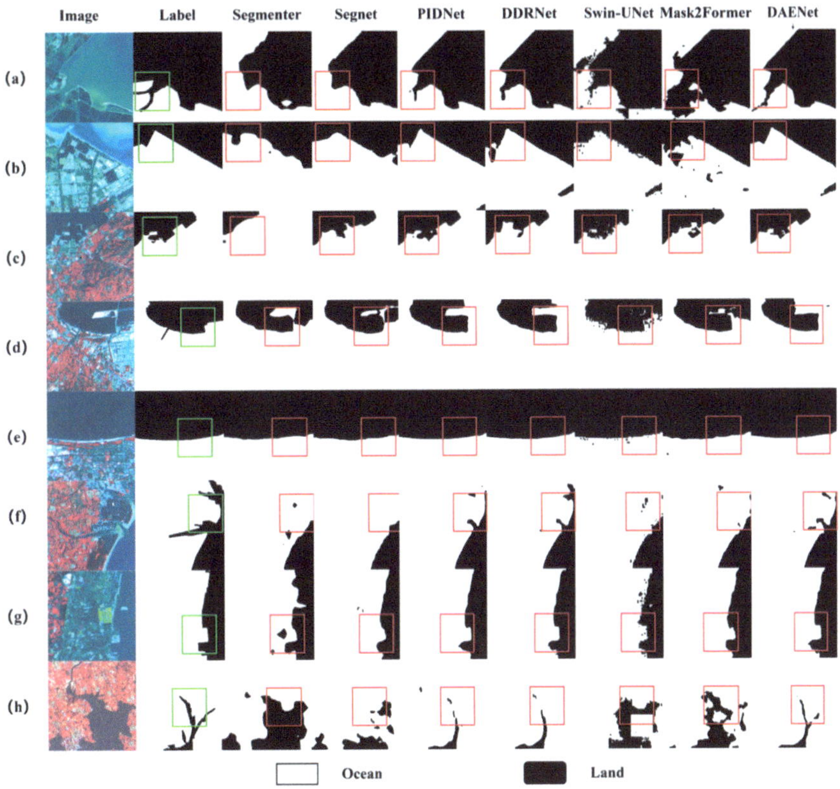

Figure 6. Examples of semantic segmentation results on the Shandong Province dataset. (**a–d**,**f–h**) artificial coastline; (**e**) sandy coastline.

With regard to the straightforward, efficient, and durable semantic segmentation model, Segmenter, our assessments have indicated that its functionality is marginally inadequate. As depicted in Figure 6 (Segmenter. e), although it capably handles simple linear coastlines with minimal discrepancies, its proficiency significantly decreases when confronting complex feature patterns, as exemplified in sections a, f, and g of the same Figures 6–8. In comparison, DAENet's performance remains superior for dense and intricate terrestrial entities, as evident in Figures 6–8.

(2) Conducting analysis across diverse regions to assess the adaptability of the model:

First, China's coastlines can be primarily classified into three types based on geological and geomorphic characteristics: sandy coasts, bedrock coasts, and artificial coasts. As depicted in Figure 6b,c for artificial coasts and Figure 7 for sandy and bedrock coasts, the coastline morphologies extracted by our DAEnet demonstrate superior performance compared to other methods.

Secondly, we also analyzed results from other types of coastlines. During the generation of estuary labels, especially in manual labeling processes, if narrow rivers extend over a considerable distance, our usual approach involves capturing either a segment of the estuary shoreline or rectifying the shoreline directly. Nevertheless, when confronted with images containing incomplete labeling (see Figure 7h) or featuring coastlines marked by aquaculture ponds (see Figure 6a,h), as well as other complex coastal types, we are nonetheless able to accurately discern the correct shoreline.

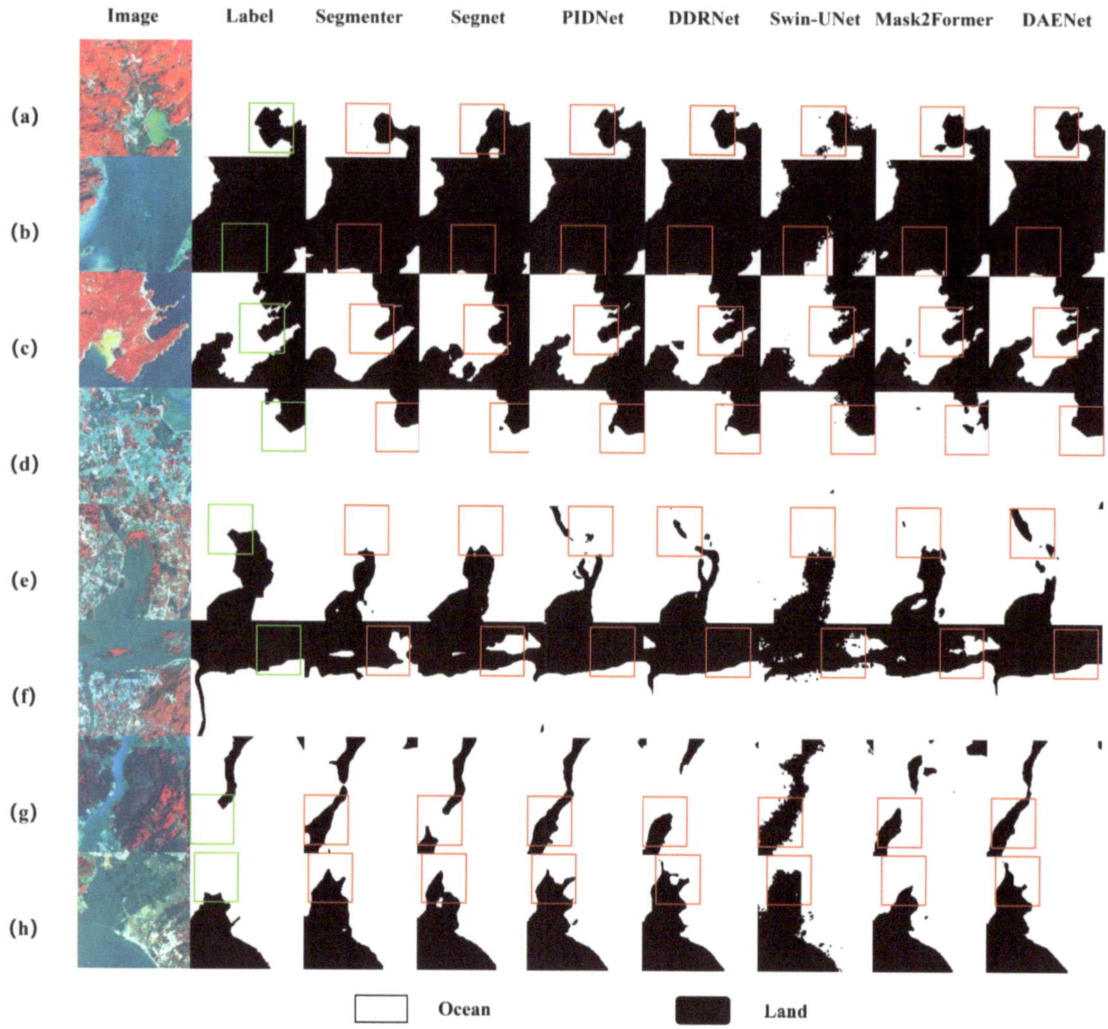

Figure 7. Examples of semantic segmentation results on the Fujian Province dataset. (**a**,**g**) artificial coastline & bedrock coastline; (**b**) bedrock coastline; (**c**) bedrock coastline & sandy coastline; (**e**) artificial coastline & biogenic coastline; (**d**,**f**,**h**) artificial coastline.

Lastly, as illustrated in Figure 7, the coastline of Fujian Province is notably intricate, winding its way in an exceptionally complex manner, making it the most complex coastline in mainland China. Despite this challenging topography, our DAENet consistently performs at a high level, with all performance metrics exceeding 84%, establishing it as a leading method in the industry. Additionally, we predicted the coastline of Guangxi Province using the DAENet model, with extracted results demonstrating superior performance compared to other models, confirming the reliability of DAENet. With sufficient data, we will be fully equipped to extract coastlines from across the country and monitor their dynamic changes.

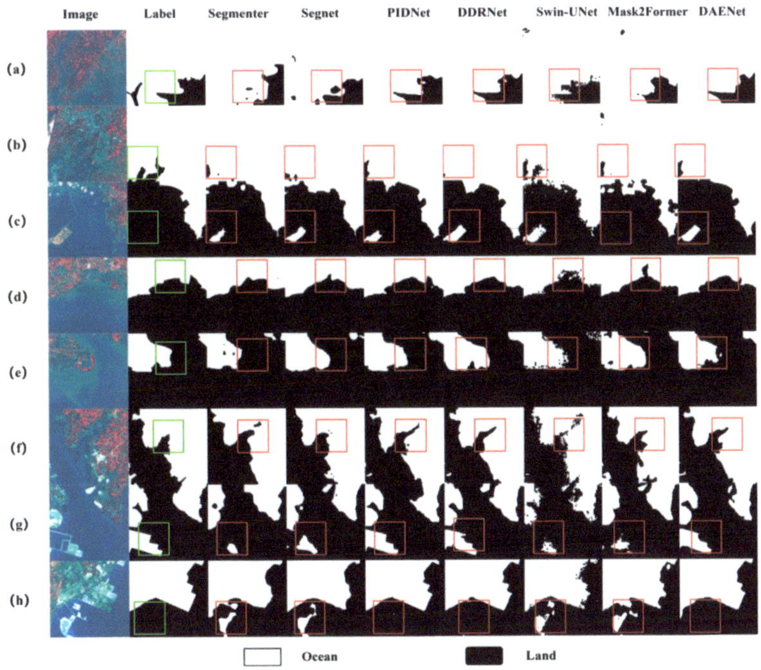

Figure 8. Examples of semantic segmentation results on the Guangxi Province dataset. (**a**,**b**) silty coastline; (**c**,**d**) artificial coastline & silty coastline; (**e**–**h**) artificial coastline.

4.2. Ablation Study

To evaluate the performance of the proposed network architecture and its three crucial modules, we utilized UNet as the foundational network for conducting an ablation study on the dataset. Furthermore, we explored the impact of the loss function on the proposed network. The subsequent comparison results indicate that the integration of the proposed MST, LAMBA, and DAT modules yields significant performance enhancements in detection (Figure 9).

(1) Impact of the Deformable Self-Attention Module: As presented in Table 5, introducing the Deformable Attention Transformer (DAT) effectively augments the segmentation performance of the UNet structure. There is an improvement ranging from 5.69% to 6.45% in the IoU accuracy metric relative to the original baseline model. The enhancement in accuracy ranges from 9.42% to 10.28%, providing substantial evidence for the effectiveness of integrating DAT. This enhancement is attributed to the feature maps first undergoing processing via window-based local attention, facilitating local information aggregation. Subsequently, the Deformable Attention block models the global relationships among the locally enhanced tokens. This alternative attention block design, equipped with local and global receptive fields, aids the model's learning process.

(2) The incorporation of the multi-scale transformation module (MST) into the Swin-UNet + LAMBA and Dat-UNet + LAMBA models yields significant results. As illustrated in Table 5, the integration of the multi-scale deformable edge detection module significantly enhances the segmentation performance of the UNet architecture. Relative to the original model, there is an improvement of 1.28% to 1.93% in the IoU accuracy metric. For the accuracy metric, the enhancement ranges from 1.01% to 1.72%. These outcomes underscore that integrating the MST module facilitates the model by capturing a richer set of feature information.

(3) Impact of the LAMBA Module: The LAMBA module was, respectively, integrated into the Swin-UNet and Dat-UNet models. As depicted in Table 5, the adaptive edge detec-

tion module's introduction bolsters the segmentation performance of the UNet structure. Compared to the original model, there is a boost of 1.03% to 2.67% in the IoU accuracy metric. Regarding the accuracy metric, the uplift spans from 1.07% to 1.45%. These findings highlight that introducing the LAMBA module amplifies the model's proficiency in edge delineation.

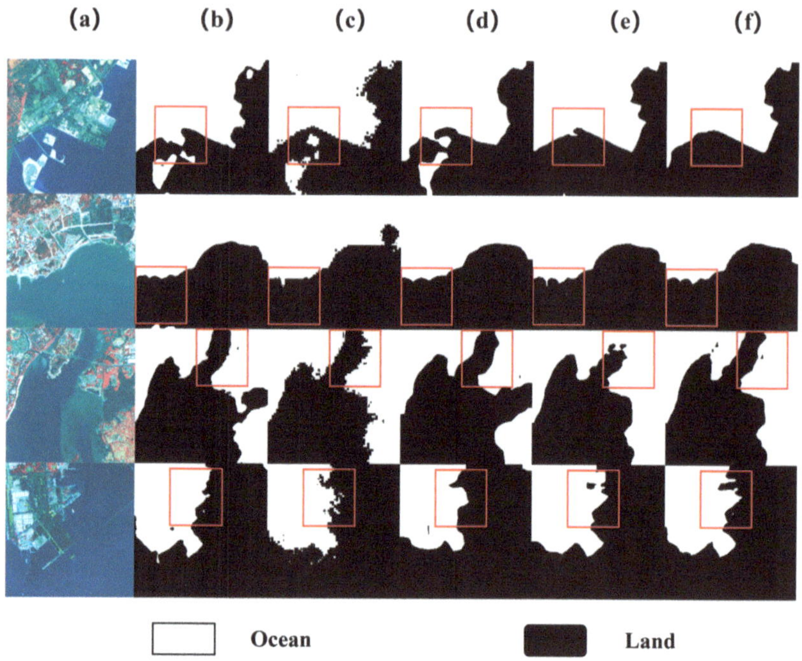

Figure 9. Examples of semantic segmentation results on the Ablation experiment: (**a**) Image, (**b**) U-Net, (**c**) Swin-UNet, (**d**) Dat-UNet, (**e**) Dat-UNet + MST, and (**f**) Dat-UNet + MST + LAMBA.

Table 5. Comparison of ablation results.

Method	IoU (%)		F_1	
	Ocean	Land	Ocean	Land
U-Net	83.21	83.87	90.84	91.23
Swin-UNet	88.59	88.33	93.95	93.80
Swin-UNet + LAMBA	88.93	88.65	94.14	93.98
Swin-UNet + MST	89.16	89.34	94.11	94.01
Swin-UNet + LAMBA + MST	90.21	90.14	94.85	94.81
Dat-UNet	89.66	89.76	94.55	94.60
Dat-UNet + LAMBA	92.33	92.41	96.01	96.06
Dat-UNet + MST	92.89	92.42	95.98	96.04
Dat-UNet + LAMBA + MST	**93.85**	**94.04**	**96.82**	**96.93**

5. Conclusions

In this study, we propose DAENet, a deep learning model that combines semantic segmentation networks with edge detection to address inaccuracies in coastline extraction and localization. To enhance the model's feature representation, we introduce the multi-scale transformation (MST) module, which incorporates canny edge detection across multiple spatial scales as input channels. By integrating MST into the U-shaped network structure, our model gains improved global modeling capability compared to traditional CNNs and patch-based Transformers. Additionally, our novel LAMBA module focuses on capturing edge features in the spatial dimension to mitigate semantic ambiguity caused by unclear object boundaries. We refine the binary Dice Loss function to expedite convergence and compute an edge-perceptive loss utilizing Canny edge maps to augment performance on edges further. Experimental results demonstrate that DAENet outperforms traditional models like Segmenter, SegNeX, PIDNet, DDRNet, Swin-UNet, and Mask2Former. Compared to the traditional model, Swin-UNet, DAENet shows a 5% improvement in MIoU.

To our knowledge, the proposed DAENet model is the first to apply the DAT block for remote sensing sea–land segmentation. It addresses the limitations of pure CNNs and enhances segmentation accuracy. The proposed network model can be effectively applied to precise positioning tasks for various complex coastal types in different regions, demonstrating its potential for coastal dynamic management and planning. Furthermore, the unique dataset created in this study allows the extracted results to approximate the actual coastline closely.

However, our model has several limitations. (1) DAENet extensively uses the Deformable Attention module, resulting in a larger parameter set and slightly longer training durations than other methods. This may limit DAENet's use in compact mobile devices, but it still offers valuable insights into the roles of Deformable Attention in remote sensing semantic segmentation. In future research, we will design more accurate geometric prior models and loss functions for SLS segmentation features or generate multi-scale features through style transfer to accelerate model convergence. (2) DAENet still requires improvements in object boundary extraction. The deficiencies mainly appear in the segmentation results, where we aim to explore advanced encoding techniques for boundary features to overcome this limitation. Augmented outcomes deviate from the actual shape of the objects and show slight noise. Additionally, we will prioritize implementing model compression methods to enhance inference efficiency. Overall, the goal is to accelerate model convergence while retaining its deformable characteristics.

Author Contributions: Conceptualization, B.K. and J.X.; Methodology, B.K., J.W. and J.X.; Software, B.K.; Validation, B.K.; Formal analysis, J.W.; Investigation, B.K.; Data curation, B.K. and J.X.; Writing—original draft, B.K.; Writing—review & editing, J.W., J.X. and C.W.; Supervision, J.W. and C.W.; Project administration, J.X.; Funding acquisition, J.X. All authors have read and agreed to the published version of the manuscript.

Funding: This research was funded by the High-Resolution Remote Sensing Applications Demonstration System for Urban Fine Management of China (Grant number 06-Y30F04-9001-20/22) and the National Key R&D Program of China (Grant number 06-Y30F04-9001-20/22).

Institutional Review Board Statement: Not applicable.

Informed Consent Statement: Not applicable.

Data Availability Statement: The original contributions presented in the study are included in the article, further inquiries can be directed to the corresponding author.

Conflicts of Interest: The authors declare no conflicts of interest.

References

1. Chen, C.; Bu, J.; Zhang, Y.; Zhuang, Y.; Chu, Y.; Hu, J.; Guo, B. The application of the tasseled cap transformation and feature knowledge for the extraction of coastline information from remote sensing images. *Adv. Space Res.* **2019**, *64*, 1780–1791. [CrossRef]
2. Gens, R. Remote sensing of coastlines: Detection, extraction and monitoring. *Int. J. Remote Sens.* **2010**, *31*, 1819–1836. [CrossRef]
3. Yang, Z.; Wang, L.; Sun, W.; Xu, W.; Tian, B.; Zhou, Y.; Yang, G.; Chen, C. A new adaptive remote sensing extraction algorithm for complex muddy coast waterline. *Remote Sens.* **2022**, *14*, 861. [CrossRef]
4. Mimura, N. Sea-level rise caused by climate change and its implications for society. *Proc. Jpn. Acad.* **2013**, *89*, 281–301. [CrossRef]
5. Small, C.; Nicholls, R.J. A global analysis of human settlement in coastal zones. *J. Coast. Res.* **2003**, *19*, 584–599.
6. Bell, P.S.; Bird, C.O.; Plater, A.J. A temporal waterline approach to mapping intertidal areas using X-band marine radar. *Coast. Eng.* **2016**, *107*, 84–101. [CrossRef]
7. Green, E.; Mumby, P.J.; Edwards, A.J.; Clark, C.D. A review of remote sensing for the assessment and management of tropical coastal resources. *Coast. Manag.* **1996**, *24*, 1–40. [CrossRef]

8. Vassilakis, E.; Papadopoulou-Vrynioti, K. Quantification of deltaic coastal zone change based on multi-temporal high resolution earth observation techniques. *ISPRS Int. J. Geo-Inf.* **2014**, *3*, 18–28. [CrossRef]
9. Zhang, Y.; Hou, X. Characteristics of coastline changes on Southeast Asia Islands from 2000 to 2015. *Remote Sens.* **2020**, *12*, 519. [CrossRef]
10. Bera, R.; Mait, R. Quantitative analysis of erosion and accretion (1975–2017) using DSAS—A study on Indian Sundarbans. *Reg. Stud. Mar. Sci.* **2019**, *28*, 100583. [CrossRef]
11. Ghosh, M.K.; Kumar, L.; Roy, C. Monitoring the coastline change of Hatiya Island in Bangladesh using remote sensing techniques. *ISPRS J. Photogramm. Remote Sens.* **2015**, *101*, 137–144. [CrossRef]
12. Konko, Y.; Okhimambe, A.; Nimon, P.; Asaana, J.; Rudant, J.P.; Kokou, K. Coastline change modelling induced by climate change using geospatial techniques in Togo (West Africa). *Adv. Remote Sens.* **2020**, *9*, 85–100. [CrossRef]
13. Hamylton, S.; Prosper, J. Development of a spatial data infrastructure for coastal management in the Amirante Islands. *Int. J. Appl. Earth Obs. Geoinf.* **2020**, *19*, 24–30. [CrossRef]
14. Wu, G.; de Leeuw, J.; Skidmore, A.K.; Liu, Y.; Prins, H.H. Performance of Landsat TM in ship detection in turbid waters. *Int. J. Appl. Earth Obs. Geoinf.* **2012**, *11*, 54–61. [CrossRef]
15. Giardino, C.; Bresciani, M.; Villa, P.; Martinelli, A. Application of remote sensing in water resource management: The case study of Lake Trasimeno, Italy. *Water Resour. Manag.* **2010**, *24*, 3885–3899. [CrossRef]
16. Qiao, G.; Mi, H.; Wang, W.; Tong, X.; Li, Z.; Li, T.; Hong, Y. 55-year (1960–2015) spatiotemporal coastline change analysis using historical DISP and Landsat time series data in Shanghai. *Int. J. Appl. Earth Obs. Geoinf.* **2018**, *68*, 238–251.
17. Sun, W.; Chen, C.; Liu, W.; Yang, G.; Meng, X.; Wang, L.; Ren, K. Coastline extraction using remote sensing: A review. *GISci. Remote Sens.* **2023**, *60*, 1780–1791. [CrossRef]
18. Chen, C.; Qin, Q.; Zhang, N.; Li, J.; Chen, L.; Wang, J.; Yang, X. Extraction of bridges over water from high-resolution optical remote-sensing images based on mathematical morphology. *Int. J. Remote Sens.* **2014**, *35*, 3664–3682. [CrossRef]
19. Chen, H.; Chen, C.; Zhang, Z.; Lu, C.; Wang, L.; He, X.; Chen, J. Changes of the spatial and temporal characteristics of land-use landscape patterns using multi-temporal Landsat satellite data: A case study of Zhoushan Island, China. *Ocean. Coast. Manag.* **2021**, *213*, 105842. [CrossRef]
20. Elkhateeb, E.; Soliman, H.; Atwan, A.; Elmogy, M.; Kwak, K.S.; Mekky, N. A novel coarse-to-Fine Sea-land segmentation technique based on Superpixel fuzzy C-means clustering and modified Chan-Vese model. *IEEE Access.* **2021**, *9*, 53902–53919. [CrossRef]
21. Tong, Q.; Shan, J.; Zhu, B.; Ge, X.; Sun, X.; Liu, Z. Object-Oriented Coastline Classification and Extraction from Remote Sensing Imagery. In Proceedings of the Remote Sensing of the Environment: 18th National Symposium on Remote Sensing of China, Beijing, China, 7–10 September 2007.
22. Wang, P.; Zhuang, Y.; Chen, H.; Chen, L.; Shi, H.; Bi, F. Pyramid integral image reconstruction algorithm for infrared remote sensing sea-land segmentation. In Proceedings of the 2017 IEEE International Geoscience and Remote Sensing Symposium (IGARSS), Fort Worth, TX, USA, 23–28 July 2017; pp. 3763–3766.
23. Wang, D.; Cui, X.; Xie, F.; Jiang, Z.; Shi, Z. Multi-feature sea–land segmentation based on pixel-wise learning for optical remote-sensing imagery. *Int. J. Remote Sens.* **2017**, *38*, 4327–4347. [CrossRef]
24. Lei, S.; Zou, Z.; Liu, D.; Xia, Z.; Shi, Z. Sea-land segmentation for infrared remote sensing images based on superpixels and multi-scale features. *Infrared Phys. Technol.* **2018**, *91*, 12–17. [CrossRef]
25. Chen, C.; Fu, J.; Zhang, S.; Zhao, X. Coastline information extraction based on the tasseled cap transformation of Landsat-8 OLI images. *Estuar. Coast. Shelf Sci.* **2019**, *217*, 281–291. [CrossRef]
26. Chen, C.; Liang, J.; Xie, F.; Hu, Z.; Sun, W.; Yang, G.; Zhang, Z. Temporal and Spatial Variation of Coastline Using Remote Sensing Images for Zhoushan Archipelago, China. *Int. J. Appl. Earth Obs. Geoinf.* **2022**, *107*, 102711. [CrossRef]
27. Fisher, P. The Pixel: A Snare and a Delusion. *Int. J. Remote Sens.* **1997**, *18*, 679–685. [CrossRef]
28. Liu, Y.; Zhang, M.; Xu, P.; Guo, Z. SAR ship detection using sea-land segmentation-based convolutional neural network. In Proceedings of the 2017 International Workshop on Remote Sensing with Intelligent Processing (RSIP), Shanghai, China, 18–21 May 2017.
29. Li, R.; Liu, W.; Yang, L.; Sun, S.; Hu, W.; Zhang, F.; Li, W. DeepUNet: A Deep Fully Convolutional Network for Pixel-level Sea-Land Segmentation. *IEEE J. Sel. Top. Appl. Earth Obs. Remote Sens.* **2018**, *11*, 3954–3962. [CrossRef]
30. Lin, Z.; Ji, K.; Leng, X.; Kuang, G. Squeeze and excitation rank faster R-CNN for ship detection in SAR images. *IEEE Trans. Geosci. Remote Sens. Lett.* **2019**, *16*, 751–755. [CrossRef]
31. Yang, T.; Jiang, S.; Hong, Z.; Zhang, Y.; Han, Y.; Zhou, R.; Kuc, T.Y. Sea-land segmentation using deep learning techniques for landsat-8 OLI imagery. *Mar. Geod.* **2020**, *43*, 105–133. [CrossRef]
32. Hui, G.; Xiaodong, Y.; Heng, Z.; Yiting, N.; Jiaqi, W. Multi-scale sea-land segmentation method for remote sensing images based on Res2Net. *Acta Optia Sin.* **2022**, *42*, 1828004.
33. Li, Y.; Wang, X.; Zhang, X.; Fang, J.; Zhang, X. WRBSNet: A Novel Sea–Land Segmentation Network With a Wider Range of Batch Sizes. *IEEE Geosci. Remote Sens. Lett.* **2024**, *21*, 1–5. [CrossRef]
34. Chen, C.; Zou, Z.; Sun, W.; Yang, G.; Song, Y.; Liu, Z. Mapping the distribution and dynamics of coastal aquaculture ponds using Landsat time series data based on U2-Net deep learning model. *Int. J. Digit. Earth* **2024**, *17*, 2346258. [CrossRef]
35. Ji, X.; Tang, L.; Lu, T.; Cai, C. DBENet: Dual-Branch Ensemble Network for Sea-Land Segmentation of Remote Sensing Images. *IEEE Trans. Instrum. Meas.* **2023**, *72*, 5503611. [CrossRef]

36. Sun, S.; Mu, L.; Feng, R.; Chen, Y.; Han, W. Quadtree decomposition-based Deep learning method for multiscale coastline extraction with high-resolution remote sensing imagery. *Sci. Remote Sens.* **2024**, *9*, 100112. [CrossRef]
37. Li, Z.; Cui, B.; Yang, G. Edge detection network model of coastline based on deep learning. *Comput. Engin. Sci.* **2022**, *44*, 2220–2229.
38. Yu, F.; Koltun, V. Multi-scale context aggregation by dilated convolutions. *arXiv* **2015**, arXiv:1511.07122.
39. Dong, R.; Pan, X.; Li, F. DenseU-Net-Based Semantic Segmentation of Small Objects in Urban Remote Sensing Images. *IEEE Access.* **2019**, *7*, 65347–65356. [CrossRef]
40. Chen, X.; Li, Z.; Jiang, J.; Han, Z.; Deng, S.; Li, Z.; Liu, M. Adaptive Effective Receptive Field Convolution for Semantic Segmentation of VHR Remote Sensing Images. *IEEE Geosci. Remote Sens.* **2021**, *59*, 3532–3546. [CrossRef]
41. Ding, L.; Tang, H.; Bruzzone, L. LANet: Local Attention Embedding to Improve the Semantic Segmentation of Remote Sensing Images. *IEEE Trans. Geosci. Remote Sens.* **2021**, *59*, 426–435. [CrossRef]
42. Yang, L.; Wang, X.; Zhai, J. Waterline Extraction for Artificial Coast With Vision Transformers. *Front. Environ. Sci.* **2022**, *10*, 799250. [CrossRef]
43. Zheng, S.; Lu, J.; Zhao, H.; Zhu, X.; Luo, Z.; Wang, Y.; Fu, Y.; Feng, J.; Xiang, T.; Torr, P.H.S.; et al. Rethinking Semantic Segmentation from a Sequence-To-Sequence Perspective with Transformers. *arXiv* **2021**, arXiv:2012.15840.
44. Xie, E.; Wang, W.; Yu, Z.; Anandkumar, A.; Alvarez, J.M.; Luo, P. SegFormer: Simple and efficient design for semantic segmentation with transformers. *Adv. Neural Inf. Process. Syst.* **2021**, *34*, 12077–12090.
45. Yang, Z.; Wang, G.; Feng, L.; Wang, Y.; Wang, G.; Liang, S. A Transformer Model for Coastline Prediction in Weitou Bay, China. *Remote Sens.* **2023**, *15*, 4771. [CrossRef]
46. Zhu, Y.; Wang, B.; Liu, Q.; Tan, S.; Wang, S.; Ge, W. SRMA: A Dual-Branch Parallel Multi-Scale Attention Network for Remote Sensing Images Sea-Land Segmentation. *Int. J. Remote Sens.* **2024**, *45*, 3370–3395. [CrossRef]
47. Han, K.; Wang, Y.; Chen, H.; Chen, X.; Guo, J.; Liu, Z.; Tang, Y.; Xiao, A.; Xu, C.; Xu, Y.; et al. A Survey on Vision Transformer. *IEEE Trans. Pattern Anal. Mach. Intell.* **2022**, *45*, 87–110. [CrossRef]
48. Guo, J.; Han, K.; Wu, H.; Tang, Y.; Chen, X.; Wang, Y.; Xu, C. CMT: Convolutional Neural Networks Meet Vision Transformers. In Proceedings of the IEEE/CVF Conference on Computer Vision and Pattern Recognition, New Orleans, LA, USA, 18–24 June 2022; pp. 12175–12185.
49. Guo, M.-H.; Xu, T.-X.; Liu, J.-J.; Liu, Z.-N.; Jiang, P.-T.; Mu, T.-J.; Zhang, S.-H.; Martin, R.R.; Cheng, M.-M.; Hu, S.-M. Attention Mechanisms in Computer Vision: A Survey. *Comp. Vis. Media* **2022**, *8*, 331–368. [CrossRef]
50. Liu, Z.; Lin, Y.; Cao, Y.; Hu, H.; Wei, Y.; Zhang, Z.; Lin, S.; Guo, B. Swin Transformer: Hierarchical Vision Transformer using Shifted Windows. In Proceedings of the 2021 IEEE/CVF International Conference on Computer Vision (ICCV), Montreal, QC, Canada, 10–17 October 2021; pp. 9992–10002.
51. Wang, W.; Xie, E.; Li, X.; Fan, D.-P.; Song, K.; Liang, D.; Lu, T.; Luo, P.; Shao, L. Pyramid Vision Transformer: A Versatile Backbone for Dense Prediction Without Convolutions. In Proceedings of the IEEE/CVF International Conference on Computer Vision, Montreal, QC, Canada, 10–17 October 2021.
52. Dai, J.; Qi, H.; Xiong, Y.; Li, Y.; Zhang, G.; Hu, H.; Wei, Y. Deformable Convolutional Networks. In Proceedings of the IEEE International Conference on Computer Vision, Venice, Italy, 22–29 October 2017; pp. 764–773.
53. Wang, Y.; Liu, W.; Sun, W.; Meng, X.; Yang, G.; Ren, K. A Progressive Feature Enhancement Deep Network for Large-Scale Remote Sensing Image Superresolution. *IEEE Trans. Geosci. Remote Sens.* **2023**, *61*, 5619413. [CrossRef]
54. Li, X.; Xu, F.; Liu, F.; Tong, Y.; Lyu, X.; Zhou, J. Semantic Segmentation of Remote Sensing Images by Interactive Representation Refinement and Geometric Prior-Guided Inference. *IEEE Trans. Geosci. Remote Sens.* **2024**, *62*, 5400318. [CrossRef]
55. Xia, Z.; Pan, X.; Song, S.; Li, L.E.; Huang, G. Vision Transformer with Deformable Attention. In Proceedings of the IEEE/CVF Conference on Computer Vision and Pattern Recognition, New Orleans, LA, USA, 18–24 June 2022; pp. 4784–4793.
56. Hou, X.; Wu, T.; Hou, W.; Chen, Q.; Wang, Y.; Yu, L. Characteristics of coastline changes in mainland China since the early 1940s. *Sci. China Earth Sci.* **2016**, *59*, 1791–1802. [CrossRef]
57. Wu, T.; Hou, X.; Xu, X. Spatio-temporal characteristics of the mainland coastline utilization degree over the last 70 years in China. *Ocean. Coast. Manag.* **2014**, *98*, 150–157. [CrossRef]
58. He, X.; Zhou, Y.; Zhao, J.; Zhang, D.; Yao, R.; Xue, Y. Swin transformer embedding UNet for remote sensing image semantic segmentation. *IEEE Trans. Geosci. Remote Sens.* **2022**, *60*, 4408715. [CrossRef]
59. Wang, X.; Liu, Y.; Ling, F.; Liu, Y.; Fang, F. Spatio-temporal change detection of Ningbo coastline using Landsat time-series images during 1976–2015. *ISPRS Int. J. Geo-Inf.* **2017**, *6*, 68. [CrossRef]
60. Meyer, E.L.; Matzke, N.J.; Williams, S.J. Remote sensing of intertidal habitats predicts West Indian topsnail population expansion but reveals scale-dependent bias. *J. Coast. Conserv.* **2015**, *19*, 107–118. [CrossRef]
61. Xu, J.; Zhang, Z.; Zhao, X.; Wen, Q.; Zuo, L.; Wang, X.; Yi, L. Spatial and temporal variations of coastlines in northern China (2000–2012). *Int. J. Geogr. Inf. Sci.* **2013**, *24*, 18–32. [CrossRef]
62. Guo, H. Big Earth data in support of the sustainable development goals (2019). *Bull. Chin. Acad. Sci.* **2021**, *36*, 932–939.
63. Seale, C.; Redfern, T.; Chatfield, P.; Luo, C.; Dempsey, K. Coastline detection in satellite imagery: A deep learning approach on new benchmark data. *Remote Sens. Environ.* **2022**, *278*, 113044. [CrossRef]
64. Zou, Z.; Chen, C.; Liu, Z.; Zhang, Z.; Liang, J.; Chen, H.; Wang, L. Extraction of aquaculture ponds along coastal region using u2-net deep learning model from remote sensing images. *Remote Sens.* **2022**, *14*, 4001. [CrossRef]

65. Sun, W.F.; Ma, Y.; Zhang, J.; Liu, S.W.; Ren, G.B. Study of remote sensing interpretation keys and extraction technique of different types of shoreline. *Bull. Surv. Mapp.* **2011**, 3, 41–44.
66. Strudel, R.; Garcia, R.; Laptev, I.; Schmid, C. Segmenter: Transformer for semantic segmentation. In Proceedings of the IEEE/CVF International Conference on Computer Vision, Electrical Network, Montreal, BC, Canada, 11–17 October 2021; pp. 7262–7272.
67. Guo, M.H.; Lu, C.Z.; Hou, Q.; Liu, Z.; Cheng, M.M.; Hu, S.M. Segnext: Rethinking convolutional attention design for semantic segmentation. *arXiv* **2022**, arXiv:2209.08575.
68. Xu, J.; Xiong, Z.; Bhattacharyya, S.P. PIDNet: A Real-Time Semantic Segmentation Network Inspired by PID Controllers. In Proceedings of the IEEE/CVF Conference on Computer Vision and Pattern Recognition, Vancouver, BC, Canada, 17–24 June 2023; pp. 19529–19539.
69. Hong, Y.; Pan, H.; Sun, W.; Jia, Y. Deep dual-resolution networks for real-time and accurate semantic segmentation of road scenes. *arXiv* **2021**, arXiv:2101.06085.
70. Cheng, B.; Misra, I.; Schwing, A.G.; Kirillov, A.; Girdhar, R. Masked-attention mask transformer for universal image segmentation. In Proceedings of the IEEE/CVF Conference on Computer Vision and Pattern Recognition, New Orleans, LA, USA, 18–24 June 2022; pp. 1290–1299.

remote sensing

Article

Multi-Scale Window Spatiotemporal Attention Network for Subsurface Temperature Prediction and Reconstruction

Jiawei Jiang [1,2,†], Jun Wang [3,4,†], Yiping Liu [5], Chao Huang [1], Qiufu Jiang [1], Liqiang Feng [6], Liying Wan [7,†] and Xiangguang Zhang [1,2,*]

1 Key Laboratory of Ocean Circulation and Waves, Institute of Oceanology, Chinese Academy of Sciences, Qingdao 266071, China; jjw@qdio.ac.cn (J.J.); chaohuang@qdio.ac.cn (C.H.); jiangqiufu@qdio.ac.cn (Q.J.)
2 University of Chinese Academy of Sciences, Beijing 100049, China
3 College of Meteorology and Oceanography, National University of Defense Technology, Changsha 410073, China; wwjj103@nudt.edu.cn
4 Hunan Key Laboratory for Marine Detection Technology, Changsha 410073, China
5 Yantai Center of Coastal Zone Geological Survey, China Geological Survey, Yantai 264000, China; lyiping@mail.cgs.gov.cn
6 Ocean Big Data Center, Institute of Oceanology, Chinese Academy of Sciences, Qingdao 266071, China; fenglq@qdio.ac.cn
7 Key Laboratory of Research on Marine Hazards Forecasting, National Marine Environmental Forecasting Center, Beijing 100081, China; liying.wan@nmefc.cn
* Correspondence: zxg@qdio.ac.cn
† These authors contributed equally to this work.

Abstract: In this study, we investigate the feasibility of using historical remote sensing data to predict the future three-dimensional subsurface ocean temperature structure. We also compare the performance differences between predictive models and real-time reconstruction models. Specifically, we propose a multi-scale residual spatiotemporal window ocean (MSWO) model based on a spatiotemporal attention mechanism, to predict changes in the subsurface ocean temperature structure over the next six months using satellite remote sensing data from the past 24 months. Our results indicate that predictions made using historical remote sensing data closely approximate those made using historical in situ data. This finding suggests that satellite remote sensing data can be used to predict future ocean structures without relying on valuable in situ measurements. Compared to future predictive models, real-time three-dimensional structure reconstruction models can learn more accurate inversion features from real-time satellite remote sensing data. This work provides a new perspective for the application of artificial intelligence in oceanography for ocean structure reconstruction.

Keywords: temperature structure prediction; temperature structure reconstruction; spatiotemporal window ocean; satellite observations; spatiotemporal attention mechanism

Citation: Jiang, J.; Wang, J.; Liu, Y.; Huang, C.; Jiang, Q.; Feng, L.; Wan, L.; Zhang, X. Multi-Scale Window Spatiotemporal Attention Network for Subsurface Temperature Prediction and Reconstruction. *Remote Sens.* **2024**, *16*, 2243. https://doi.org/10.3390/rs16122243

Academic Editor: Wenfang Lu

Received: 26 May 2024
Revised: 13 June 2024
Accepted: 16 June 2024
Published: 20 June 2024

1. Introduction

The ocean plays a critical role in regulating the stability of the Earth's system by absorbing a significant portion of global heat. Temperature, as one of the most fundamental marine physical quantities, is intricately linked to the density structure of the ocean [1,2]. This relationship influences not only the flow field but also biological activities and chemical reactions within the marine environment [3]. Recent studies have demonstrated a notable upward trend in the heat content of the upper ocean over the past few decades [4]; such increases in ocean temperature are associated with the potential for meteorological disasters, including typhoons and storm surges [5]. Consequently, investigating and understanding changes in ocean temperature structure are essential for promoting marine environmental awareness, ecological protection, and disaster prevention.

Numerical simulation is a traditional method employed to obtain the three-dimensional structure of the ocean and predict its dynamic processes. Various models exhibit distinct advantages, for instance, the Hybrid Coordinate Ocean Model (HYCOM) [6] features a significant vertical hierarchical structure, establishing a practical hybrid vertical coordinate system. The Finite-Volume Coastal Model (FVCOM) [7] accurately fits the coastline boundary and seabed topography using specific conservation equations. Despite their utility, these models rely on dynamic simplification processes and settings of various model parameters, such as bottom friction coefficient and eddy viscosity coefficients. They thus may not capture the full complexity and variability of the real ocean. As a result, the accuracy of numerical simulation data can be limited [8]. Additionally, the high computational overhead associated with these simulations poses a significant challenge. To address these limitations, numerous research institutions employ various instruments such as subsurface mooring, large buoys, drifting buoys, and gliders for real-time observation of the three-dimensional ocean structure [9–11]. However, field measurement methods face inherent challenges, including high sampling costs, difficulty in data acquisition, and low spatiotemporal resolution [12,13]. Satellite remote sensing offers high resolution, large-scale coverage, and long-term continuity but is restricted to detecting surface information due to transmission media limitations [14,15]. To overcome this, researchers have proposed data-driven deep learning strategies to reconstruct subsurface ocean structures from satellite data, a process termed deep ocean remote sensing (DORS) [16,17]. Recent years have seen DORS emerge as an efficient and innovative approach to obtaining the three-dimensional structure of the ocean.

There are currently four major schemes for DORS work (Figure 1). Figure 1a shows the use of known satellite remote sensing data of the ocean surface to predict future changes at the ocean surface. For instance, Zhang et al. [18] employed the Long Short-Term Memory (LSTM) method to forecast sea surface temperature in offshore China, He et al. [19] developed the DSL method for predicting sea surface temperature, Zhang et al. [20] devised a U-Net method for time series prediction of sea surface salinity in the Western Pacific Ocean, Xu et al. [21] combined Memory in Memory (MIM) and Variational Mode Decomposition (VMD) to propose the VMD-MIM model, which further enhances the prediction performance of sea surface temperature. These approaches primarily focus on predicting surface changes, whereas subsurface data are often more valuable to researchers [22].

Figure 1b depicts the use of historical subsurface three-dimensional structure data to forecast future subsurface changes. For example, Sun et al. [23] proposed a 3D U-Net model, utilizing the past 12 months of subsurface data to predict the subsurface structure for the next 12 months. Yue et al. [24] introduced the SA-PredRNN model, based on SA-ConvLSTM and PredRNN, to achieve similar predictions using historical subsurface data. Although these predictive models are valuable, they rely on extensive measured data as input, which poses practical limitations due to the data's scarcity and acquisition challenges.

Figure 1c represents the most extensively studied approach, focusing on the reconstruction of subsurface ocean structures using remote sensing data. Su et al. [25] explored the reconstruction of subsurface temperature and salinity anomalies using machine learning methods such as Random Forest (RF) and Extreme Gradient Boosting (XGBoost), while Meng et al. [26] achieved high-resolution subsurface reconstruction through Convolutional Neural Networks (CNNs). Xie et al. [27] implemented subsurface reconstruction in the South China Sea region using a U-Net model based on convolutional neural networks. Recently, the introduction of the Transformer framework and attention mechanisms has facilitated the efficient reconstruction of subsurface structures by learning from historical data. For instance, Zhang et al. [28] developed a discrete point historical time series model to invert subsurface structures by capturing temporal changes, while Su et al. [29] utilized the Transformer framework to sample and extract spatial features from images of different scales, reducing computational costs and enabling efficient reconstruction of subsurface density fields. These methods, however, are focused on reconstruction and do not predict future subsurface structures.

Figure 1d illustrates the relatively novel approach of using historical remote sensing data to predict future subsurface changes in the ocean. In 2024, Liu et al. [22] introduced an artificial intelligence model that employed historical remote sensing images to forecast future subsurface ocean structures, training the model with satellite remote sensing and numerical simulation data, and achieving notable results in the South China Sea region.

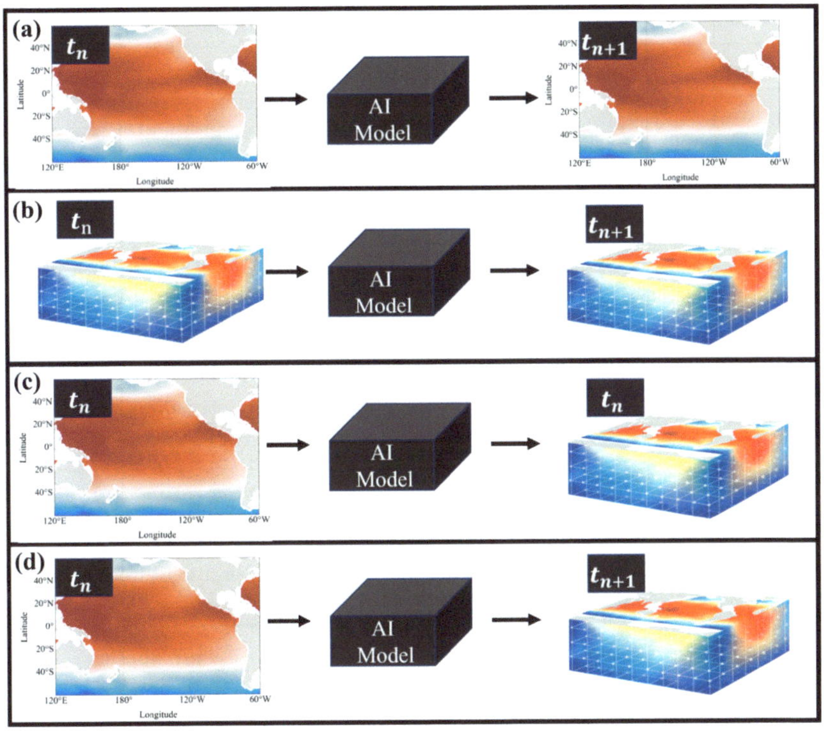

Figure 1. (**a**) Surface ocean prediction using remote sensing images. (**b**) Subsurface ocean prediction using measured profile data. (**c**) Subsurface ocean reconstruction using satellite remote sensing images. (**d**) Subsurface ocean prediction using satellite remote sensing images.

By comparing these DORS methods, we identify several pertinent questions for further exploration: (1) Is there a performance difference between using historical remote sensing data to predict future subsurface ocean profile structures and using historical subsurface ocean profile structures for the same purpose? (2) What is the relationship between the reconstruction of subsurface structures that combines historical and real-time data, and the prediction of subsurface structures using only historical data?

In this paper, we propose the multiscale spatiotemporal window ocean (MSWO) model, which combines the features of the spatiotemporal series network SimVP [30] and the computer vision Swin Transformer [31] network. The MSWO model employs a window attention mechanism in the spatiotemporal dimension to achieve low computational-load attention effects. Additionally, it extracts the nonlinear spatiotemporal relationships in the data through multi-scale residuals. To better represent the spatial and temporal characteristics of the data, we introduce global location coding during information extraction and use the channel attention mechanism to extract key feature information from remote sensing images. The model was validated against measured Argo grid data in the Central South Pacific. Historical remote sensing data from the past 24 months were used to predict the subsurface temperature structure trends in the Central South Pacific for the next 6 months. Furthermore, we explored the effect of using measured profile structures from the past 24 months to predict the subsurface structure for the next 6 months, comparing the reconstructed results with the predicted results on the test dataset. The results indicate that our model achieves the highest inversion accuracy in both reconstruction and prediction processes. This method presents a promising new approach for transparent ocean observation.

2. Related Work

Both the reconstruction of ocean subsurface structures and the prediction of future subsurface changes can be accomplished using spatiotemporal series models. With the development of artificial intelligence, more and more remote sensing images are used to reconstruct ocean subsurface structures and begin to invert historical data as one of the input factors [28,29].

Spatiotemporal series models have their origins in single-element time series prediction models, with the most traditional method being the recurrent neural networks (RNNs). The LSTM [32] subsequently improved the model's efficiency in learning temporal patterns. In 2015, Shi et al. [33] advanced this field by introducing the ConvLSTM model, which replaced the fully connected layer with a convolutional layer in the FC-LSTM, marking a transition from time series models to spatiotemporal series models. This innovation extended LSTM input data to multidimensional images, maintaining the fundamental principles of LSTM and other recurrent neural networks. The prediction for the next time point is achieved by iterating through the forgetting gate, output gate, and memory cell. In 2017, Wang et al. [34] proposed PredRNN, based on ConvLSTM, which introduced a long-term memory module to enhance the accuracy of spatiotemporal series prediction tasks.

The introduction of the attention mechanism and the Transformer framework [35] in 2017 revolutionized natural language-processing research. Researchers discovered that the attention mechanism's ability to capture global changes was superior to that of convolutional neural networks, leading to its application in computer vision. Various Transformer framework variants [31,36–38] have since achieved remarkable success in object detection, image classification, and semantic segmentation. The attention mechanism's capability to capture global information has also been applied to time series prediction tasks [39,40]. In 2020, Lin et al. [41] incorporated the self-attention module into the ConvLSTM model, resulting in the SA-ConvLSTM model, which can capture global spatiotemporal correlations. However, with higher-resolution image inputs, spatiotemporal series prediction models faced similar computational cost challenges as those in computer vision. The Swin Transformer [31] improved model performance while significantly reducing computational complexity. Consequently, in 2023, Tang et al. [42] replaced the convolutional layer in ConvLSTM with the Swin Transformer, achieving optimal results on video prediction datasets such as Moving MINST.

However, the iterative RNN-based spatiotemporal series models have inherent flaws. As illustrated in Figure 2a, the prediction for the current time step in an iterative model is influenced by the previous time step's prediction, leading to error accumulation over time. Additionally, this process tends to make the model overly reliant on immediate past context information, hindering long-term information learning [43]. To address this issue, various generative models have been developed to generate all prediction results by capturing global information in a single step [39,44], as shown in Figure 2b. In 2023, Tan et al. [30] proposed the SimVP model, which uses an Encoder–Decoder architecture for feature extraction and large-kernel convolution to simulate a global attention mechanism, achieving excellent results in spatiotemporal prediction. With the advent of a new generation of satellites, remote sensing images now feature larger scales and higher resolutions. Consequently, the primary objective of this paper is to design an efficient artificial intelligence model that supports high-resolution images to predict the future three-dimensional structure of the ocean.

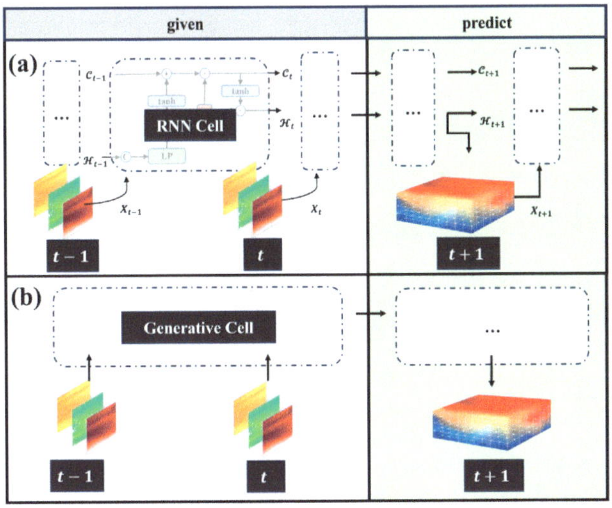

Figure 2. (**a**) Flow chart of RNN iterative framework on spatiotemporal prediction tasks. (**b**) Flow chart of the generative framework for spatiotemporal prediction tasks.

3. Method

3.1. Overall Architecture

The MSWO model aims to design an autoencoder architecture suitable for spatiotemporal sequence prediction tasks, and predicts future images based on the learning of past temporal related frames. Different from the traditional iterative RNN, the MSWO model uses a generative strategy. The traditional automatic encoder generates a single step frame in static time and minimizes the difference between the true probability distribution P_x and the predicted probability distribution $P'_x = \mathcal{G}_\theta(\mathcal{X})$ by mapping $\mathcal{G}_\theta : \mathcal{X} \rightarrow \hat{\mathcal{Y}}$. The optimal parametric neural network is as follows:

$$\theta^* = \underset{\theta}{argmin} Div(P_x, P'_x), \tag{1}$$

where *Div* represents a specific loss function. Similarly, MSWO encodes the frames of the entire historical time dimension, decodes the future prediction time dimension, and minimizes the difference between the predicted and true probability distributions by mapping $\mathcal{F}_\theta : \mathcal{X}^{t,T} \rightarrow \hat{\mathcal{Y}}^{t+1,T'}$. The optimal parametric neural network is:

$$\theta^* = \underset{\theta}{argmin} \sum_{t+1}^{t+1+T'} Div(P_{x^i}, P'_{x^i}) \tag{2}$$

In the experiment, we choose the L_2 loss function to minimize the error between this mapping:

$$\theta^* = \underset{\theta}{argmin} \sum_{t+1}^{t+1+T'} \| \mathcal{Y}^i - \mathcal{F}_\theta\left(\mathcal{X}^i\right) \|^2 \tag{3}$$

3.2. MSWO Module

Assuming that the size of the input remote sensing image is $H \times W \times C$ in T times, H represents the length of the image, W represents the width of the image, and C represents the number of channels of the image, the entire input can be represented by a tensor $\mathcal{X} \in \mathbb{R}^{T \times H \times W \times C}$. As shown in Figure 3, the input shape is $B \times T \times C \times H \times W$, where B represents batch size. The whole MSWO model consists of three parts: spatial encoder, window attention, and spatial decoder.

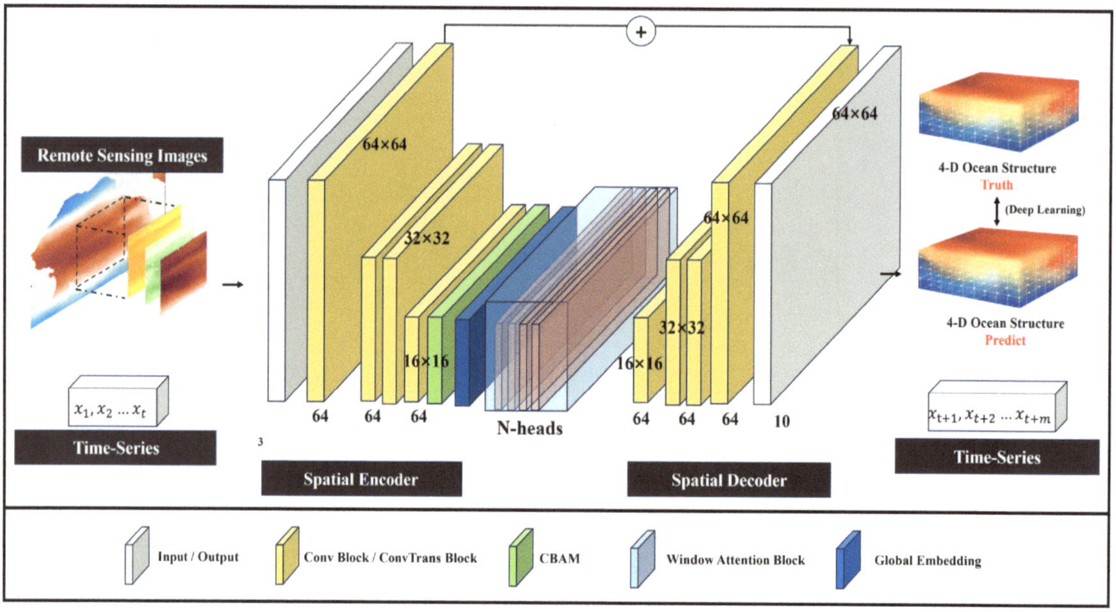

Figure 3. Flow diagram of MSWO using satellite remote sensing data to predict subsurface ocean temperature structure. 4-D represents the xyz axis in the time dimension and space.

The spatial encoder consists of two subsampled convolution layers, the convolutional block attention module (CBAM) [45], global attention coding [46], and residual concatenation. By stacking convolution layers, high-dimensional satellite remote sensing images are encoded into the low-dimensional potential space as follows:

$$z = \delta(Norm(Conv(z))), \tag{4}$$

where δ represents the activation function used to extract nonlinear variations from features, and *Norm* is the normalization layer. CBAM includes a channel attention module and a spatial attention module, which use average pooling and max pooling methods to extract useful channel information and spatial information, helping the model to better learn critical information and enhance sensitivity, thereby improving the overall performance of the model.

After spatial encoder, we reshape the dimensions of the tensor as $B \times (T \times C) \times H^* \times W^*$. The purpose of this is to stack the single frame images at different times along the time axis and update the channel of the image to the combination of the doped time dimension. The tensor is then entered into the window attention module. For large-scale and high-resolution spatiotemporal series prediction tasks, the computational complexity of global attention is generally $\Omega(MSA)_1 = 4HWCT^2 + 2(HWC)^2T$, and the spatial complexity is $O\left((HWC)^2\right)$. Such square-level computational complexity causes a very large load on intensive prediction tasks [36]. By stacking time dimension and channel dimension, MSWO controls the matrix multiplication operation in smaller dimensions. The overall computational complexity of the model is $\Omega(MSA)_2 = 4HW(CT)^2 + 2(HW)^2CT$, and the space complexity becomes $O\left((HW)^2\right)$, which greatly reduces the overhead of the model. In addition, inspired by the Swin Transformer [35] and SwinLSTM [40] models, we designed SWO and MSWO options in the window attention section, as shown in Figure 4c,d. SWO includes general window segmentation and a sliding window module. By dividing the whole image and performing attention calculation in the window, the computing overhead is further reduced. At this time, the whole image is divided into

$\frac{H}{M} \times \frac{W}{M}$ window matrices of $M \times M$ size, and the computational complexity is reduced to $\Omega(MSA)_2 = \frac{HW}{M^2} \times \left(4M^2\left(CT\right)^2 + 2M^4CT\right) = 4HW(CT)^2 + 2M^2HWCT$, and the space complexity is $O\left(M^4\right)$. In MSWO, we use patch merging to acquire multi-scale image data, and through the residual joining operation, we can focus on a larger sensory field in the $M \times M$ size window.

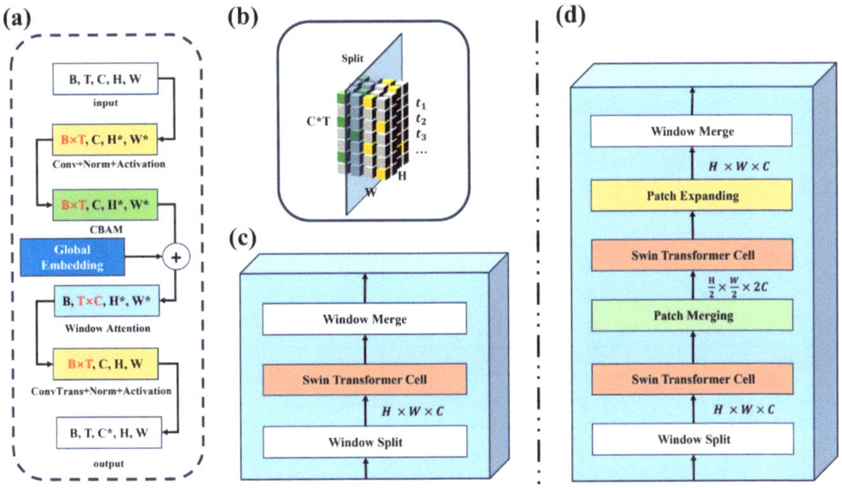

Figure 4. Details in the MSWO process. (**a**) The process of tensor dimension change in the whole process. (**b**) Window segmentation and attention calculation diagram. (**c**) SWO model flow chart. (**d**) MSWO model flow chart.

The Swin transformer cell is divided into two modules. The first module captures the token relationship inside the window through the window-multi-head attention mechanism (W-MHA); the second module is the same as the first module, only the W-MHA is replaced by the shift window-multi-head attention (SW-MHA) to capture the token relationship between adjacent windows. Finally, the decoding process also needs to reshape the tensor shape into $(B \times T) \times C \times H^* \times W^*$ to extract the information of a single image, and map the subsampled spatiotemporal data back to the original size through a deconvolution operation as follows:

$$z = \delta(Norm(ConvTrans(z))) \quad (5)$$

4. Experiments

In the experimental part, we introduce the whole experimental design of the model, including data sources, processing methods, parameters of the training model, evaluation indexes of the model results, and the design of the ablation experiment.

4.1. Data

In this study, we utilized globally available datasets, including sea surface temperature (SST), absolute dynamic topography (ADT), sea surface salinity (SSS), and Argo-measured datasets. The temporal resolution for all data is monthly averages, and the spatial resolution was interpolated to $1^\circ \times 1^\circ$ using interpolation methods. During the training and testing phases, satellite remote sensing data were integrated as input variables, and Argo profile data served as output variables. As detailed in Table 1, SST data were sourced from the National Oceanic and Atmospheric Administration (NOAA)'s OISST dataset, ADT data from Aviso, and SSS data from SMOS satellite's Level 3 data. The study area encompassed the Central South Pacific, spanning from 44.5°S to 18.5°N and 116.5°W to 179.5°W. The input dimensions for the remote sensing images were 64×64 pixels. The temporal sequence of the dataset ranged from January 2011 to December 2019, encompassing 108 months. The time series data were divided into training, validation, and testing sets in a 7:1:1 ratio, employing a sliding window method [39]. The input time series covered 24 months, with predictions extending 6 months into the future. Argo depth selection focused on the upper 250 m of the ocean, divided into 10 depth layers (10, 20, 30, 50, 75, 100, 125, 150, 200, and 250 m). Outlier data were processed and optimally interpolated, and the following method was used for data normalization:

$$\mathcal{X}_{norm} = \frac{\mathcal{X} - \mathcal{X}_{min}}{\mathcal{X}_{max} - \mathcal{X}_{min}} \tag{6}$$

The MSWO model was implemented using the PyTorch framework. The training process utilized the L_2 loss function and the Adam optimizer with a learning rate of 0.001. The model was trained for 1000 epochs, with an early stopping criterion set at 50 epochs, and a batch size of 2. To ensure robustness of the final results, all experiments were repeated three times on an Nvidia Tesla V100 GPU, with the mean and standard deviation of these repetitions reported as the final results.

Table 1. Data sources and resolutions used in present study.

Data	Data Source	Spatial Resolution
Argo	http://apdrc.soest.hawaii.edu/projects/Argo/ (accessed on 1 May 2024)	1°, monthly
ADT	https://www.aviso.altimetry.fr/en/data/products/ (accessed on 1 May 2024)	0.25°, monthly
SST	https://climatedataguide.ucar.edu/climate-data/sst-data-noaa-high-resolution-025x025-blended-analysis-daily-sst-and-ice-oisstv2 (accessed on 1 May 2024)	0.25°, monthly
SSS	https://www.catds.fr/Products/Catalogue-CPDC/Catds-products-from-Sextant#/metadata/0f02fc28-cb86-4c44-89f3-ee7df6177e7b (accessed on 1 May 2024)	25 km, monthly

4.2. Evaluation Indicator

We used three evaluation metrics to evaluate the performance of the predicted results against the measured results, which are mean square error (MSE), root mean square error (RMSE), and mean absolute error (MAE). These three indicators are used to measure the correlation between two data vectors, and the calculation process is as follows:

$$MSE = \frac{1}{m}\sum_{i=1}^{m}(y_i - \hat{y}_i)^2, \tag{7}$$

$$RMSE = \sqrt{MSE}, \tag{8}$$

$$MAE = \frac{1}{m}\sum_{i=1}^{m}|(y_i - \hat{y}_i)|, \tag{9}$$

where, y_i and $\hat{y}_i$ represent the true value and reconstructed value, respectively, and m represents the number of samples. The three indicators approaching zero means that the predicted results are closer to the measured results.

5. Result and Discussion

In this study, we primarily explored the performance of the MSWO model and its capability to predict future changes in ocean subsurface structures using remote sensing satellite images. Under limited training data conditions, the MSWO achieved optimal prediction accuracy. Furthermore, we compared the prediction model with the ocean subsurface structure reconstruction model. Over the 12 months of the test dataset, the prediction and reconstruction models exhibited complementary trends in model accuracy, which may offer new insights for future research.

5.1. Results of Ablation Experiment

The selection of different modules or methods can significantly impact the performance of the MSWO model. In the ablation study section, we investigated the performance of the SWO and MSWO models when incorporating global encoding and the shift window mechanism. Compared to the SWO model, the MSWO model employs the same down-sampling and up-sampling mechanisms as the SwinLSTM model, capturing larger-scale spatial processes by resizing the entire image. Additionally, unlike the spatial relative position encoding used in the window attention mechanism, the global encoding is introduced before the window attention layer. This global encoding allows elements within each small window to not only focus on elements within the same window but also to consider the global context. The shift window mechanism supplements the attention relationships between adjacent independent windows, aligning with the approach proposed by the Swin Transformer. The results of all ablation experiments are shown in Table 2. Overall, using down sampling to capture large-scale spatial information and introducing global spatiotemporal encoding enhances model accuracy. The best SWO model achieved an average MSE of 0.661, RMSE of 0.757, and MAE of 0.528 on the test dataset, while the best SWO-D model achieved an average MSE of 0.648, RMSE of 0.749, and MAE of 0.516 on the test dataset. For convenience, the SWO and MSWO models referred to in the following sections are the best-performing models.

Table 2. The results of ablation experiments are recorded, and the Model includes SWO and MSWO; Global represents whether global encoding of space-time was introduced, and Shifted represents whether a shifted window mechanism was used. The bold part represents the group with the best performance in the SWO and MSWO models.

Model	Global	Shifted	MSE	RMSE	MAE
SWO	×	×	0.842 ± 0.106	0.835 ± 0.049	0.567 ± 0.026
SWO	√	×	$\mathbf{0.661 \pm 0.045}$	$\mathbf{0.757 \pm 0.027}$	$\mathbf{0.528 \pm 0.023}$
SWO	×	√	0.737 ± 0.051	0.798 ± 0.033	0.547 ± 0.031
SWO	√	√	0.788 ± 0.104	0.820 ± 0.051	0.557 ± 0.027
MSWO	×	×	0.737 ± 0.074	0.803 ± 0.047	0.558 ± 0.035
MSWO	√	×	$\mathbf{0.648 \pm 0.047}$	$\mathbf{0.749 \pm 0.026}$	$\mathbf{0.516 \pm 0.012}$
MSWO	×	√	0.804 ± 0.104	0.838 ± 0.058	0.595 ± 0.032
MSWO	√	√	0.676 ± 0.075	0.766 ± 0.044	0.523 ± 0.023

5.2. Model Comparison

In the field of spatiotemporal prediction, numerous advanced artificial intelligence models have been proposed for tasks such as video prediction and weather forecasting. Similar to these works, we utilized these established spatiotemporal sequence models as baseline models to evaluate the performance of the MSWO model. The baseline models include ConvLSTM, PredRNN, SwinLSTM, EarthFormer, SA-ConvLSTM, and SimVP. The input and output formats for MSWO were kept consistent with these baseline models, and the accuracy of the predictions was evaluated using MSE, RMSE, and MAE metrics. To further reduce the likelihood of random events, all experiments were repeated three times, with the mean and the standard deviation recorded for each experiment. Each training and testing session was conducted on an Nvidia Tesla V100 GPU. The average results obtained from all baseline models are presented in Table 3. The overall mean MSE for ConvLSTM, PredRNN, SwinLSTM, EarthFormer, SA-ConvLSTM, SimVP, and our MSWO model was 0.829, 1.007, 1.074, 0.789, 0.935, 0.768, 0.661, and 0.648, respectively. The overall mean RMSE was 0.858, 0.925, 0.963, 0.834, 0.891, 0.813, 0.757, and 0.749, respectively. The overall mean MAE was 0.580, 0.634, 0.648, 0.596, 0.610, 0.575, 0.528, and 0.516, respectively. The MSWO model achieved the best predictive accuracy among all the compared models, indicating its advantage in predicting changes in ocean subsurface structures. This demonstrates that the MSWO model has superior predictive performance in this domain.

Table 3. Comprehensive prediction comparison results between the SWO model and other baseline models on the test dataset, where Size represents the size of the image in the calculation process, and EarthFormer and MSWO perform multi-resolution sampling operations, respectively. The bold part represents the group with the best performance in different model experiments.

Method	Size (H = W)	MSE	RMSE	MAE	Cite
ConvLSTM	16	0.829 ± 0.105	0.858 ± 0.060	0.580 ± 0.048	[33]
PredRNN	16	1.007 ± 0.143	0.925 ± 0.062	0.634 ± 0.043	[34]
SwinLSTM	16	1.074 ± 0.156	0.963 ± 0.064	0.648 ± 0.031	[42]
EarthFormer	16/8	0.789 ± 0.091	0.834 ± 0.054	0.596 ± 0.046	[46]
SA-ConvLSTM	16	0.935 ± 0.105	0.891 ± 0.054	0.610 ± 0.034	[41]
SimVP	16	0.768 ± 0.067	0.813 ± 0.035	0.575 ± 0.023	[30]
SWO (ours)	16	0.661 ± 0.045	0.757 ± 0.027	0.528 ± 0.023	-
MSWO (ours)	16/8	**0.648 ± 0.047**	**0.749 ± 0.026**	**0.516 ± 0.012**	-

To illustrate the variations in predictive accuracy over space and time, we have plotted Figures 5 and 6. Figure 5 presents the performance metrics of different models at various depths within the upper 250 m of the ocean, while Figure 6 shows the performance metrics across different months. All models exhibited a trend where the prediction error initially increases and then decreases, with the maximum error occurring around the 100 m depth. This corresponds to the thermocline, a transitional layer between warmer surface water and cooler deep water, characterized by a steep vertical temperature gradient. The thermocline significantly affects the ocean's density and acoustic fields. Numerous previous studies [22,29,47,48] have highlighted the challenges artificial intelligence models face in accurately reconstructing subsurface structures due to the presence of the thermocline, posing a considerable challenge for all predictive tasks. For the study area, another challenge is the irregular variations in ocean structure caused by the El Niño–Southern Oscillation (ENSO) phenomenon. The limited availability of measured data hampers the ability of AI models to learn long-term decadal variations, thus obstructing accurate future predictions. Figure 6 shows how model errors change over time, with almost all models displaying an increase in error as the prediction horizon extends. This error increase is due to both the declining correlation between the output results and the real satellite data inputs, as well as the accumulation of errors from the recursive process. Overall, the MSWO model achieved the best predictive performance among all compared models, with the lowest error increase over time. This demonstrates that the MSWO model has broad application potential in the field of spatiotemporal sequence prediction.

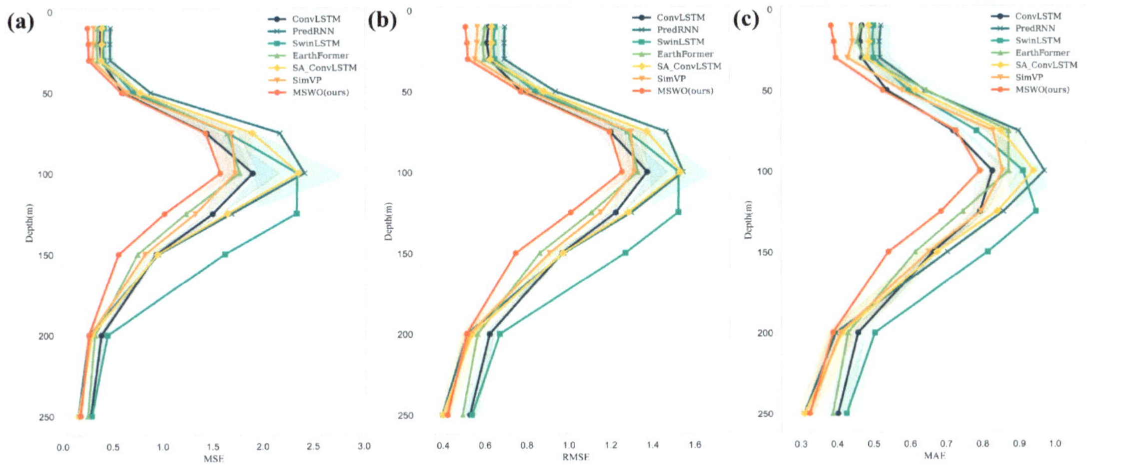

Figure 5. Evaluation index results of different models at different depths. (**a**) MSE, (**b**) RMSE, (**c**) MAE.

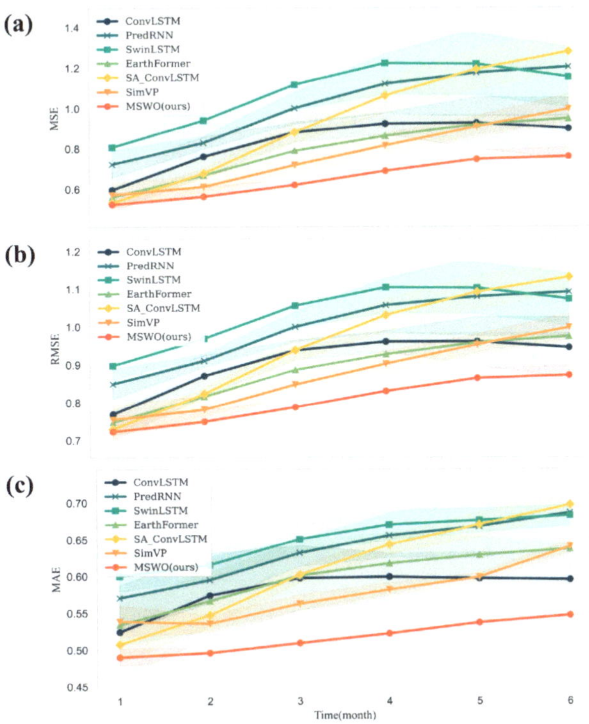

Figure 6. Evaluation index results of different models in the predicted 6-month period. (**a**) MSE, (**b**) RMSE, (**c**) MAE.

5.3. Compare with P2P Schemes

In this study, we introduce a method for predicting future subsurface ocean temperature structures using historical satellite remote sensing images. However, we pose a new inquiry: How does the performance of using satellite remote sensing imagery as input for prediction compare to using historical profile data? Employing satellite remote sensing image for predicting future ocean structures is more convenient due to its lower cost of acquisition, wider coverage, and longer available period. Hence, we replaced historical satellite remote sensing data with historical measured Argo data as input, employing the MSWO model for training and testing. The experiments were replicated three times on an Nvidia Tesla V100 GPU, utilizing MSE, RMSE, and MAE as evaluation metrics. The performance of the model across depths is depicted in Figure 7.

In Figure 7, S2P (surface to profile) and P2P (profile to profile) represent predictions of future profile structures using historical satellite remote sensing and historical profile data, respectively. Unexpectedly, compared to using solely satellite remote sensing data, employing larger volumes of historical profile data did not result in a significant increase in predictive accuracy; the predictive errors of P2P remained largely comparable to those of S2P. A comparison between the S2P and P2P modes reveals the superior performance of S2P in the upper layers of the ocean and the relatively better results around 100 m depth for P2P. This suggests that variations captured by satellite remote sensing data exhibit more relevant features in the prediction process for upper ocean layers. As the inversion depth increases, the introduction of historical profile data may lead to improved inversion outcomes. Overall, the comprehensive performance of S2P and P2P modes in this experiment did not demonstrate significant differences. This finding may be attributed to the fact that changes in the upper 250 m of ocean structure are largely influenced by surface ocean dynamics, and the variations captured by historical satellite remote sensing data are already sufficiently comprehensive. Despite being influenced by dataset limitations and variations related to ENSO phenomena in the study area, predictive models still have considerable room for improvement in accuracy. Nevertheless, this discovery underscores the efficacy of utilizing historical satellite remote sensing data for predicting future subsurface ocean structures, with an accuracy comparable to using historically measured data for forecasting future ocean profiles. This facilitates the direct utilization of satellite remote sensing imagery for large-scale, long-term prediction forecasts in practical applications.

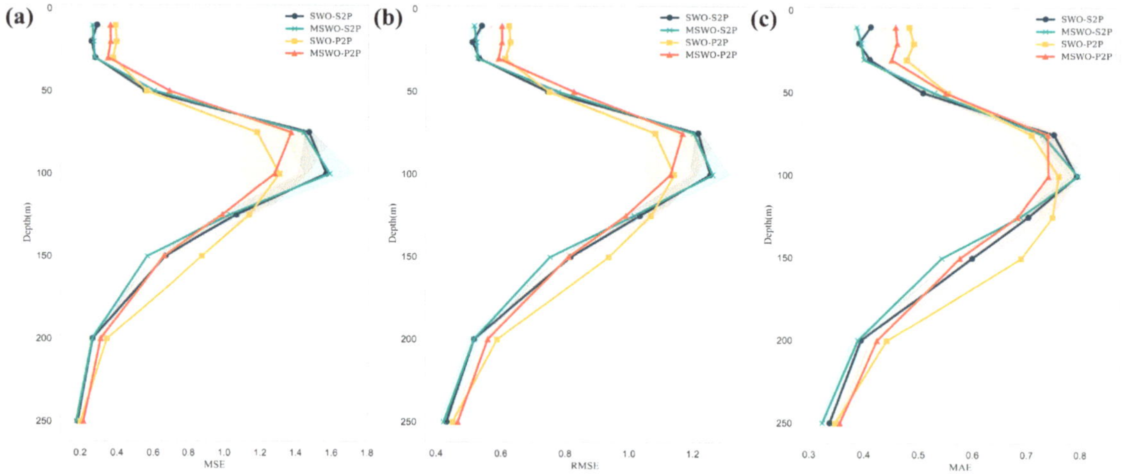

Figure 7. Evaluation index results of historical satellite remote sensing predicting future profile structure (S2P) and historical profile predicting future profile structure (P2P) models at different depths. (**a**) MSE, (**b**) RMSE, (**c**) MAE.

The task of predicting the next 6 months using the preceding 24 months differs from the reconstruction task, where the input comprises the preceding 12 months' data alongside the known 6 months' data to reconstruct the subsurface ocean temperature structure. In contrast to future prediction tasks, reconstruction tasks benefit from the utilization of real-time satellite remote sensing data, enabling them to better capture real-time changes while learning historical patterns. To examine the similarities and differences in the inversion results between the two approaches, we conducted a comparative analysis between the prediction and reconstruction results. Specifically, we divided the test set into 12 months, which were then categorized into four seasons based on the Northern Hemisphere's months (with December to February categorized as spring, March to May as summer, June to August as autumn, and September to November as winter), as depicted in Figure 8. Figure 8a represents the prediction results for the 12 months of 2019 obtained by different models using the prediction method, with the line graph illustrating the RMSE between predicted and actual values. Figure 8c depicts the reconstruction results for the 12 months of 2019 obtained by different models using the reconstruction method, with the line graph showing the RMSE between reconstructed and actual values. Figure 8b,d showcase the seasonal variation in errors across different models.

From Figure 8, it is evident that the prediction models generally exhibit higher errors compared to the reconstruction models. Interestingly, we observe an inverse trend in inversion errors between the prediction and reconstruction models on the test dataset (highlighted in the yellow box in the figure). We attribute this phenomenon to the prediction models primarily learning regularities on the temporal scale, lacking input of real-time spatial change data. In contrast, the reconstruction task for subsurface ocean structure better addresses this gap, optimizing inversion results further by incorporating real-time satellite remote sensing data onto the foundation of spatiotemporal sequence prediction. This holds practical significance, particularly for real-time forecasting endeavors.

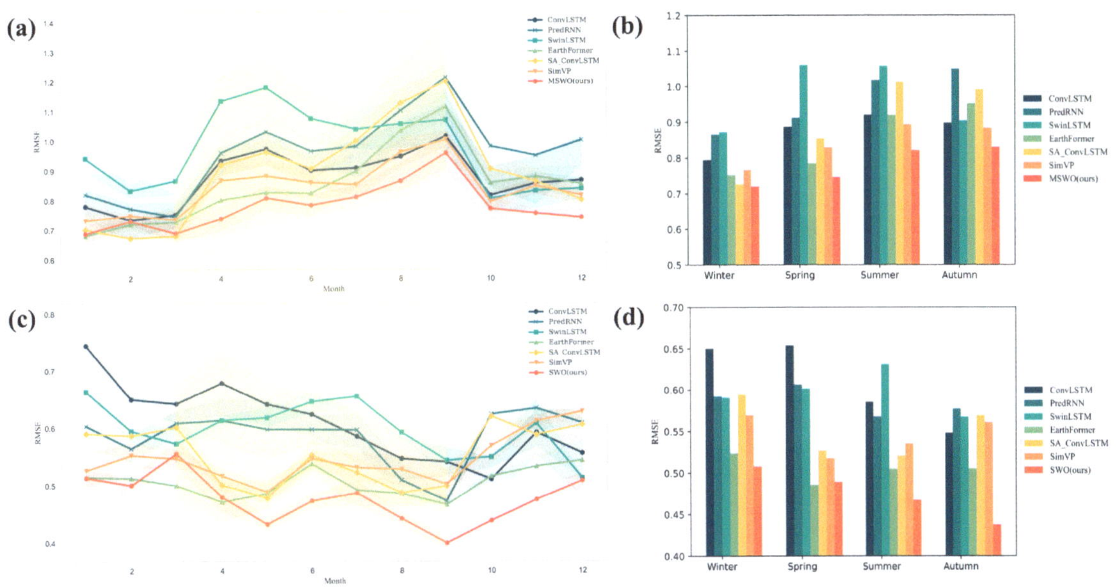

Figure 8. (**a**) RMSE changes of different baseline models on the 12-month prediction task of the test dataset. (**b**) Seasonal RMSE bar chart of different benchmark models on the prediction task of the test dataset. (**c**) RMSE changes for different baseline models on 12-month reconstruction tasks of the test dataset. (**d**) Seasonal RMSE histogram for different benchmark models on the reconstruction task of the test dataset.

5.4. Section Cutting Comparison

To further validate the consistency between the three-dimensional subsurface ocean structure predicted by MSWO and the measured data, we selected four latitudinal and three longitudinal transects for cross-sectional plotting. As shown in Figure 9, along the latitude direction, we selected four transects at 15.5°N (A1), 0.5°N (A2), 15.5°S (A3), and

40.5°S (A4), while along the longitude direction, we chose three transects at 169.5°W (B1), 139.5°W (B2), and 119.5°W (B3). We present the measured Argo profiles, predicted profiles, and their differences plotted along these transects using the MSWO model on the test dataset for January 2019, as illustrated in Figures 10 and 11.

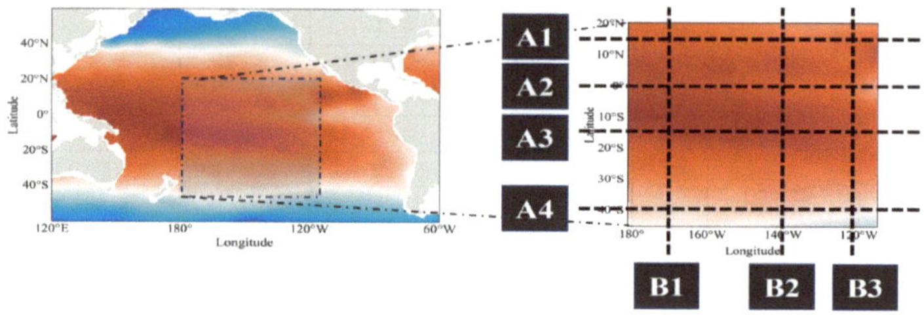

Figure 9. Section selection diagram.

Observing the relationship between the measured and true profiles in Figure 10, it is evident that near the equator, warm water driven by the easterly winds converges into the western Pacific warm pool before sinking, leading to a gradient distribution in subsurface temperatures and shallower thermoclines in the eastern Pacific and deeper thermoclines in the western Pacific. Compared to the measured data, the predicted results from the MSWO model also reflect these gradient changes. Similarly, examining the relationship between the measured profiles in Figure 11, it is observed that the subsurface temperatures in the warm pool region near the equator are higher than those in the eastern Pacific, and subsurface temperatures in the low latitudes of the South Pacific are higher than those in the North Pacific. This distribution correlates strongly with the South Equatorial Current (SEC), Equatorial Under Current (EUC), North Equatorial Counter Current (NECC), and North Equatorial Current (NEC) [49], indicating that MSWO can effectively predict changes in subsurface water masses within the ocean.

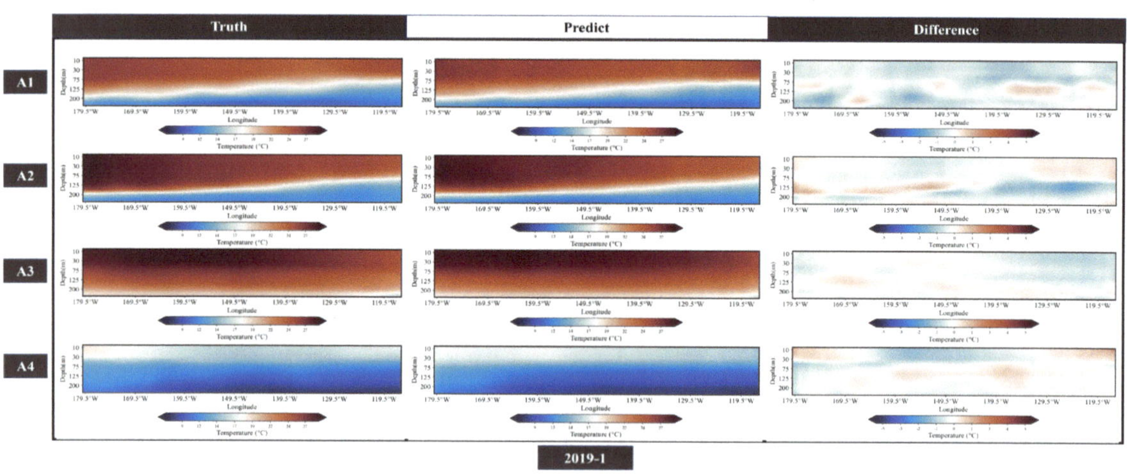

Figure 10. Profile diagram of the January 2019 test dataset drawn at A1–A4 cross sections.

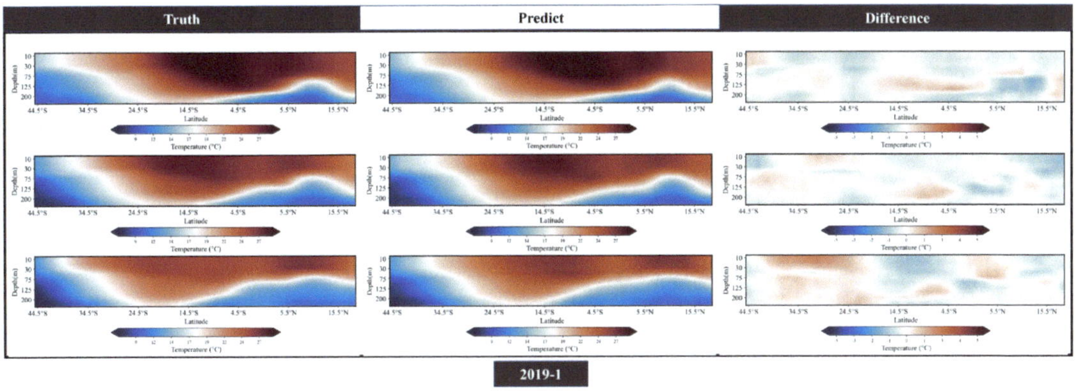

Figure 11. Profile diagram of the January 2019 test dataset drawn at B1–B3 cross sections.

5.5. Planar Error and Density Scatter

The planar error facilitates an understanding of the distribution of errors between the predicted results and the measured data, while the density scatter plot evaluates the correlation between the measured values and the model's predicted results, as well as the robustness of the model's performance. Thus, we generated planar error maps and density scatter plots for all baseline models in July 2019, focusing on the upper 100 m of the ocean. As depicted in Figure 12, significant errors were observed near the equator across all models, with the eastern Pacific displaying underestimations and the western Pacific showing overestimations, a pattern closely associated with the 2019 El Niño event. Overall, MSWO exhibited favorable performance advantages in predicting the development and changes in subsurface oceanic structures. Furthermore, compared to other baseline models, MSWO demonstrated the best robustness and most concentrated inversion effects in the density scatter plot.

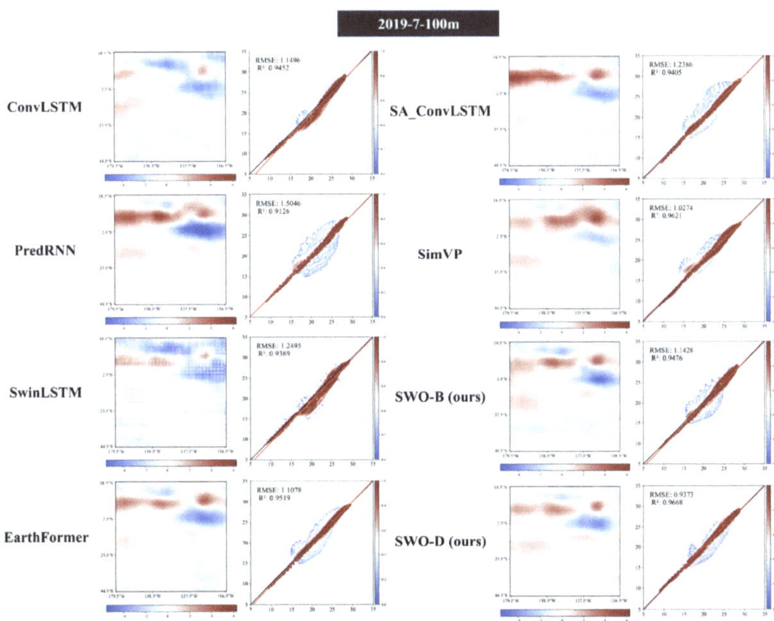

Figure 12. All baseline models were compared with the MSWO model for planar error plots and density scatter plots.

6. Conclusions

With the advancement of human science, the development of artificial intelligence technology has gradually provided new methods and perspectives for realizing the strategy of transparent oceans. In this paper, we proposed the MSWO model and employed it to predict the subsurface temperature structure for the next six months in the Central South Pacific region using satellite remote sensing data from the past 24 months. MSWO introduces a global positional encoding to enrich the spatiotemporal relationship features among the data and utilizes channel attention mechanisms to extract crucial information. Subsequently, MSWO employs multiscale residual operations and window attention mechanisms to extract spatiotemporal correlations within the input data, facilitating predictions of future subsurface oceanic structures. In comparative experiments, MSWO achieved the best predictive performance, benefiting from the attention mechanism's extraction of global spatiotemporal information and its efficient utilization. Additionally, we explored the impact of using either satellite remote sensing data alone or profile data alone as inputs on future predictions. Surprisingly, the results from both strategies were similar, indicating that we can directly predict and forecast future oceanic changes using satellite remote sensing methods, which are more practically applicable compared to profile prediction models that require scarce measured data. Furthermore, we compared the performance of MSWO in prediction tasks and reconstruction tasks. The experimental results show that the error trends of prediction models and reconstruction models exhibited complementary characteristics across the 12 months of the test dataset. The incorporation of remote sensing images as inputs in reconstruction models further complements the real-time features missing in prediction models, thereby improving inversion accuracy.

However, there are still many opportunities for further development in this research. For instance, the lack of measured datasets limits the model's ability to capture long-term decadal changes, and the monthly and spatial resolutions lead to the loss of many small and medium-scale change processes. The impact of the ENSO process on the model cannot be avoided, and the extent of its influence is currently unclear. The design of the MSWO model supports high-resolution satellite remote sensing images and large-scale spatiotemporal process forecasting. In future work, we will explore higher-resolution ocean reconstruction and forecasting.

Author Contributions: Conceptualization, L.W. and X.Z.; Data curation, Y.L.; Formal analysis, J.W.; Investigation, J.J.; Methodology, J.J.; Resources, X.Z.; Supervision, X.Z.; Validation, J.J., C.H. and Q.J.; Visualization, L.F. All authors have read and agreed to the published version of the manuscript.

Funding: This work was supported by the Technology Support Talent Program of the Chinese Academy of Sciences (E4KY31), the Chinese Academy of Sciences pilot project (XDB42000000), the Major Science and Technology Infrastructure Maintenance and Reconstruction Project of the Chinese Academy of Sciences (DSS-WXGZ-2022), the National Key Research and Development Program (2021YFC3101504), the National Natural Science Foundation of China (42176030) and High Level Innovative Talent Project of NUDT.

Data Availability Statement: The Argo data can be downloaded from http://apdrc.soest.hawaii.edu/projects/Argo/ (accessed on 1 May 2024); the ADT data can be downloaded from https://www.aviso.altimetry.fr/en/data/products/ (accessed on 1 May 2024); the SST data can be downloaded from https://climatedataguide.ucar.edu/climate-data/sst-data-noaa-high-resolution-025x025-blended-analysis-daily-sst-and-ice-oisstv2 (accessed on 1 May 2024); and the SSS data can be downloaded from https://www.catds.fr/Products/Catalogue-CPDC/Catds-products-from-Sextant#/metadata/0f02fc28-cb86-4c44-89f3-ee7df6177e7b (accessed on 1 May 2024).

Conflicts of Interest: The authors declare no conflicts of interest.

References

1. Behrenfeld, M.J.; O'Malley, R.T.; Siegel, D.A.; McClain, C.R.; Sarmiento, J.L.; Feldman, G.C.; Milligan, A.J.; Falkowski, P.G.; Letelier, R.M.; Boss, E.S. Climate-Driven Trends in Contemporary Ocean Productivity. *Nature* **2006**, *444*, 752–755. [CrossRef] [PubMed]

2. Johnson, G.C.; Lyman, J.M. Warming Trends Increasingly Dominate Global Ocean. *Nat. Clim. Change* **2020**, *10*, 757–761. [CrossRef]
3. Lowman, H.E.; Emery, K.A.; Dugan, J.E.; Miller, R.J. Nutritional Quality of Giant Kelp Declines Due to Warming Ocean Temperatures. *Oikos* **2022**, *2022*. [CrossRef]
4. Cheng, L.; Abraham, J.; Zhu, J.; Trenberth, K.E.; Fasullo, J.; Boyer, T.; Locarnini, R.; Zhang, B.; Yu, F.; Wan, L.; et al. Record-Setting Ocean Warmth Continued in 2019. *Adv. Atmos. Sci.* **2020**, *37*, 137–142. [CrossRef]
5. Li, J.; Sun, L.; Yang, Y.; Yan, H.; Liu, S. Upper Ocean Responses to Binary Typhoons in the Nearshore and Offshore Areas of Northern South China Sea: A Comparison Study. *Coas* **2020**, *99*, 115–125. [CrossRef]
6. Chassignet, E.P.; Hurlburt, H.E.; Smedstad, O.M.; Halliwell, G.R.; Hogan, P.J.; Wallcraft, A.J.; Baraille, R.; Bleck, R. The HYCOM (HYbrid Coordinate Ocean Model) Data Assimilative System. *J. Mar. Syst.* **2007**, *65*, 60–83. [CrossRef]
7. Chen, C.; Liu, H.; Beardsley, R.C. An Unstructured Grid, Finite-Volume, Three-Dimensional, Primitive Equations Ocean Model: Application to Coastal Ocean and Estuaries. *J. Atmos. Ocean. Technol.* **2003**, *20*, 159–186. [CrossRef]
8. Meng, Y.; Rigall, E.; Chen, X.; Gao, F.; Dong, J.; Chen, S. Physics-Guided Generative Adversarial Networks for Sea Subsurface Temperature Prediction. *IEEE Trans. Neural Netw. Learn. Syst.* **2023**, *34*, 3357–3370. [CrossRef] [PubMed]
9. Wang, J.; Ma, Q.; Wang, F.; Lu, Y.; Pratt, L.J. Seasonal Variation of the Deep Limb of the Pacific Meridional Overturning Circulation at Yap-Mariana Junction. *JGR Ocean.* **2020**, *125*, e2019JC016017. [CrossRef]
10. Liu, C.; Feng, L.; Köhl, A.; Liu, Z.; Wang, F. Wave, Vortex and Wave-Vortex Dipole (Instability Wave): Three Flavors of the Intra-Seasonal Variability of the North Equatorial Undercurrent. *Geophys. Res. Lett.* **2022**, *49*, e2021GL097239. [CrossRef]
11. Shu, Y.; Chen, J.; Li, S.; Wang, Q.; Yu, J.; Wang, D. Field-Observation for an Anticyclonic Mesoscale Eddy Consisted of Twelve Gliders and Sixty-Two Expendable Probes in the Northern South China Sea during Summer 2017. *Sci. China Earth Sci.* **2019**, *62*, 451–458. [CrossRef]
12. Tian, T.; Leng, H.; Wang, G.; Li, G.; Song, J.; Zhu, J.; An, Y. Comparison of Machine Learning Approaches for Reconstructing Sea Subsurface Salinity Using Synthetic Data. *Remote Sens.* **2022**, *14*, 5650. [CrossRef]
13. Zhou, G.; Han, G.; Li, W.; Wang, X.; Wu, X.; Cao, L.; Li, C. High-Resolution Gridded Temperature and Salinity Fields from Argo Floats Based on a Spatiotemporal Four-Dimensional Multigrid Analysis Method. *JGR Ocean.* **2023**, *128*, e2022JC019386. [CrossRef]
14. Klemas, V.; Yan, X.-H. Subsurface and Deeper Ocean Remote Sensing from Satellites: An Overview and New Results. *Prog. Oceanogr.* **2014**, *122*, 1–9. [CrossRef]
15. Li, X.; Liu, B.; Zheng, G.; Ren, Y.; Zhang, S.; Liu, Y.; Gao, L.; Liu, Y.; Zhang, B.; Wang, F. Deep-Learning-Based Information Mining from Ocean Remote-Sensing Imagery. *Natl. Sci. Rev.* **2020**, *7*, 1584–1605. [CrossRef]
16. Yan, X.-H.; Schubel, J.R.; Pritchard, D.W. Oceanic Upper Mixed Layer Depth Determination by the Use of Satellite Data. *Remote Sens. Environ.* **1990**, *32*, 55–74. [CrossRef]
17. Khedouri, E.; Szczechowski, C.; Cheney, R. Potential Oceanographic Applications Of Satellite Altimetry For Inferring Subsurface Thermal Structure. In Proceedings of the Proceedings OCEANS '83, San Francisco, CA, USA, 29 August–1 September 1983; pp. 274–280.
18. Zhang, Q.; Wang, H.; Dong, J.; Zhong, G.; Sun, X. Prediction of Sea Surface Temperature Using Long Short-Term Memory. *IEEE Geosci. Remote Sens. Lett.* **2017**, *14*, 1745–1749. [CrossRef]
19. He, Q.; Zha, C.; Song, W.; Hao, Z.; Du, Y.; Liotta, A.; Perra, C. Improved Particle Swarm Optimization for Sea Surface Temperature Prediction. *Energies* **2020**, *13*, 1369. [CrossRef]
20. Zhang, X.; Zhao, N.; Han, Z. A Modified U-Net Model for Predicting the Sea Surface Salinity over the Western Pacific Ocean. *Remote Sens.* **2023**, *15*, 1684. [CrossRef]
21. Xu, S.; Dai, D.; Cui, X.; Yin, X.; Jiang, S.; Pan, H.; Wang, G. A Deep Learning Approach to Predict Sea Surface Temperature Based on Multiple Modes. *Ocean Model.* **2023**, *181*, 102158. [CrossRef]
22. Liu, Y.; Zhang, L.; Hao, W.; Zhang, L.; Huang, L. Predicting Temporal and Spatial 4-D Ocean Temperature Using Satellite Data Based on a Novel Deep Learning Model. *Ocean Model.* **2024**, *188*, 102333. [CrossRef]
23. Sun, N.; Zhou, Z.; Li, Q.; Zhou, X. Spatiotemporal Prediction of Monthly Sea Subsurface Temperature Fields Using a 3D U-Net-Based Model. *Remote Sens.* **2022**, *14*, 4890. [CrossRef]
24. Yue, W.; Xu, Y.; Xiang, L.; Zhu, S.; Huang, C.; Zhang, Q.; Zhang, L.; Zhang, X. Prediction of 3-D Ocean Temperature Based on Self-Attention and Predictive RNN. *IEEE Geosci. Remote Sens. Lett.* **2024**, *21*, 1–5. [CrossRef]
25. Su, H.; Lu, W.; Wang, A.; Tianyi, Z. AI-Based Subsurface Thermohaline Structure Retrieval from Remote Sensing Observations. In *Artificial Intelligence Oceanography*; Springer Nature: Singapore, 2023; pp. 105–123, ISBN 978-981-19637-4-2.
26. Meng, L.; Yan, C.; Zhuang, W.; Zhang, W.; Geng, X.; Yan, X.-H. Reconstructing High-Resolution Ocean Subsurface and Interior Temperature and Salinity Anomalies from Satellite Observations. *IEEE Trans. Geosci. Remote Sens.* **2022**, *60*, 1–14. [CrossRef]
27. Xie, H.; Xu, Q.; Cheng, Y.; Yin, X.; Fan, K. Reconstructing Three-Dimensional Salinity Field of the South China Sea from Satellite Observations. *Front. Mar. Sci.* **2023**, *10*. [CrossRef]
28. Zhang, S.; Deng, Y.; Niu, Q.; Zhang, Z.; Che, Z.; Jia, S.; Mu, L. Multivariate Temporal Self-Attention Network for Subsurface Thermohaline Structure Reconstruction. *IEEE Trans. Geosci. Remote Sens.* **2023**, *61*, 1–16. [CrossRef]
29. Su, H.; Qiu, J.; Tang, Z.; Huang, Z.; Yan, X.-H. Retrieving Global Ocean Subsurface Density by Combining Remote Sensing Observations and Multiscale Mixed Residual Transformer. *IEEE Trans. Geosci. Remote Sens.* **2024**, *62*, 1. [CrossRef]
30. Tan, C.; Gao, Z.; Li, S.; Li, S.Z. SimVP: Towards Simple yet Powerful Spatiotemporal Predictive Learning. *arXiv* **2023**, arXiv:2211.12509.

31. Liu, Z.; Lin, Y.; Cao, Y.; Hu, H.; Wei, Y.; Zhang, Z.; Lin, S.; Guo, B. Swin Transformer: Hierarchical Vision Transformer Using Shifted Windows. *arXiv* **2021**, arXiv:2103.14030. [CrossRef]
32. Hochreiter, S.; Schmidhuber, J. Long Short-Term Memory. *Neural Comput.* **1997**, *9*, 1735–1780. [CrossRef]
33. Shi, X.; Chen, Z.; Wang, H.; Yeung, D.-Y.; Wong, W.; Woo, W. Convolutional LSTM Network: A Machine Learning Approach for Precipitation Nowcasting. *Adv. Neural Inf. Process. Syst.* **2015**, *28*.
34. Wang, Y.; Long, M.; Wang, J.; Gao, Z.; Yu, P.S. PredRNN: Recurrent Neural Networks for Predictive Learning Using Spatiotemporal LSTMs. In Proceedings of the 31st International Conference on Neural Information Processing Systems, Long Beach, CA, USA, 4–9 December 2017; Curran Associates Inc.: Red Hook, NY, USA, 2017; pp. 879–888.
35. Vaswani, A.; Shazeer, N.; Parmar, N.; Uszkoreit, J.; Jones, L.; Gomez, A.N.; Kaiser, Ł. Illia Polosukhin Attention Is All You Need. In Proceedings of the Advances in Neural Information Processing Systems, Long Beach, CA, USA, 4–9 December 2017; Curran Associates Inc.: Red Hook, NY, USA, 2017; Volume 30.
36. Dosovitskiy, A.; Beyer, L.; Kolesnikov, A.; Weissenborn, D.; Zhai, X.; Unterthiner, T.; Dehghani, M.; Minderer, M.; Heigold, G.; Gelly, S.; et al. An Image Is Worth 16x16 Words: Transformers for Image Recognition at Scale. *arXiv* **2021**, arXiv:2010.11929. [CrossRef]
37. d'Ascoli, S.; Touvron, H.; Leavitt, M.; Morcos, A.; Biroli, G.; Sagun, L. ConViT: Improving Vision Transformers with Soft Convolutional Inductive Biases. *J. Stat. Mech.* **2022**, *2022*, 114005. [CrossRef]
38. Wang, W.; Xie, E.; Li, X.; Fan, D.-P.; Song, K.; Liang, D.; Lu, T.; Luo, P.; Shao, L. Pyramid Vision Transformer: A Versatile Backbone for Dense Prediction without Convolutions. In Proceedings of the IEEE/CVF International Conference on Computer Vision, Montreal, BC, Canada, 11–17 October 2021; pp. 568–578.
39. Zhou, H.; Zhang, S.; Peng, J.; Zhang, S.; Li, J.; Xiong, H.; Zhang, W. Informer: Beyond Efficient Transformer for Long Sequence Time-Series Forecasting. In Proceedings of the AAAI Conference on Artificial Intelligence, Virtual, 2–9 February 2021.
40. Lim, B.; Arık, S.Ö.; Loeff, N.; Pfister, T. Temporal Fusion Transformers for Interpretable Multi-Horizon Time Series Forecasting. *Int. J. Forecast.* **2021**, *37*, 1748–1764. [CrossRef]
41. Lin, Z.; Li, M.; Zheng, Z.; Cheng, Y.; Yuan, C. Self-Attention ConvLSTM for Spatiotemporal Prediction. *Proc. AAAI Conf. Artif. Intell.* **2020**, *34*, 11531–11538. [CrossRef]
42. Tang, S.; Li, C.; Zhang, P.; Tang, R. SwinLSTM: Improving Spatiotemporal Prediction Accuracy Using Swin Transformer and LSTM. *arXiv* **2023**, arXiv:2308.09891.
43. Wang, Y.; Wu, H.; Zhang, J.; Gao, Z.; Wang, J.; Yu, P.S.; Long, M. PredRNN: A Recurrent Neural Network for Spatiotemporal Predictive Learning. *IEEE Trans. Pattern Anal. Mach. Intell.* **2022**, *45*, 2208–2225. [CrossRef] [PubMed]
44. Zou, R.; Duan, Y.; Wang, Y.; Pang, J.; Liu, F.; Sheikh, S.R. A Novel Convolutional Informer Network for Deterministic and Probabilistic State-of-Charge Estimation of Lithium-Ion Batteries. *J. Energy Storage* **2023**, *57*, 106298. [CrossRef]
45. Woo, S.; Park, J.; Lee, J.-Y.; Kweon, I.S. CBAM: Convolutional Block Attention Module. In Proceedings of the European Conference on Computer Vision (ECCV), Munich, Germany, 8–14 September 2018.
46. Gao, Z.; Shi, X.; Wang, H.; Zhu, Y.; Wang, Y.; Li, M.; Yeung, D.-Y. Earthformer: Exploring Space-Time Transformers for Earth System Forecasting. *Adv. Neural Inf. Process. Syst.* **2022**, *35*, 25390–25403.
47. Meng, L.; Yan, X.-H. Remote Sensing for Subsurface and Deeper Oceans: An Overview and a Future Outlook. *IEEE Geosci. Remote Sens. Mag.* **2022**, *10*, 72–92. [CrossRef]
48. Zhu, Y.; Zhang, R.-H.; Moum, J.N.; Wang, F.; Li, X.; Li, D. Physics-Informed Deep-Learning Parameterization of Ocean Vertical Mixing Improves Climate Simulations. *Natl. Sci. Rev.* **2022**, *9*, nwac044. [CrossRef] [PubMed]
49. Ménesguen, C.; Delpech, A.; Marin, F.; Cravatte, S.; Schopp, R.; Morel, Y. Observations and Mechanisms for the Formation of Deep Equatorial and Tropical Circulation. *Earth Space Sci.* **2019**, *6*, 370–386. [CrossRef]

remote sensing

MDPI

Article

Improving Short-Term Prediction of Ocean Fog Using Numerical Weather Forecasts and Geostationary Satellite-Derived Ocean Fog Data Based on AutoML

Seongmun Sim [1,2], Jungho Im [1,3,4,*], Sihun Jung [1] and Daehyeon Han [1]

1 Department of Civil, Urban, Earth, and Environmental Engineering, Ulsan National Institute of Science and Technology, Ulsan 44919, Republic of Korea; smsim@unist.ac.kr (S.S.); jsihunh@unist.ac.kr (S.J.); daehyeon.han@unist.ac.kr (D.H.)
2 NARA Space Technology, Seoul 07245, Republic of Korea
3 Graduate School of Carbon Neutrality, Ulsan National Institute of Science and Technology, Ulsan 44919, Republic of Korea
4 Artificial Intelligence Graduate School, Ulsan National Institute of Science and Technology, Ulsan 44919, Republic of Korea
* Correspondence: ersgis@unist.ac.kr

Abstract: Ocean fog, a meteorological phenomenon characterized by reduced visibility due to tiny water droplets or ice particles, poses significant safety risks for maritime activities and coastal regions. Accurate prediction of ocean fog is crucial but challenging due to its complex formation mechanisms and variability. This study proposes an advanced ocean fog prediction model for the Yellow Sea region, leveraging satellite-based detection and high-performance data-driven methods. We used Himawari-8 satellite data to obtain a lot of spatiotemporal ocean fog references and employed AutoML to integrate numerical weather prediction (NWP) outputs and sea surface temperature (SST)-related variables. The model demonstrated superior performance compared to traditional NWP-based methods, achieving high performance in both quantitative—probability of detection of 81.6%, false alarm ratio of 24.4%, f1 score of 75%, and proportion correct of 79.8%—and qualitative evaluations for 1 to 6 h lead times. Key contributing variables included relative humidity, accumulated shortwave radiation, and atmospheric pressure, indicating the importance of integrating diverse data sources. The study emphasizes the potential of using satellite-derived data to improve ocean fog prediction, while also addressing the challenges of overfitting and the need for more comprehensive reference data.

Keywords: data-driven; Himawari-8; LDAPS; CALIPSO; ASOS; shortwave radiation; variable contribution

Citation: Sim, S.; Im, J.; Jung, S.; Han, D. Improving Short-Term Prediction of Ocean Fog Using Numerical Weather Forecasts and Geostationary Satellite-Derived Ocean Fog Data Based on AutoML. *Remote Sens.* **2024**, *16*, 2348. https://doi.org/10.3390/rs16132348

Academic Editors: Xiao-Hai Yan, Hua Su and Wenfang Lu

Received: 19 May 2024
Revised: 24 June 2024
Accepted: 25 June 2024
Published: 27 June 2024

1. Introduction

Ocean fog, also known as sea fog or marine fog, is a meteorological phenomenon that causes fog to form over the ocean. Ocean fog consists of tiny water droplets or ice particles formed by the condensation of water vapor [1–3]. Due to the Mie scattering process, the presence of these tiny particles causes a substantial reduction in visibility to an extent of 1 km or less. Low visibility raises safety concerns not only for shipping, fishing, and maritime activities but also for traffic controls in coastal regions when ocean fog extends inland [4–6]. Such ocean-fog-caused accidents often lead to socio-economic losses, including human fatalities, and thus, it is crucial to predict ocean fog in a timely manner.

To predict fog over the land, including ocean fog intrusion, ground observation time series have been frequently used. Various approaches have been adopted to predict low visibility at ground stations, including ordinary classification [7] and long short-term memory networks [8]. Nevertheless, relying solely on field observations is not a practical method for directly predicting ocean fog, as they are inherently aspatial, resulting in poor expandability to areas without in situ data.

Given the complex interaction between the ocean and atmosphere, people frequently use numerical weather prediction (NWP) to predict the occurrence of ocean fog because it provides visibility predictions for both the land and the ocean. However, the accuracy of NWP visibility forecasts over the ocean is relatively low because the optimization of microphysics and boundary layers for simulating ocean fog varies greatly with time and location [9–11]. Several studies have attempted to enhance NWP forecasts in order to accurately simulate ocean fog by coupling multiple models or adopting additional parameters [12,13]. However, the intricate nature of ocean fog phenomena poses a significant challenge to accurately forecasting ocean fog.

Data from satellite scatterometers are consistently assimilated with NWP wind vectors over the ocean, leading to relatively high accuracy in wind forecasts, potentially enabling the use of satellite-detected ocean fog for predicting its movement [14]. Although this method accurately predicted the centroids of ocean fog patches, it failed to simulate the shapes and sizes of ocean fog due to the assumption that the initial shape of the detected ocean fog must remain constant over time.

Other NWP outputs, such as pressure (P), air temperature (Ta), and relative humidity (RH), have systematic errors due to the difficulties of obtaining observational data for data assimilation over the ocean [15–17]. These systematic errors limit the effectiveness of NWP models in directly predicting ocean fog. Therefore, to overcome these limitations, the use of data-driven modeling techniques, such as machine learning, could be an attractive solution to predict ocean fog from NWP forecast outputs. Furthermore, using data related to sea surface temperatures (SSTs), such as a simulated SST product or accumulated incoming solar radiation, which is strongly linked to the formation of ocean fog [18,19], can improve the performance of ocean fog prediction when combined with NWP data [12,20,21].

To use such data-driven techniques, it is necessary to have diverse ocean fog reference cases on various spatial and temporal domains. While in situ data are typically obtained at limited locations (generally on the land), satellite-based ocean fog detection may be an attractive approach for collecting a wide range of ocean fog samples in terms of space and time. As ocean fog has distinct optical characteristics, many algorithms for detecting daytime ocean fog were proposed using satellite visible channels as input features [22–27]. However, at night, ocean fog also exhibits distinct radiative characteristics in the short and long infrared wavelength channels, leading to the proposal of such channel-based algorithms for nighttime ocean fog detection [28,29]. Daytime and nighttime ocean fog detection models have been combined to produce operational, continuous ocean fog detection. However, inconsistent detection results have often been observed during the transitional period, such as at dawn and dusk, when solar influence changes. To mitigate such an inconsistency, Sim and Im (2023) suggested an algorithm that detects ocean fog continuously, regardless of the time of day, by solely utilizing infrared brightness temperature from geostationary satellite data, resulting in good performance even during transitional periods [6]. As a result, using works of Sim and Im (2023) allows for spatiotemporally diverse and highly reliable ocean fog detection examples [6]. It can also reflect more diverse characteristics of ocean fog than in situ observations, which are limited in both space and time, and help develop generalized algorithms.

Therefore, this study proposed an ocean fog prediction model that considered the spatial and temporal diversity of ocean fog in the Yellow Sea region. Specifically, the aims were to achieve the following: (1) collect ocean fog samples from various locations using satellite-based ocean fog detection; (2) overcome systematic errors in NWP using AutoML, a high-performance data-driven method; and (3) confirm high performance by utilizing SST-related variables with atmospheric variables.

2. Study Area and Data

2.1. Study Area

The study area is the Yellow Sea in Northeast Asia, located between the Korean Peninsula and China. The shallow depth of the Yellow Sea allows the mixing layer to expand downward until it reaches the cold-water layer at the bottom, resulting in abnormal SST drops during warm seasons. This leads to the formation of cold SST zones, even in the summer. The predominant type of fog over the ocean is advection fog, which is primarily formed when warm air passes over a cold surface. Therefore, we selected the Yellow Sea as the study region of focus due to the frequent reports of ocean fog occurrences there [30]. However, the coverage area of the Local Data Assimilation and Prediction System (LDAPS) used in this study is restricted to the Korean peninsula and its surrounding seas. Therefore, the study area was set on the eastern part of the Yellow Sea, extending from 35–40°N to 121.5–127.5°E (Figure 1).

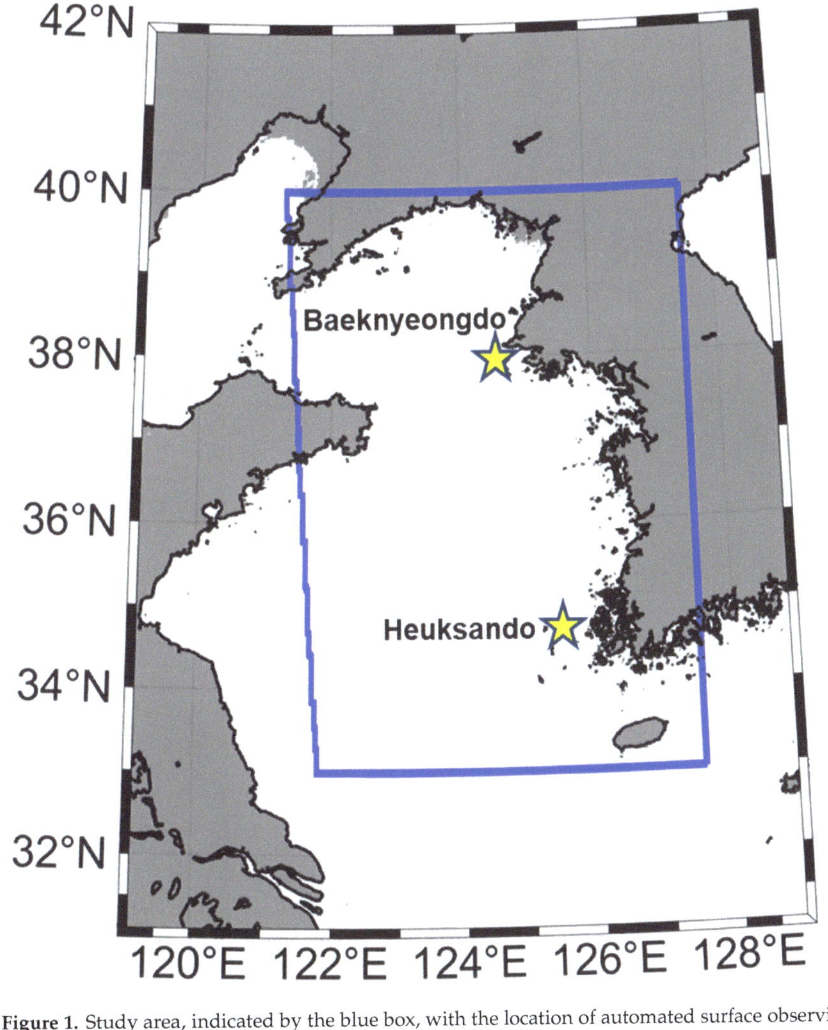

Figure 1. Study area, indicated by the blue box, with the location of automated surface observing system stations located in Baeknyeongdo and Heuksando, which measure various meteorological variables including visibility.

2.2. Himawari-8

Himawari-8 is a geostationary meteorological satellite equipped with a multispectral sensor called the Advanced Himawari Imager (AHI), operated by the Japan Meteorological Agency [31]. This sensor collects data on 16 distinct wavelength channels every 10 min, covering the region of East Asia. The ocean fog detection model developed by Sim and Im (2023) utilizes the infrared channels of Himawari-8 to obtain spatially extensive ocean fog references [6]. Furthermore, when the ocean is unobstructed by clouds or fog, it absorbs solar radiation and accumulates thermal energy. We can estimate the total accumulated solar energy that reaches the ocean by taking into account factors such as the sun's angle and cloud obstructions. Incoming solar radiation raises SSTs, which might restrict the formation and/or maintenance of ocean fog [32–34]. Therefore, we used features that accumulated the hourly level-3 downward shortwave radiation (SWR) product of Himawari-8 from the previous 24 h, 12 h, 9 h to 6 h from the targeting time, the previous day's total accumulation, and cooling hours as input variables for estimating the short-term heat content of the water mass (Table 1) [18,35].

Table 1. Summary of input variables used in the ocean fog prediction model.

Source	Variable Name	Description
Himawari-8	SWR_6to9h	Accumulated shortwave radiation from −9 to −6 h
	SWR_6to12h	Accumulated shortwave radiation from −12 to −6 h
	SWR_6to24h	Accumulated shortwave radiation from −24 to −6 h
	SWR_preday	Accumulated shortwave radiation in previous day
	cooling_H	Cooling hours without shortwave radiation (hours)
LDAPS	Ta	Air temperature (°C)
	RH	Relative humidity (%)
	U	u-vector wind (m/s)
	V	v-vector wind (m/s)
	WS	Wind speed (m/s)
	P	Pressure (Pa)
	VIS	Visibility (m)
HYCOM	SST	Sea surface temperature (°C)
LDAPS & HYCOM	TD	Temperature difference between sea surface and air (°C)

2.3. *Field Reference Data*

In order to evaluate predicted ocean fog, ocean fog occurrence data from the cloud aerosol lidar and infrared pathfinder satellite observation (CALIPSO) and the automated surface observing system (ASOS) were used. CALIPSO is a satellite with a sun-synchronous orbit equipped with a cloud-aerosol lidar and an orthogonal polarization sensor. As CALIPSO provides data across the ocean with a wide spatial range, it could be utilized to assess the spatial reliability of predicted ocean fog [36]. CALIPSO uses two distinct laser beams with wavelengths of 532 and 1064 nm to analyze the vertical distribution of atmospheric particle components. The vertical feature mask (VFM) product provides the categorized class profile, which consists of 545 vertical layers. While there is no specific classification for ocean fog, clouds that are located close to the ocean surface or on unusually high ocean surfaces can be considered to be ocean fog. Consequently, the instances of ocean fog were utilized as reference cases after conducting the quality assessment procedure described in Wu et al. (2015) [5].

ASOS is a meteorological and weather measuring system consisting of field-installed measurements on the land. Currently, 103 ASOS stations are operational in South Korea, providing meteorological and weather information on an hourly basis. Among the ASOS data, visibility information of less or equal to 1 km, which has undergone quality checks based on cloud amount and weather information, can be utilized as ocean fog reference data for coastal locations. Baekneoung-do and Heuksan-do stations, respectively located in 38°N to 124.7°E and 34.4°N to 125.3°E, were selected for this study based on their closeness to the coast and the frequency that the ocean fog was reported. Unfortunately, as cloud and weather information are determined through visual inspection by the administrator, it is essential to use the data after careful quality control.

2.4. LDAPS

LDAPS is an NWP model developed by the Korea Meteorological Administrator (KMA) based on the unified model (UM) of the United Kingdom Met Office [37]. LDAPS assimilates surface and upper air observation data via quality assurance processes, resulting in reduced forecasting errors. LDAPS provides data on 70 vertical profiles with a spatial resolution of 1.5 km around the Korean peninsula eight times per day. The forecast data for a time span of up to 36 h, referred to as 'anal6h', are produced at 00, 06, 12, and 18 UTC. Similarly, the forecast data for a time span of up to 3 h, designated as 'anal3h', along with analysis data, are produced at 03, 09, 15, and 21 UTC. Among LDAPS outputs, surface products highly related to ocean fog occurrence, i.e., air temperature (Ta), relative humidity (RH), pressure (P), u-vector (U), v-vector (V), wind speed (WS), and visibility (VIS), were used as input variables for the ocean fog prediction model [17,38] (Table 1). LDAPS identifies ocean fog when VIS is less than or equal to 1km, called 'V1KM', and we used it as a control model.

2.5. HYCOM SST

The hybrid coordinate ocean model (HYCOM) is a comprehensive global ocean reanalysis system that incorporates various data sources, including field observations from instruments like Argo floats, buoys, and ship-based equipment, as well as satellite observations of SST and ocean surface winds. The HYCOM generates outputs of 40 vertical layers that extend to a depth of 5000 m with a horizontal resolution of $1/12° \times 1/12°$. HYCOM has been extensively studied to assess its reliability in measuring various ocean parameters, including surface salinity and ocean currents [39,40]. The literature has consistently demonstrated that HYCOM performs well in capturing mesoscale ocean phenomena. Regrettably, as HYCOM does not provide a repository of forecast data, the analysis data of the global ocean forecasting system 3.1 version, serving data at 3 h intervals were used under the assumption that the forecast data were closely aligned with the analysis data. Therefore, in this study, the HYCOM SST analysis product was used as an input variable for the ocean fog prediction model after the hourly interpolation of the product (Table 1). We used the temperature difference between SST and LDAPS-derived air temperature (TD) as an input variable in addition to SST, which theoretically indicates the condensation potential of the water vapor [41].

3. Methodology

This study proposed an ocean fog prediction model focusing on the Yellow Sea region, which used Himawari-derived ocean fog data [6], LDAPS outputs, and SWR accumulations with machine learning (Figure 2). Because ocean fog mechanisms and behaviors are complex and LDAPS outputs contain systematic errors, an advanced ensemble-based machine learning model known as automatic machine learning (AutoML) was used. Among the 2019–2022 study period, samples from 2020, which had stable and large ocean fog patches suitable for qualitative assessment, were used for testing the model, including quantitative and qualitative assessments, while samples from the other periods were used to train the model. In addition to validation, the variable contributions of input variables to the AutoML model were investigated. CALIPSO and ASOS data, as well as Himawari-derived ocean fog data, were used in the assessment.

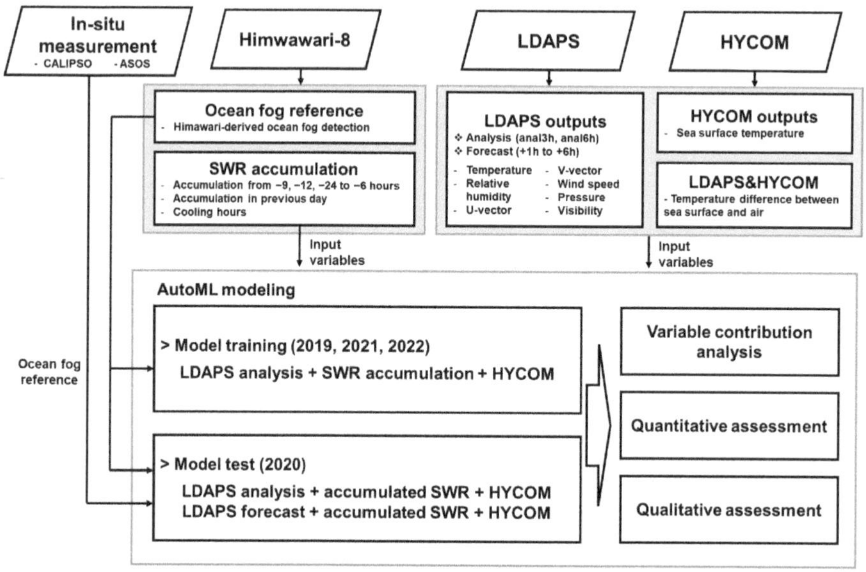

Figure 2. Process flow diagram proposed in this study.

3.1. Extraction of Ocean Fog Reference Data

Satellite-based ocean fog detection is highly useful, but its accuracy may be restricted to specific circumstances due to variations in the optical thickness of atmospheric obstructions, leading to potential missed or misclassified ocean fog occurrences [6,23,24]. Therefore, in order to identify ocean fog locations that are consistently stable and reliable, areas where ocean fog has been observed for more than three hours were designated as highly reliable ocean fog areas. Similarly, we designated areas with consistently clear skies for more than three hours as high-reliability clear sky references. We used the ocean fog detection algorithm by Sim and Im (2023), which has been optimized to identify ocean fog in the study area, the Yellow Sea [6] (Table 2). This algorithm demonstrated consistent performance regardless of time and space; thus, it guaranteed stable spatial references by properly filtering out ocean fog and clear skies.

Table 2. Number of ocean fog reference cases derived from Himawari-8. The number of clear sky reference cases is equal to that of the ocean fog cases.

Purpose	Year	Data Type	Number of Ocean Fog Cases
Training	2019	Analysis	3001
	2021	Analysis	1987
	2022	Analysis	912
Test	2020	Analysis	2187
		Forecast +1 h	2170
		Forecast +2 h	2300
		Forecast +3 h	2008
		Forecast +4 h	624
		Forecast +5 h	767
		Forecast +6 h	1111

3.2. Modeling

In order to predict the occurrence and persistence of ocean fog in complex and diverse instances with numerical weather data and satellite-based SWR accumulation data, AutoML, a massive modeling technique in the form of an ensemble, was used. Specifically, this study used AutoGluon, an open-source AutoML package developed by Amazon Web Services [42]. AutoGluon uses the sequential stacking of multiple machine learning models with repeating n-fold cross validation to achieve the best performance while mitigating overfitting issues. It also optimizes computational resources through hyperparameter sharing, allowing for highly accurate ocean fog prediction results with limited computing resources [43,44]. It is a model that automatically optimizes for optimal performance and requires no hyperparameters other than the number of sequences, folds to cross-validate, and machine learning tasks to run.

In this study, AutoGluon with four sequences was employed (Figure 3). In the first sequence, known as a base layer, various types of machine learning models (e.g., random forest, extremely randomized trees, k-nearest neighbors, light gradient boosting, catboost, extreme gradient boost, and neural network) were utilized to predict ocean fog using the original input variables. In the second sequence, the same machine learning models used in the first layer were trained to predict ocean fog using the base layer's results and the original input variables. Following that, the third layer used the same machine learning models as the first layer to predict ocean fog, but only with the second layer's results. To reduce computing costs, the hyperparameters of each model were shared across these sequences. Finally, in the last sequence, the meta-learning model was trained by concatenating the results from the previous sequence. AutoGluon tuned and fit the hyperparameters and models by repeating the procedure until the user-specified time limit. We used 10-fold cross validation with a time limit of 2 h. The random permutation-based contribution of the input variables used in the modeling was also provided in the form of variable importance, and it was used to estimate how input variables contribute to the model [44,45]. This study utilized samples from LDAPS analysis data for model training and forecast data for model test due to their consistent performance when compared to forecast data. The study period spans from 2019 to 2022; the samples of 2020 were utilized to test the model for both quantitative and qualitative evaluations and the remaining data were used to train the model.

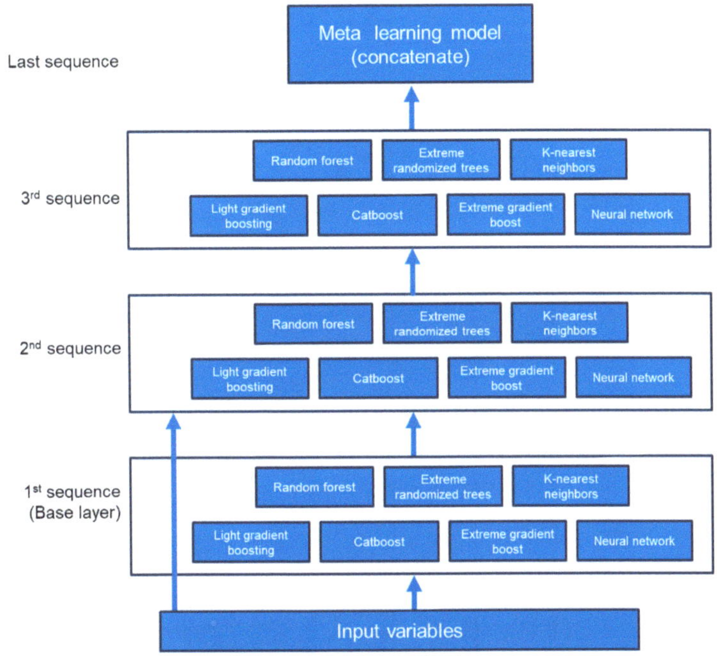

Figure 3. The structure of AutoGluon used in this study.

3.3. Evaluation

The model produced binary output, which was either the presence or absence of ocean fog. The model output can be arranged into a 2 × 2 contingency table using the reference data (Table 3). Four accuracy metrics—probability of detection (POD), false alarm ratio (FAR), F1 score (F1), and proportion correct (PC)—were calculated from the table. POD indicates ocean fog prediction performance, which is the proportion of actual ocean fog cases that are correctly classified (Equation (1)), whereas FAR indicates the proportion of cases predicted to be ocean fog that are incorrectly classified (Equation (2)). The F1 reflects how well POD and FAR perform in all aspects (Equation (3)), whereas PC is the proportion of correctly classified cases out of all cases, which represents overall accuracy (Equation (4)). The evaluation was carried out on a case-by-case basis, with each ocean fog and non-fog patch case containing a collection of multiple pixel samples divided into classes based on the majority of the classified area.

$$POD = \frac{TT}{TT + TF} \times 100\% \quad (1)$$

$$FAR = \frac{FT}{TT + FT} \times 100\% \quad (2)$$

$$F1 = 2 \times \frac{POD \times (100\% - FAR)}{POD + (100\% - FAR)} \quad (3)$$

$$PC = \frac{TT + FF}{TT + FT + TF + FF} \times 100\% \quad (4)$$

Table 3. Contingency table for ocean fog classification. A capitalized "T" means true (ocean fog), and "F" means false (non-fog), with the label of the reference coming first and the label of the predicted result coming last.

		Reference	
		Ocean fog	**Non-fog**
Predicted	**Ocean fog**	TT	FT
	Non-fog	TF	FF

In addition to the quantitative evaluation, a qualitative evaluation based on the spatial distribution of predicted ocean fog was conducted. The spatial reliability of the prediction was assessed using the CALIPSO data, which provide extensive spatial references for ocean fog. We also evaluated the temporal reliability of the model using the ASOS data, which provide temporally continuous ocean fog references.

4. Results and Discussion

4.1. Quantitative Assessment

The proposed AutoGluon model performed well, resulting in a POD of 80.0%, FAR of 23.5%, F1 of 78.2%, and PC of 80.9% for the analysis data, whereas V1KM had a POD of 14.3%, FAR of 1.6%, F1 of 24.9%, and PC of 63.1% (Figure 4). AutoGluon outperformed V1KM for most accuracy metrics, but V1KM exhibited outstanding performance for FAR (Figure 4). It was because the V1KM classified the majority of the samples as non-fog, resulting in a lower FAR, as evidenced by the significantly higher PC compared to the lower F1. Prior research utilizing LDAPS data has demonstrated a tendency for overestimating the visibility of LDAPS [37,46,47], suggesting that the V1KM's accuracy in predicting ocean fog may be compromised due to this factor. Both Autogluon and V1KM exhibited comparable results in terms of prediction accuracy when compared to the analysis data. As the lead time increased to 6 h, performance metrics decreased slightly, but AutoGluon still outperformed V1KM (Figure 4). It is notable that AutoGluon demonstrates superior performance in predicting ocean fog with a lead time of 4 and 5 h (Figure 4). By examining the local time (KST: UTC+9 h, CST: UTC+8 h) of those particular moments, it becomes evident that they correspond to the periods of dawn and dusk. When there is thin ocean fog during that time of day, the detection result may be inconsistent because of solar radiation. However, ocean fog samples that were consistently detected for more than 3 h can be considered to be relatively intense ocean fog cases. The selection of only intense ocean fog cases can be deduced for the purpose of improving the quality of accuracy assessment. While the performance of this study was not validated using the same geographical area and data, the performance metrics from the present study were higher than Kamangir et al. (2021) that had a POD of 65% and FAR of 40% based on deep learning methods with operational NWP results [17] (Figure 4).

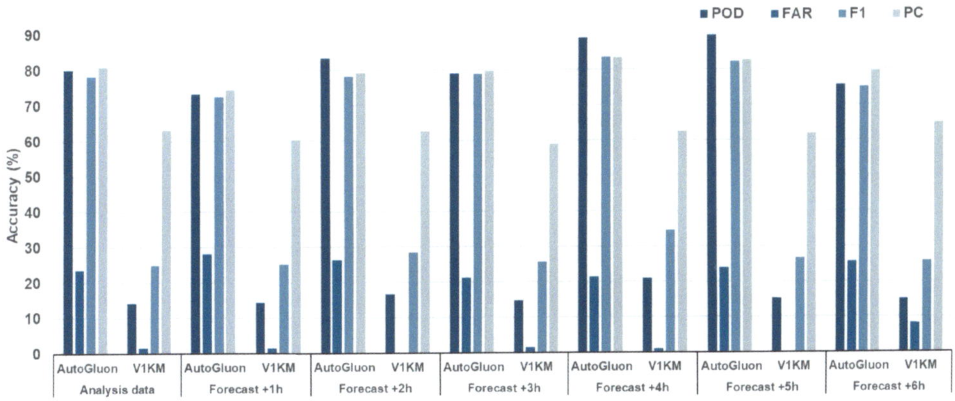

Figure 4. Quantitative performances of AutoGluon and LDAPS V1KM models for hindcast samples of analysis and forecast data with lead times ranging from +1 to +6 h in 2020. Performance metrics such as the probability of detection, false alarm ratio, F1, and proportion correct are displayed in order.

4.2. Variable Contribution Analysis

The relative contribution of the input variables was assessed using the random permutation method (Figure 5). RH was the most contributing variable among the 14 input variables, followed by SWR_preday, P, VIS, cooling_H, and SWR_6to9h. RH is used to compute the aerosol extinction coefficient and the dry air aerosol mass-mixing ratio [37]. VIS is determined by the inverse of the total of the extinction coefficients of clear air and aerosol [37]. In other words, RH is more useful than VIS for simulating ocean fog, dominated by condensed droplets, because VIS simulations use other variables [4,48,49]. Even though there was a high correlation with visibility, the value distributions for ocean fog and non-fog samples were similar, with median values of 89.0% and 88.1%, respectively. This suggests that RH, while making a valuable contribution, did not solely determine the prediction of ocean fog, but interacted with other variables.

While SST and TD were anticipated to be essential components in the ocean fog formation mechanism [20], those ranked 9th and 13th because the SWR-related variables (i.e., SWR_preday, cooling_H, and SWR_6to9h) posed more contributing variables (Figure 5). The sun's shortwave radiation heats the ocean's surface, but when there are clouds or a sunset, the shortwave warming is blocked and the ocean's longwave radiation becomes the primary driver, resulting in the cooling of the ocean surface [32,50–52]. Simply put, a reduction in SWR accumulation results in greater cooling of ocean water, which promotes the formation and persistence of ocean fog [12,44]. Hence, the notable contribution of SWR_preday in this model suggests that it could be valuable for predicting ocean fog when there is a low amount of overall solar energy accumulated in the preceding day. The variable cooling_H denotes the importance of determining the precise timing of persistent cooling in order to predict ocean fog. Moreover, the variable SWR_6to9h serves as an indicator that the recent utilization of solar energy distribution can be valuable in predicting ocean fog. This is trustworthy because the variable SWR_6to9h varies over time, enabling us to pinpoint the exact location of either ocean fog or non-fog.

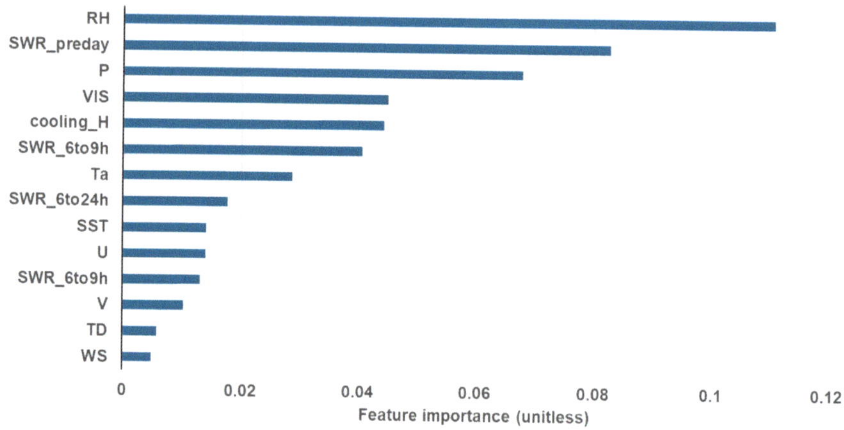

Figure 5. Variable contributions of input variables identified by the AutoGluon model.

Wind is known to significantly influence the development of ocean fog [53,54]. Specifically, gentle breezes are considered crucial for the formation of ocean fog, with advection fog serving as the primary mechanism. However, the evaluation of the ocean fog prediction model revealed minimal influence of U, V, and WS. This can be inferred from the characteristics of the Himawari-derived ocean fog sample used for training; in order to enhance the accuracy of detecting ocean fog, only cases of ocean fog lasting for more than 3 h were chosen, indicating a preference for stable ocean fog instances that have already formed rather than those in the early stages of development. In short, this refers to a situation in which ocean fog samples have been predominantly used at a point where the influence of wind has become insignificant. Furthermore, the P parameter, which serves to indicate a state of macroscopic atmospheric stability, exhibits a high value of model contribution. At lower P, the stability of the atmosphere decreases, leading to an increase in turbulent exchange, a decrease in the stratification of moisture, and a decrease in the presence of liquid water [9,55–57]. These conditions are not favorable for the existence of ocean fog. Conversely, high levels of P are advantageous for the formation of both ocean fog and clear skies. It will be incorporated with other factors in ocean fog prediction models.

4.3. Evaluation of Spatial Distribution

The spatial distribution of the predicted ocean fog was compared to the CALIPSO data in the region characterized by abundant clear skies and ocean fog areas (Figure 6). According to the CALIPSO observation, ocean fog with the unknown class was most prevalent from 34°N to 36°N, clear skies from 36°N to 40°N, and clouds partially with ocean fog from 32°N to 34°N (Figure 7).

We first conducted a comparison between the spatial distribution of the Himawari-derived ocean fog and the predicted ocean fog. According to the results from CALIPSO, the Himawari-derived ocean fog exhibited a mixture of ocean fog and clouds or unknowns below latitudes of 36°N, while the skies were clear in the higher latitudes. All ocean fog prediction models accurately predicted the absence of ocean fog in the clear sky region (Figures 6 and 7). In the ocean fog-dominant region, both AutoGluon using anal3h and anal6h predicted the presence of ocean fog. However, the predicted region was smaller than the detected ocean fog, with the model using anal3h, classifying a smaller area as ocean fog compared to anal6h. Conversely, the results obtained by V1KM indicated a complete lack of prediction for ocean fog in the given region. Despite the presence of clouds, the AutoGluon models predicted the occurrence of ocean fog in the region from 32°N to 34°N. Even though we assume that there is ocean fog beneath clouds, there is no consistent pattern in the distribution of detected and predicted ocean fog. Therefore, we compared the spatial distribution of the input variables that contributed to the prediction of ocean fog in AutoGluon.

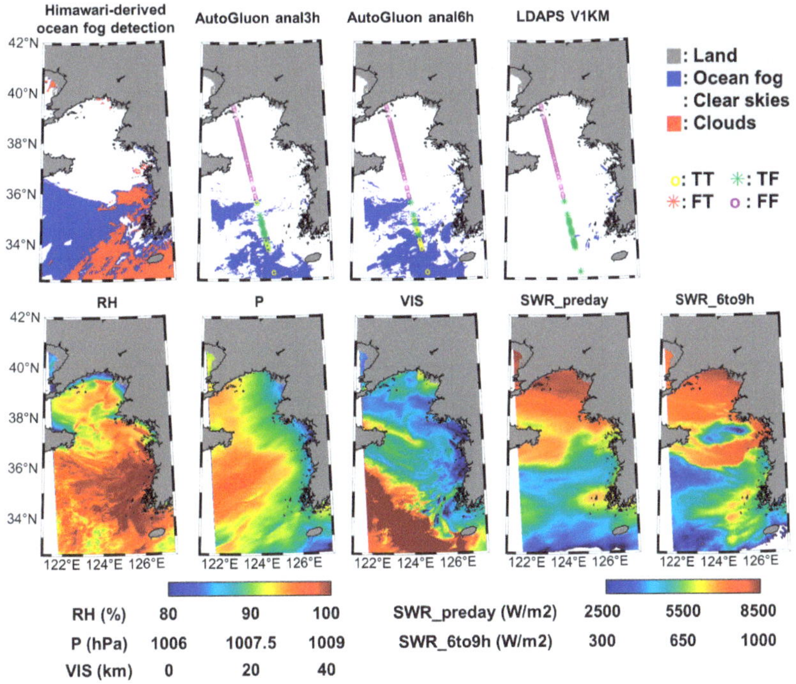

Figure 6. Detected and predicted ocean fog maps and highly contributing input variables for the AutoGluon model in the Yellow Sea at 06 June 2020 05:00 UTC with CALIPSO-based ocean fog observations acquired at 06 June 2020 05:20 UTC. AutoGluon anal3h and anal6h indicate the ocean fog prediction results using the forecast data produced at 03UTC and 00UTC as input, respectively. TT indicates correctly classified ocean fog, TF indicates missed ocean fog, FT indicates falsely classified ocean fog, and FF indicates correctly classified non-fog (refer to Section 3.3).

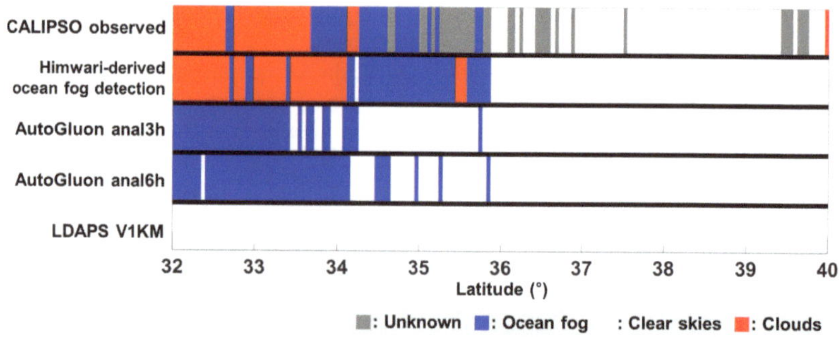

Figure 7. Ocean fog results from the detection and prediction models along with CALIPSO observations on 6 June 2020, at 05:20 UTC. The unknown class includes cases with two or more of the following composite characteristics: ocean fog, clear sky, and cloud.

The RH values were predominantly low in the non-fog areas and high in the areas where significant ocean fog was detected (Figure 6). When comparing the RH to the ocean fog prediction based on AutoGluon, it was found that some areas with a high RH (~100%) were predicted to have ocean fog. Further, Figure 6 shows a majority of agreement between areas predicted to have non-fog and those with a low RH value (<95%). This indicates that RH plays a crucial role in categorizing non-fog conditions. Regarding variable P, even though it exhibits lower spatial resolution compared to other variables, the distribution of high p values and detected ocean fog locations were found to be similar, as reported in the literature [9,47]. A high level of P, located between 35°N and 36°N and 122°E and 124°E, corresponds to the predicted location of ocean fog (Figure 6). This confirms that P is a valuable factor for identifying the presence of ocean fog on a large scale.

The coastal regions located between 34°N–37°N and 126°E–127°E demonstrate reduced VIS levels that correspond to the detected presence of ocean fog. However, in other regions where ocean fog was detected, there was a consistent tendency to overestimate VIS levels > 20 km. Hence, it is clear that VIS has drastically decreased in the area between 37°N and 40°N, where there were clear skies, compared to the area between 32°N–36°N and 122°E–125°E, where there was ocean fog. After analyzing the patterns of the predicted ocean fog based on AutoGluon and VIS, it was determined that there were no similarities (Figure 6). Therefore, it can be concluded that VIS was not used to predict ocean fog in this particular ocean fog case.

While SWR_preday and SWR_6to9h were not provided for the target time, interestingly, these variables showed notable similarities to the detected ocean fog distribution based on their historical accumulation data. The SWR_6to9h data exhibited low values (<500 W/m^2) that closely corresponded to the detected pattern of ocean fog at locations 35°N–36°N and 122°N–124°N (Figure 6). Furthermore, it was discovered that the SWR_6to9h data exhibited similar patterns to the predicted ocean fog area from AutoGluon (Figure 6). AutoGluon identified the region between 32°N–34°N and 122°E–126°E, which has low values of SWR_preday, as an area of ocean fog. However, because the region was classified as cloudy by the ocean fog detection model, a reliable verification could not be performed. Nevertheless, the SWR_6to9h variable in the region of 37°N–38°N, 123°E–126°E exhibits a low value, despite being classified as clear skies by the ocean fog detection model. This suggests that the prediction of ocean fog is not exclusively influenced by the SWR_6to9h variable but rather by a combination of several rules.

Specifically, we compared the CAPLIPSO-based ocean fog case to the ocean fog detection and prediction results (Figure 7). At 32–34°N, where there was a mixture of ocean fog and clouds; the ocean fog detection results were similar to CALIPSO, but the AutoGluon predicted the area as mostly ocean fog. Even though clouds can obscure the presence of ocean fog, the area identified as a mixture of ocean fog and clouds by both

CALIPSO and ocean fog detection indicated a high likelihood of ocean fog beneath clouds. In the 34–36°N region, which was a mixture of ocean fog and the unknown class, the ocean fog detection model classified the region mostly as ocean fog, while the AutoGluon model predicted it as a mixture of ocean fog and non-fog (Figure 7). RH and P were favorable for ocean fog in this area, but the SWR accumulation variables, which were the primary influences in this case, did not show strong favorable patterns for this location, leading to the speculation that only a small area was predicted to be ocean fog. V1KM made no predictions for ocean fog in this region. The 36–40°N region, which had clear skies, was identified as non-fog in both ocean fog detection and prediction models (Figure 7). These results confirm that the AutoGluon-based ocean fog prediction model can predict ocean fog over cloudy areas and be sensitive to the distribution of the SWR accumulation variable.

4.4. Evaluation of Spatiotemporal Distribution

To assess the spatiotemporal continuity and stability of ocean fog prediction models, we chose cases with low cloud cover. First, we investigated the case of ocean fog detected at 37–39°N 124–126°E on 20 June 2020, at 12:00 UTC (Figures 8 and 9). The detected ocean fog was large and persistent, with no cloud contamination, movement, or size variation. However, the ocean fog prediction results demonstrated a different trend, with AutoGluon not classifying any ocean fog at 12:00 UTC, then predicting an ocean fog patch, which grew until the lead time reached 3 h, after which it consistently predicted a similar size patch for the 37–39°N 124–126°E location until the lead time reached 6 h. V1KM did not show noticeable ocean fog areas until the 2 h lead time, and it predicted an ocean fog patch at the 3 h lead time, with the size of the patch increasing until the 6 h lead time. Because AutoGluon as well as V1KM went from undetected to detected and expanded in size for the ocean fog patch, which did not change in size or location, it was assumed that the contribution of LDAPS input variables to AutoGluon's prediction in this case was significant. To investigate the reasons for the prediction trend of ocean fog patches, we examined the spatial distribution of variables with high contributions in AutoGluon (Figure 9). The time series distribution of the input variables demonstrated that RH and VIS followed a similar pattern for the detected ocean fog patches, as did SWR_preday starting at the lead time of 3 h.

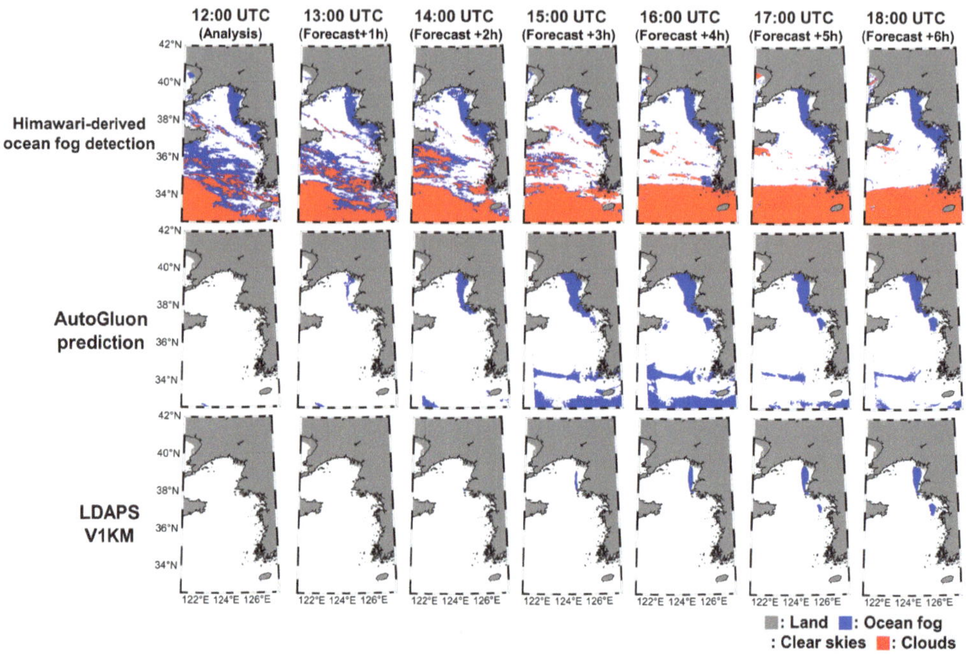

Figure 8. Timeseries mapping results of ocean fog detection, prediction, and LDAPS V1KM on 20 June 2020, from 12:00 UTC to 18:00 UTC. Analysis indicates the use of analysis data for input variables, and forecast indicates predicted results with lead times.

In terms of RH, areas with higher levels seemed to be similar to ocean fog areas determined by Himawari satellite data throughout the observation period. The AutoGluon and Himawari-derived ocean fog distributions exhibited high similarity after the lead time of 3 h, but the earlier time periods showed low similarity, even with extremely high RH values. Specifically, the highest RH value recorded at 12:00 UTC was only at 95%, indicating that the underestimated RH in LDAPS during the analysis and early lead time periods may have contributed an impact in the delayed prediction of ocean fog by AutoGluon. Vis has a similar distribution of low values to RH, but Vis below 1 km becomes noticeable at 3 h lead time, explaining V1KM's failure to predict ocean fog at 2 h lead time. Furthermore, Vis's low value area is very small in comparison to the AutoGluon-predicted patches of ocean fog, implying that Vis played no significant role in predicting ocean fog in this case. The SWR_preday, which represents the previous day's SWR accumulation, is used as the same value by all predictions of AutoGluon on that day. Although low values favor the presence of ocean fog, the spatial distribution differed considerably across all lead times compared to detected ocean fog cases found at 37–39°N 124–126°E. However, there is a line-shaped region at 34.5°N with low SWR_preday values, which AutoGluon predicted as ocean fog with a 2 h lead time. Although it was expected that it would not be used to contribute to the detailed spatial distribution because it is a static variable, it was discovered to have a high contribution at certain moments and is used to predict the detailed spatial distribution of ocean fog based on changes in the contribution degree as the value of other input variables changes.

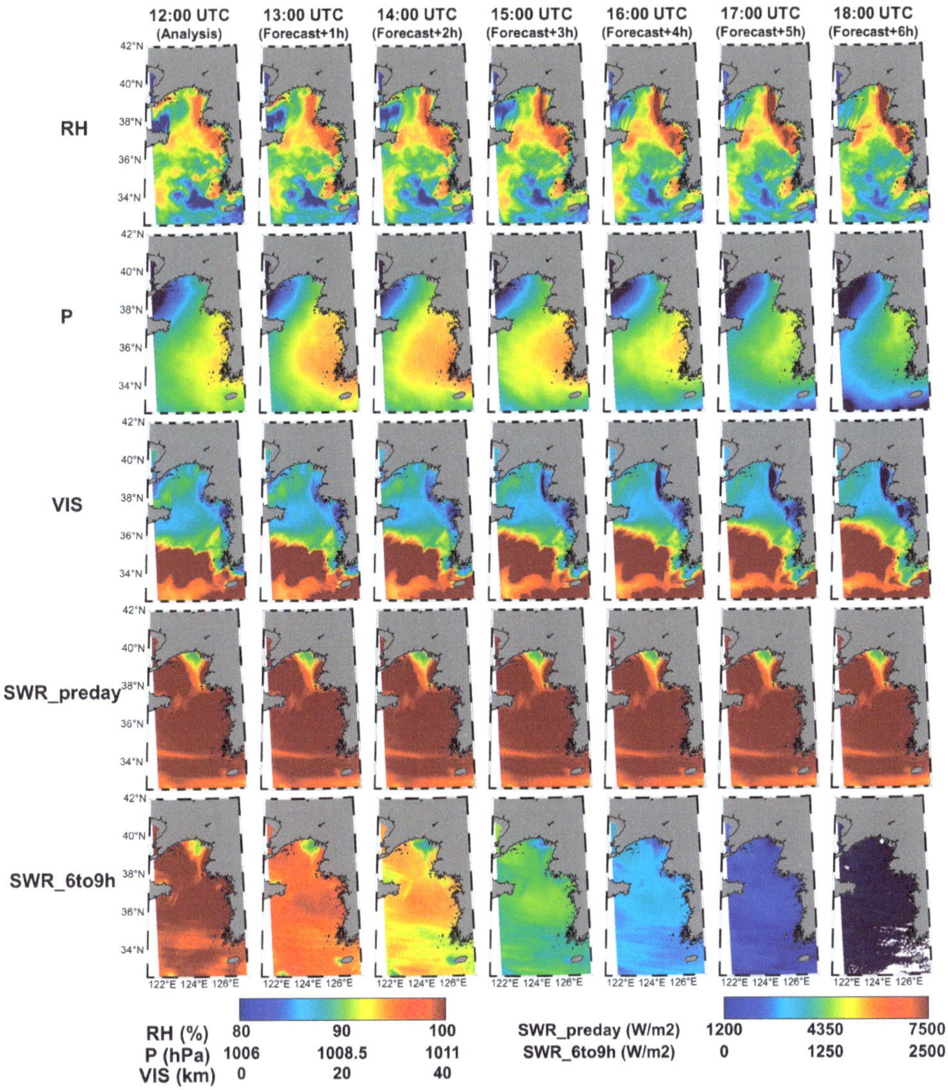

Figure 9. Timeseries mapping results of relative humidity, pressure, visibility, accumulative shortwave radiance of previous day and from −6 h to −9 h on 20 June 2020, from 12:00 UTC to 18:00 UTC. Analysis indicates the use of analysis data for input variables, and forecast indicates the data with the lead times.

We examined trends in ocean fog detection and predictions at Baekneoung-do ASOS location from 20 June 2020, 13:00 UTC to 21 June 2020, 05:00 UTC, including the period covered in Figure 8 when ocean fog was reported (Figures 1 and 10). Ocean fog was observed continuously from 20 June 2020, 13:00 UTC to 21 June 2020, 00:00 UTC, followed by a 2 h increase in measured VIS, and then non-fog was observed beginning 21 June 2020, 03:00 UTC. More than 80% of the area around Baekneoung-do ASOS was correctly classified as ocean fog until 20 June 2020, 20:00 UTC, but the Himawari-derived ocean fog results were unstable between 20 June 2020, 22:00 UTC and 21 June 2020, 00:00 UTC. This is due to cloud contamination, and as the clouds cleared at 01:00 UTC on 21 June 2020, the percentage of ocean fog detections decreased in favor of non-fog, consistent with the observations. At the time, the AutoGluon prediction model performed well in classifying the majority of the area around Baekneoung-do ASOS as ocean fog until 21 June 2020, 01:00 UTC in both anal3h and anal6h, after which the percentage of space predicted as ocean fog gradually declined. However, the dissipation of ocean fog was delayed by about 3 h compared to the ASOS results, indicating a gap in the different views of ocean fog in the assessment values. This is an unavoidable error when comparing the observed presence of ocean fog in one ASOS region to the percentage of ocean fog in a 100 km^2 area. Similar to the results in Figure 8, the predicted area of ocean fog gradually increased from 0% to 100% at 19:00 UTC on 20 June 2020, and then decreased, resulting in a non-fog forecast around the Baekneoung-do station beginning at 22:00 UTC on 20 June 2020. These results show that the AutoGluon-based ocean fog prediction model is stable and performs well regardless of cloud cover.

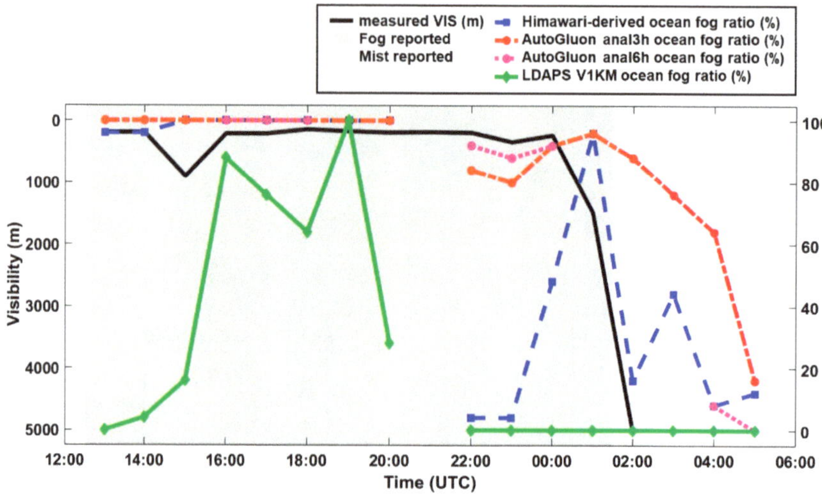

Figure 10. Timeseries of measured visibility, weather report, and ocean fog detection and prediction from 20 June 2020, 13:00 UTC to 21 June 2020, 05:00 UTC at the Baekneoung-do ASOS station. The ocean fog ratio is the proportion of ocean fog coverage within a 100 km^2 surrounding area of the station.

We investigated the case of ocean fog detected at 34–39°N on 17 August 2020, at 00:00 UTC (Figures 11 and 12). This ocean fog was stationary, unobstructed, yet changing in size. The temporal progression of the Himawari-derived ocean fog indicates a reduction in the overall extent of the foggy conditions between 00:00 UTC and 06:00 UTC, with particular emphasis on the fog along the coastline at 34–37°N and 125–126°E. It is shortly after sunrise in the local time zone (KST: UTC+9h, CST: UTC+8h); therefore, the fog slowly disperses from the eastern direction as the sun ascends [6]. Unlike the decreasing fog that was observed, the AutoGluon prediction indicates a sudden and significant decrease in fog area between 02:00 UTC and 03:00 UTC. Additionally, the V1KM prediction fails to predict any ocean fog at all. In order to determine the cause of the discontinuity and lack of predicted ocean fog, the spatial distribution of the variables that have a significant impact on AutoGluon was examined. Following 03:00 UTC, when the ocean fog patch predicted by AutoGluon has a consistently organized appearance, ocean fog areas resemble the distribution of regions with high values of RH. In addition, the predicted fog's areas at that time overlaps with the high P regions, demonstrating adherence to the theoretical basis, and low values of SWR_6to9h. The low VIS values correspond to the predicted ocean fog locations, but the area is relatively small and concentrated along the coast. Furthermore, VIS consistently predicts visibility over 40 km regardless of the time of day, implying that

VIS overestimates offshore areas and, thus, V1KM may not be a good predictor of ocean fog offshore.

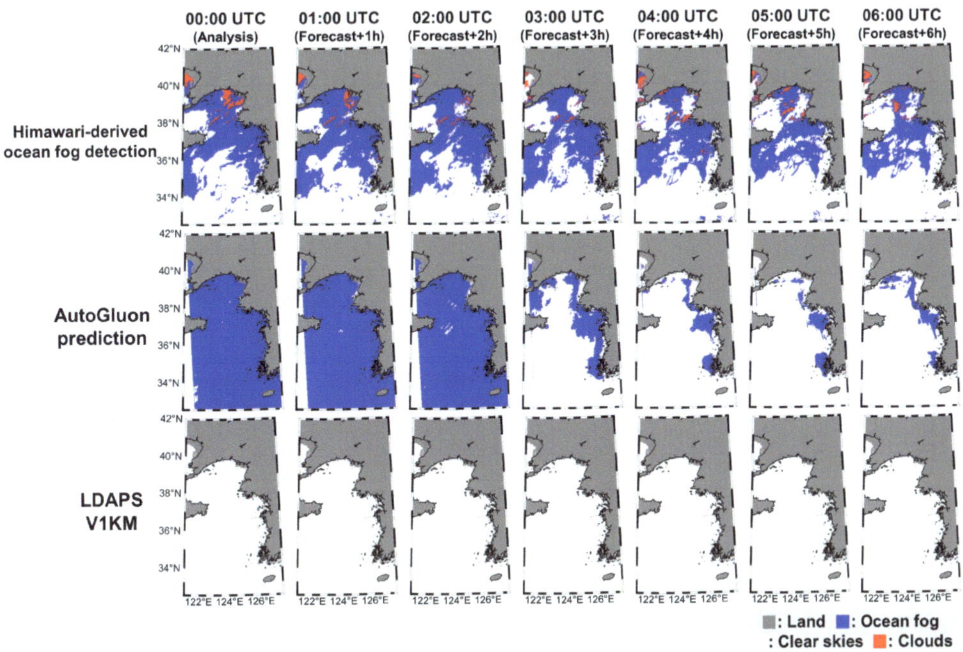

Figure 11. Timeseries mapping results of ocean fog detection, prediction, and LDAPS V1KM on 17 August 2020, from 00:00 UTC to 06:00 UTC. Analysis indicates the use of analysis data for input variables, and forecast indicates predictions with the lead times.

In a whole periodic view, since this period is shortly after sunrise, the time series distribution of SWR_6to9h, where the accumulated SWR value is 0 until 02:00 UTC, exhibits a comparable pattern to the time series discontinuity observed in the AutoGluon prediction. It appears that the AutoGluon model excessively depends on the SWR_6to9h variable, leading to a discontinuity in the time series. This outcome alone might give the perception that the model is overfitting at a specific time. However, areas with persistent ocean fog or clouds lasting more than 3 h also have lower values (~0) for SWR_6to9h. This is reasonable because the cooling effect of longwave radiation creates an optimal condition for the formation of ocean fog [43,50,51]. Nevertheless, the issue of time series discontinuity can pose a challenge; it can be mitigated by increasing the number of training cases in subsequent iterations.

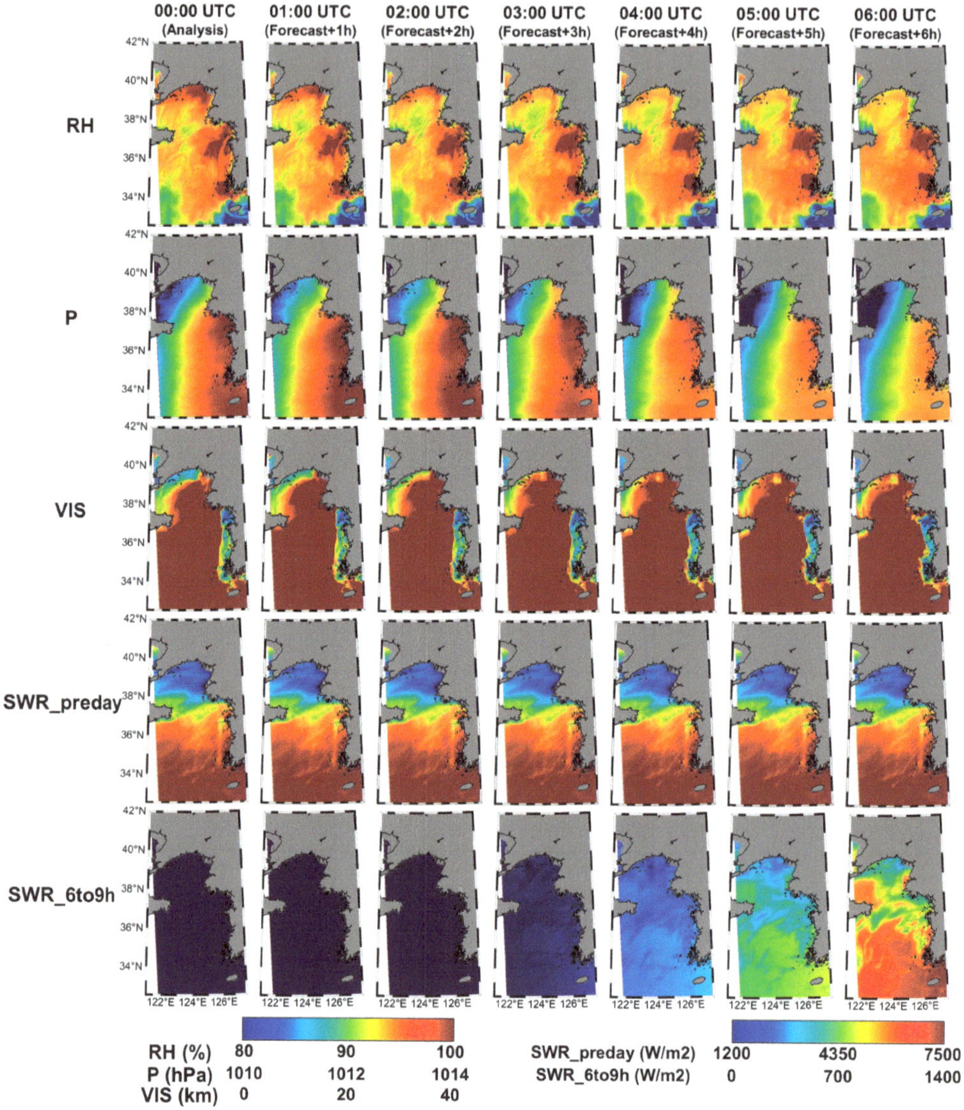

Figure 12. Timeseries mapping results of relative humidity, pressure, visibility, accumulative short-wave radiance of previous day and from −6 h to −9 h on 17 August 2020, from 00:00 UTC to 06:00 UTC. Analysis indicates the use of analysis data for input variables, and forecast indicates the data with the lead times.

As the presence of ocean fog was also recorded at the ASOS station situated on the Heuksan-do, the time period from 16 August 2020, 19:00 UTC to 17 August 2020, 12:00 UTC was examined (Figures 1 and 13). This period included the dissipation and formation of ocean fog. At the Heuksan-do ASOS station, the ocean fog lasted until 16 August 2020, 22:00 UTC, when it transitioned from ocean fog to mist to non-fog with a significant increase in visibility (Figure 13). Subsequently, the weather conditions remained unchanged, and there was a significant decrease in measured VIS starting on 17 August at 09:00 UTC (Figure 13).

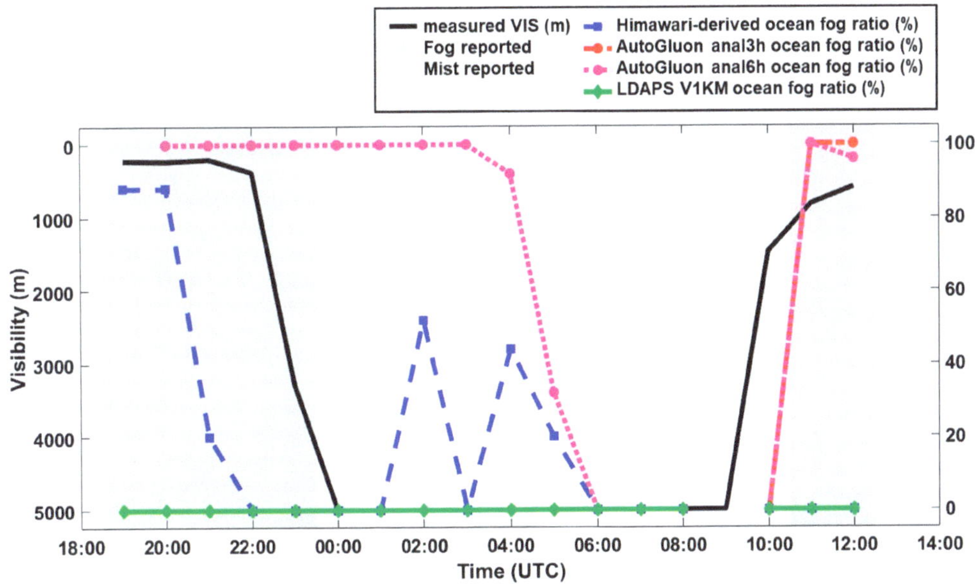

Figure 13. Temporal ocean fog related results of measured visibility, weather report, and ocean fog detection and prediction from 16 August 2020, 19:00 UTC to 17 August 2020, 13:00 UTC at the Heuksan-do ASOS station. The ocean fog ratio is the proportion of ocean fog coverage within a 100 km^2 surrounding area of the station.

The AutoGluon ocean fog predictions indicate that the dissipation occurs after 03:00 UTC on 17 August from both the anal3h and anal6h data, which is likely caused by the overfitting of the SWR_6to9h variable on the map (Figure 13). Subsequently, the ocean fog was rediscovered at 11:00 UTC, and prediction for the ocean fog aligned with the ASOS observations and persisted until 12:00 UTC. At that moment, the lead time for predicting the occurrence of ocean fog was 5 to 6 h, indicating that the AutoGluon model is effective in accurately predicting ocean fog conditions within a 6 h timeframe.

5. Conclusions

Although predicting ocean fog is important, it remains a challenging subject for numerical simulation due to the complexity of the favorable environment. Consequently, data-driven approaches have been utilized for predicting ocean fog, but they have shown limited performance and generality due to a lack of field data that reflected spatial and temporal variability. In this study, we constructed ocean fog cases based on reliable ocean fog detection results reflecting spatial and temporal variability for the Yellow Sea region and used automated machine learning to predict ocean fog with a lead time of up to 6 h. Based on quantitative and qualitative evaluations, the proposed approach outperformed the operational numerical forecasting model's visibility-based ocean fog prediction results. Even though it only occurred in a few cases, the proposed model accurately predicted ocean fog under cloud cover. This demonstrated its future potential for removing cloud contamination from ocean fog detection results. According to the theoretical basis, sea surface temperature and the difference between it and air temperature were considered important input variables, but it was confirmed that the cumulative values of past shortwave radiance contributed more to the prediction of ocean fog, demonstrating the utility of satellite observation data for predicting ocean fog. However, there was a tendency to overfit satellite-derived variables or numerical model outputs during certain periods, which deserves to be further explored in the future. We eliminated all winter ocean fog events as they did not meet the criteria, i.e., lasting more than three hours. In the future, if the quality control of satellite-based ocean fog sample extraction is further improved or differentiated by season, more generalized ocean fog predictions may be possible. In addition, a more accurate sampling of ocean fog cases can be obtained by conducting a thorough analysis of the uncertainty and bias of satellite data. Furthermore, training the model to distinguish not only advection fog but also ocean fog cases caused by cloud lowering is expected to result in more precise and interpretable ocean fog forecasts, which will contribute to operational ocean fog forecasts.

Author Contributions: Conceptualization, S.S. and J.I.; methodology, S.S., S.J. and D.H.; writing—original draft preparation, S.S.; writing—review and editing, J.I.; supervision, J.I.; writing—review and editing, J.I.; funding acquisition, J.I. All authors have read and agreed to the published version of the manuscript.

Funding: This research was funded by Korea Institute of Marine Science & Technology (KIMST) funded by the Ministry of Oceans and Fisheries (RS-2023-00256330, Development of risk managing technology tackling ocean and fisheries crisis around Korean Peninsula by Kuroshio Current) and (20210046, Development of technology using analysis of ocean satellite images), and by Institute of Information & communications Technology Planning & Evaluation (IITP) grant funded by the Korea government (MSIT) (No. 2020-0-01336, Artificial Intelligence Graduate School Program (UNIST)).

Data Availability Statement: The data used in this study are freely available as follows.

- Himawari-8: ftp.ptree.jaxa.jp
  The access account is required following the completion of the registration.
- ASOS: https://data.kma.go.kr/data/grnd/selectAsosRltmList.do?pgmNo=36
- LDAPS: https://data.kma.go.kr/data/rmt/rmtList.do?code=340&pgmNo=65
- HYCOM: https://www.hycom.org/dataserver

Conflicts of Interest: The authors declare no conflicts of interest.

References

1. Duo, Y.; Ritchie, H.; Desjardins, S.; Pearson, G.; MacAfee, A.; Gultepe, I. High-Resolution GEM-LAM Application in Marine Fog Prediction: Evaluation and Diagnosis. *Weather Forecast.* **2010**, *25*, 727–748. [CrossRef]
2. Heo, K.Y.; Ha, K.J. A Coupled Model Study on the Formation and Dissipation of Sea Fogs. *Mon. Weather Rev.* **2010**, *138*, 1186–1205. [CrossRef]
3. He, J.; Ren, X.; Wang, H.; Shi, Z.; Zhang, F.; Hu, L.; Zeng, Q.; Jin, X. Analysis of the Microphysical Structure and Evolution Characteristics of a Typical Sea Fog Weather Event in the Eastern Sea of China. *Remote Sens.* **2022**, *14*, 5604. [CrossRef]
4. Gultepe, I.; Müller, M.D.; Boybeyi, Z. A New Visibility Parameterization for Warm-Fog Applications in Numerical Weather Prediction Models. *J. Appl. Meteorol. Climatol.* **2006**, *45*, 1469–1480. [CrossRef]
5. Wu, D.; Lu, B.; Zhang, T.; Yan, F. A Method of Detecting Sea Fogs Using CALIOP Data and Its Application to Improve MODIS-Based Sea Fog Detection. *J. Quant. Spectrosc. Radiat. Transf.* **2015**, *153*, 88–94. [CrossRef]
6. Sim, S.; Im, J. Improved Ocean-Fog Monitoring Using Himawari-8 Geostationary Satellite Data Based on Machine Learning With SHAP-Based Model Interpretation. *IEEE J. Sel. Top. Appl. Earth Obs. Remote Sens.* **2023**, *16*, 7819–7837. [CrossRef]
7. Guijo-Rubio, D.; Gutiérrez, P.A.; Casanova-Mateo, C.; Sanz-Justo, J.; Salcedo-Sanz, S.; Hervás-Martínez, C. Prediction of Low-Visibility Events Due to Fog Using Ordinal Classification. *Atmos. Res.* **2018**, *214*, 64–73. [CrossRef]
8. Zang, Z.; Bao, X.; Li, Y.; Qu, Y.; Niu, D.; Liu, N.; Chen, X. A Modified RNN-Based Deep Learning Method for Prediction of Atmospheric Visibility. *Remote Sens.* **2023**, *15*, 553. [CrossRef]
9. Koračin, D.; Businger, J.A.; Dorman, C.E.; Lewis, J.M. Formation, Evolution, and Dissipation of Coastal Sea Fog. *Bound.-Layer Meteorol.* **2005**, *117*, 447–478. [CrossRef]
10. Chaouch, N.; Temimi, M.; Weston, M.; Ghedira, H. Sensitivity of the Meteorological Model WRF-ARW to Planetary Boundary Layer Schemes during Fog Conditions in a Coastal Arid Region. *Atmos. Res.* **2017**, *187*, 106–127. [CrossRef]
11. Wang, X.; Qiu, X.; Wu, B.; Dong, Q.; Liu, L.; Li, Y.; Tian, M. Analysis of the Different Influence between Initial/Boundary and Physical Perturbation during Ensemble Forecast of Fog. *Meteorol. Atmos. Phys.* **2023**, *135*, 44. [CrossRef]
12. Lin, C.; Zhang, Z.; Pu, Z.; Wang, F. Numerical Simulations of an Advection Fog Event over Shanghai Pudong International Airport with the WRF Model. *J. Meteorol. Res.* **2017**, *31*, 874–889. [CrossRef]
13. Richter, D.H.; MacMillan, T.; Wainwright, C. A Lagrangian Cloud Model for the Study of Marine Fog. *Bound.-Layer Meteorol.* **2021**, *181*, 523–542. [CrossRef]
14. Harun-Al-Rashid, A.; Yang, C.S. A Simple Sea Fog Prediction Approach Using GOCI Observations and Sea Surface Winds. *Remote Sens. Lett.* **2018**, *9*, 21–30. [CrossRef]
15. Hacker, J.P.; Rife, D.L. A Practical Approach to Sequential Estimation of Systematic Error on Near-Surface Mesoscale Grids. *Weather Forecast.* **2007**, *22*, 1257–1273. [CrossRef]

16. Warner, T.T. *Numerical Weather and Climate Prediction*; Cambridge University Press: Cambridge, UK, 2010.
17. Kamangir, H.; Collins, W.; Tissot, P.; King, S.A.; Dinh, H.T.H.; Durham, N.; Rizzo, J. FogNet: A Multiscale 3D CNN with Double-Branch Dense Block and Attention Mechanism for Fog Prediction. *Mach. Learn. Appl.* **2021**, *5*, 100038. [CrossRef]
18. Feng, Y.; Gao, Z.; Xiao, H.; Yang, X.; Song, Z. Predicting the Tropical Sea Surface Temperature Diurnal Cycle Amplitude Using an Improved XGBoost Algorithm. *J. Mar. Sci. Eng.* **2022**, *10*, 1686. [CrossRef]
19. Dorman, C.E. *Marine Fog: Challenges and Modeling, and in Observations*; Springer: Berlin/Heidelberg, Germany, 2017; ISBN 9783319452272.
20. Fallmann, J.; Lewis, H.; Sanchez, J.C.; Lock, A. Impact of High-Resolution Ocean–Atmosphere Coupling on Fog Formation over the North Sea. *Q. J. R. Meteorol. Soc.* **2019**, *145*, 1180–1201. [CrossRef]
21. Liu, S.; Tian, L.; Lu, Z.; Sai, H.; Liu, C.; Li, P. The Boundary Layer Characteristics and Development Mechanism of a Warm Advective Fog Event over the Yellow Sea. *J. Phys. Conf. Ser.* **2023**, *2486*, 012004. [CrossRef]
22. Jeon, J.-Y.; Kim, S.-H.; Yang, C.-S. Fundamental Research on Spring Season Daytime Sea Fog Detection Using MODIS in the Yellow Sea. *Korean J. Remote Sens.* **2016**, *32*, 339–351. [CrossRef]
23. Yuan, Y.; Qiu, Z.; Sun, D.; Wang, S.; Yue, X. Daytime Sea Fog Retrieval Based on GOCI Data: A Case Study over the Yellow Sea. *Opt. Express* **2016**, *24*, 787. [CrossRef] [PubMed]
24. Han, J.H.; Suh, M.S.; Yu, H.Y.; Roh, N.Y. Development of Fog Detection Algorithm Using GK2A/AMI and Ground Data. *Remote Sens.* **2020**, *12*, 3181. [CrossRef]
25. Kim, D.; Park, M.-S.; Park, Y.-J.; Kim, W. Geostationary Ocean Color Imager (GOCI) Marine Fog Detection in Combination with Himawari-8 Based on the Decision Tree. *Remote Sens.* **2020**, *12*, 149. [CrossRef]
26. Ryu, H.S.; Hong, S. Sea Fog Detection Based on Normalized Difference Snow Index Using Advanced Himawari Imager Observations. *Remote Sens.* **2020**, *12*, 1521. [CrossRef]
27. Mahdavi, S.; Amani, M.; Bullock, T.; Beale, S. A Probability-Based Daytime Algorithm for Sea Fog Detection Using GOES-16 Imagery. *IEEE J. Sel. Top. Appl. Earth Obs. Remote Sens.* **2021**, *14*, 1363–1373. [CrossRef]
28. Shin, D.; Kim, J.H. A New Application of Unsupervised Learning to Nighttime Sea Fog Detection. *Asia-Pac. J. Atmos. Sci.* **2018**, *54*, 527–544. [CrossRef] [PubMed]
29. Amani, M.; Mahdavi, S.; Bullock, T.; Beale, S. Automatic Nighttime Sea Fog Detection Using GOES-16 Imagery. *Atmos. Res.* **2020**, *238*, 104712. [CrossRef]
30. Yu, H.Y.; Suh, M.S. Development of High-Resolution Fog Detection Algorithm for Daytime by Fusing GK2A/AMI and GK2B/GOCI-II Data. *Korean J. Remote Sens.* **2023**, *39*, 1779–1790. [CrossRef]
31. Lee, S.; Kang, Y.; Sung, T.; Im, J. Efficient Deep Learning Approaches for Active Fire Detection Using Himawari-8 Geostationary Satellite Images. *Korean J. Remote Sens.* **2023**, *39*, 979–995. [CrossRef]
32. Larson, K.; Hartmann, D.L. Interactions among Cloud, Water Vapor, Radiation, and Large-Scale Circulation in the Tropical Climate. Part II: Sensitivity to Spatial Gradients of Sea Surface Temperature. *J. Clim.* **2003**, *16*, 1441–1455. [CrossRef]
33. Wu, R.; Kinter, J.L. Shortwave Radiation-SST Relationship over the Mid-Latitude North Pacific during Boreal Summer in Climate Models. *Clim. Dyn.* **2011**, *36*, 2251–2264. [CrossRef]
34. Huang, Y.; Guo, B.; Subrahmanyam, M.V. Shortwave Radiation and Sea Surface Temperature Variations over East and West Tropical Pacific Ocean. *Open Access Libr. J.* **2018**, *5*, 1–9. [CrossRef]
35. Kuang, W.; Liu, A.; Dou, Y.; Li, G.; Lu, D. Examining the Impacts of Urbanization on Surface Radiation Using Landsat Imagery. *GIScience Remote Sens.* **2019**, *56*, 462–484. [CrossRef]
36. Hong, J.; Kim, J.; Jung, Y.; Kim, W.; Lim, H.; Jeong, S.; Lee, S. Potential Improvement of XCO2 Retrieval of the OCO-2 by Having Aerosol Information from the A-Train Satellites. *GIScience Remote Sens.* **2023**, *60*, 2209968. [CrossRef]
37. Clark, P.A.; Harcourt, S.A.; Macpherson, B.; Mathison, C.T.; Cusack, S.; Naylor, M. Prediction of Visibility and Aerosol within the Operational Met Office Unified Model I: Model Formulation and Variational Assimilation. *Q. J. R. Meteorol. Soc.* **2008**, *134*, 1801–1816. [CrossRef]
38. Zhen, M.; Yi, M.; Luo, T.; Wang, F.; Yang, K.; Ma, X.; Cui, S.; Li, X. Application of a Fusion Model Based on Machine Learning in Visibility Prediction. *Remote Sens.* **2023**, *15*, 1450. [CrossRef]
39. Jung, S.; Kim, Y.J.; Park, S.; Im, J. Prediction of Sea Surface Temperature and Detection of Ocean Heat Wave in the South Sea of Korea Using Time-Series Deep-Learning Approaches. *Korean J. Remote Sens.* **2020**, *36*, 1077–1093. [CrossRef]
40. Kim, Y.J.; Han, D.; Jang, E.; Im, J.; Sung, T. Remote Sensing of Sea Surface Salinity: Challenges and Research Directions. *GIScience Remote Sens.* **2023**, *60*, 2166377. [CrossRef]
41. Koračin, D.; Dorman, C.E.; Lewis, J.M.; Hudson, J.G.; Wilcox, E.M.; Torregrosa, A. Marine Fog: A Review. *Atmos. Res.* **2014**, *143*, 142–175. [CrossRef]
42. AutoGluon. AutoGluon Official Web Page. Available online: https://auto.gluon.ai/stable/index.html (accessed on 22 June 2024).
43. Erickson, N.; Mueller, J.; Shirkov, A.; Zhang, H.; Larroy, P.; Li, M.; Smola, A. Autogluon-tabular: Robust and accurate automl for structured data. *arXiv* **2020**, arXiv:2003.06505.
44. Raj, R.; Kannath, S.K.; Mathew, J.; Sylaja, P.N. AutoML Accurately Predicts Endovascular Mechanical Thrombectomy in Acute Large Vessel Ischemic Stroke. *Front. Neurol.* **2023**, *14*, 1259958. [CrossRef] [PubMed]
45. Song, Y.; Xu, Y.; Chen, B.; He, Q.; Tu, Y.; Wang, F.; Cai, J. Dynamic Population Mapping with AutoGluon. *Urban Inform.* **2022**, *1*, 13. [CrossRef]

46. Kim, D.J.; Kang, G.; Kim, D.Y.; Kim, J.J. Characteristics of LDAPS-Predicted Surface Wind Speed and Temperature at Automated Weather Stations with Different Surrounding Land Cover and Topography in Korea. *Atmosphere* **2020**, *11*, 1224. [CrossRef]
47. Kim, B.Y.; Cha, J.W.; Chang, K.H.; Lee, C. Visibility Prediction over South Korea Based on Random Forest. *Atmosphere* **2021**, *12*, 552. [CrossRef]
48. Gultepe, I.; Milbrandt, J.; Zhou, B.B. Visibility parameterization for forecasting model applications. In Proceedings of the 5th International Conference on Fog, Fog Collection and Dew, Münster, Germany, 25–30 July 2010.
49. Elias, T.; Dupont, J.C.; Hammer, E.; Hoyle, C.R.; Haeffelin, M.; Burnet, F.; Jolivet, D. Enhanced Extinction of Visible Radiation Due to Hydrated Aerosols in Mist and Fog. *Atmos. Chem. Phys.* **2015**, *15*, 6605–6623. [CrossRef]
50. Nakanishi, M. Large-Eddy Simulation of Radiation Fog. *Bound.-Layer Meteorol.* **2000**, *94*, 461–493. [CrossRef]
51. Wærsted, E.G.; Haeffelin, M.; Dupont, J.C.; Delanoë, J.; Dubuisson, P. Radiation in Fog: Quantification of the Impact on Fog Liquid Water Based on Ground-Based Remote Sensing. *Atmos. Chem. Phys.* **2017**, *17*, 10811–10835. [CrossRef]
52. Guo, L.; Guo, X.; Luan, T.; Zhu, S.; Lyu, K. Radiative Effects of Clouds and Fog on Long-Lasting Heavy Fog Events in Northern China. *Atmos. Res.* **2021**, *252*, 105444. [CrossRef]
53. da Rocha, R.P.; Gonçalves, F.L.T.; Segalin, B. Fog Events and Local Atmospheric Features Simulated by Regional Climate Model for the Metropolitan Area of São Paulo, Brazil. *Atmos. Res.* **2015**, *151*, 176–188. [CrossRef]
54. Penov, N.; Stoycheva, A.; Guerova, G. Fog in Sofia 2010–2019: Objective Circulation Classification and Fog Indices. *Atmosphere* **2023**, *14*, 773. [CrossRef]
55. Guo, L.J.; Guo, X.L.; Fang, C.G.; Zhu, S.C. Observation Analysis on Characteristics of Formation, Evolution and Transition of a Long-Lasting Severe Fog and Haze Episode in North China. *Sci. China Earth Sci.* **2015**, *58*, 329–344. [CrossRef]
56. Li, X.N.; Huang, J.; Shen, S.H.; Liu, S.D.; Lu, W.H. Evolution of liquid water content in a sea fog controlled by a high-pressure pattern. *J. Trop. Meteorol.* **2010**, *16*, 409.
57. Yang, L.; Liu, J.W.; Ren, Z.P.; Xie, S.P.; Zhang, S.P.; Gao, S.H. Atmospheric Conditions for Advection-Radiation Fog Over the Western Yellow Sea. *J. Geophys. Res. Atmos.* **2018**, *123*, 5455–5468. [CrossRef]

remote sensing

MDPI

Article

Spatial Downscaling of Sea Surface Temperature Using Diffusion Model

Shuo Wang †, Xiaoyan Li †, Xueming Zhu *, Jiandong Li and Shaojing Guo

Southern Marine Science and Engineering Guangdong Laboratory (Zhuhai), School of Marine Sciences, Sun Yat-sen University, Zhuhai 519082, China; wangsh557@mail2.sysu.edu.cn (S.W.); lixiaoyan@sml-zhuhai.cn (X.L.); lijd36@mail2.sysu.edu.cn (J.L.); guoshj9@mail2.sysu.edu.cn (S.G.)

* Correspondence: zhuxueming@sml-zhuhai.cn

† These authors contributed equally to this work and should be regarded as co-first authors.

Abstract: In recent years, advancements in high-resolution digital twin platforms or artificial intelligence marine forecasting have led to the increased requirements of high-resolution oceanic data. However, existing sea surface temperature (SST) products from observations often fail to meet researchers' resolution requirements. Deep learning models serve as practical techniques for improving the spatial resolution of SST data. In particular, diffusion models (DMs) have attracted widespread attention due to their ability to generate more vivid and realistic results than other neural networks. Despite DMs' potential, their application in SST spatial downscaling remains largely unexplored. Hence we propose a novel DM-based spatial downscaling model, called DIFFDS, designed to obtain a high-resolution version of the input SST and to restore most of the meso scale processes. Experimental results indicate that DIFFDS is more effective and accurate than baseline neural networks, its downscaled high-resolution SST data are also visually comparable to the ground truth. The DIFFDS achieves an average root-mean-square error of 0.1074 °C and a peak signal-to-noise ratio of 50.48 dB in the 4× scale downscaling task, which shows its accuracy.

Keywords: spatial downscaling; diffusion model; sea surface temperature; deep learning

1. Introduction

The sea surface temperature (SST) is a crucial climate variable that contributes to Earth's climate [1]. SST influences marine ecosystems, ocean–atmosphere interactions, and oceanic currents. Recent advances in high-resolution digital twin platforms or artificial intelligence marine forecasting, such as Earth-2 (with kilometer-scale resolution), have led to an increased demand for high-resolution oceanic data. High-resolution SST data can reveal more meso- or small-scale dynamic processes, enabling neural networks to learn more complex patterns.

However, due to the limitations of current observation technology, the resolution of existing satellite remote sensing SST products often fails to meet researchers' needs. This severely restricts the potential applications of SST data in various fields, like deep learning-based oceanographic models.

To mitigate this issue, oceanographers have begun to use spatial downscaling techniques to obtain higher-resolution SST datasets. By establishing mapping relationships between low- and high-resolution data, spatial downscaling can generate high-resolution versions of input low-resolution SST. The downscaled high-resolution outcomes can reveal more detailed marine dynamic features, providing a higher spatial resolution for subsequent applications.

Downscaling techniques can be categorized into dynamic downscaling, statistical downscaling, and deep learning-based downscaling methods. Dynamic downscaling is conducted by nesting regional models into low-resolution global models to produce high-resolution information. For instance, Huang et al. [2] used a variable-resolution option

Citation: Wang, S.; Li, X.; Zhu, X.; Li, J.; Guo, S. Spatial Downscaling of Sea Surface Temperature Using Diffusion Model. *Remote Sens.* **2024**, *16*, 3843. https://doi.org/10.3390/rs16203843

Academic Editor: Andrea Storto

Received: 30 August 2024
Revised: 14 October 2024
Accepted: 14 October 2024
Published: 16 October 2024

within the community earth system model to simulate California's climate, demonstrating competitive utility for studying high-resolution regional climatology compared to the weather research and forecast model. Dynamic downscaling can produce reliable results because it uses physical equations to describe dynamic processes, but it often entails high spatial-temporal complexity. In addition to dynamic downscaling, statistical downscaling methods are also widely employed to establish empirical relationships between large-scale variables and local-scale parameters to produce high-resolution data [3]. For instance, Jorge et al. [4] proposed a weather-type downscaling method for multivariate ocean wave climate based on a statistical downscaling framework. Statistical downscaling faces challenges in accurately establishing statistical relationships in areas with complex terrain or unique climate zones, and the quality of the input data affects the precision of the result.

Deep learning has shown impressive prospects in various tasks within marine science, including downscaling, prediction, and reconstruction of oceanic elements. Deep learning-based downscaling originates primarily from super-resolution (SR) technology in computer vision. Dong et al. [5] proposed the first super-resolution convolutional neural network (SRCNN). Subsequently, many CNN-based SR models were created, and their performance improved further [6–10]. The emergence of SR generative adversarial models such as enhanced SR generative adversarial network (ESRGAN) [11], and SR transformer models [12,13], also represent significant advances in SR.

Deep learning methods can generate accurate high-resolution data. They can also automatically learn feature representations from oceanic data, allowing for more effective feature extraction. These characteristics have prompted many oceanographers to explore deep-learning downscaling methods. Some researchers use interpolation algorithms, such as bicubic interpolation or nearest neighbor interpolation, to obtain a low-resolution SST from a high-resolution SST. They then employ these low-resolution SSTs as the input and the original high-resolution SST as the target to train neural networks for the spatial downscaling tasks. For instance, Aurelien and Ronan [14] utilized bicubic interpolation to generate low-resolution input from the operational sea surface temperature and sea ice analysis (OSTIA) dataset and then employed SRCNN to generate high-resolution targets. Similarly, Khoo et al. developed an SN-ESRGAN neural network to downscale low-resolution SST into high-resolution ones [15], wherein the nearest neighbor algorithm was used for generating input data from OSTIA SST. These approaches highlight the efficacy of deep learning-based spatial downscaling techniques in addressing SST downscaling challenges. In other scenarios, high-resolution infrared SST data and low-resolution microwave SST are used to train deep learning models. Izumi et al. trained a CNN-based network with 125 km resolution input and 25 km resolution ground truth, achieving high-quality results [16]. Zou et al. designed a transformer-based model to obtain a resolution output of 0.02°, using the 0.25° advanced microwave scanning radiometer 2 SST as input [17].

Recently, diffusion models (DM) have gained significant attention along with the rise of text-to-image generation models such as Imagen [18] and DALL-E2 [19]. Unlike other deep neural networks, DMs excel at producing more vivid samples and circumventing issues like mode collapse, which can be seen in GANs. These advantages have led to their broad application across various computer vision domains. In SR, diffusion model-based approaches, exemplified by works such as [20–22] have achieved remarkable results. Nevertheless, studies and practical applications focusing on DM-based SST downscaling are noticeably scarce. To explore whether the generative capabilities of DM can be harnessed to restore missing details and processes in low-resolution SSTs, we propose a novel spatial downscaling method DIFFDS based on diffusion model for image restoration (DIFFIR) [23]. In comparison to the original DIFFIR, DIFFDS redesigns the transformer block by introducing cross-attention and channel-attention mechanisms. This refinement results in fewer abnormal textures and more mesoscale details, making DIFFDS more suitable for SST spatial downscaling. We conducted $4\times$ scale downscaling experiments to reveal the superior performance of DIFFDS over several existing methods, offering a fresh perspective for the SST downscaling field.

Our contributions can be summarized as follows:

1. We extended the application of DM to address SST spatial downscaling problems. The proposed DIFFDS method leveraged the robust distribution fitting and generation capabilities of DM to reconstruct high-resolution SSTs. Experimental results demonstrate its effectiveness.
2. To ensure greater consistency between the downscaling results and the original high-resolution SSTs and to mitigate incorrect texture anomalies in the results, we restructured the transformer block in DIRformer. This enhancement allows DIFFDS to fully consider the underlying structure in low-resolution data, thereby resulting in a more reasonable reconstruction of high-resolution SST contents.
3. DIFFDS outperforms the commonly used CNN, GAN, and regression methods on most evaluation metrics and closely approximates the visual fidelity of high-resolution ground truth. This substantiates its superiority over other models.

2. Materials and Methods

2.1. Study Area and Data

2.1.1. Study Area

As shown in Figure 1, the study sea area extends from 6°N to 22°N, 107°E to 123°E. It encompasses the South China Sea, the Luzon Strait, and the Sulu Sea. This domain experiences a prevailing tropical maritime monsoon climate characterized by warm temperatures, seasonal monsoons, and significant rainfall. These conditions induce complex SST distributions and multi-scale dynamical processes, such as upwelling, mesoscale eddies, and oceanic fronts.

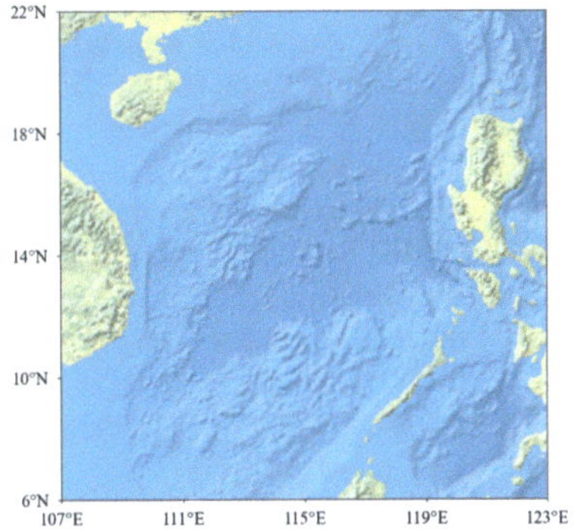

Figure 1. The study area used in this paper.

2.1.2. SST Data

The SST data employed in this study comes from operational sea surface temperature and sea ice analysis reprocessed (OSTIA-REP) [24–26] SST dataset, which is a group for high-resolution sea surface temperature (GHRSST) generated by using optimal interpolation (OI) on a global 0.05° degree grid. As a sister product to the near real-time counterpart (OSTIA-NRT), the OSTIA-REP distinguishes itself by assimilating satellite data from more than 25 distinct SST sensors, along with in situ observations sourced from drifting and moored buoys.

The original data resolution in the study region is 320 × 320 pixels (0.05°). To ensure a more acceptable training speed, we resized the original SST data by using the nearest neighbor downsampling algorithm to obtain a lower-resolution version of the data. The downsampled 48 × 48 pixels ($\frac{1}{3}$°) SST data and 192×192 pixels ($\frac{1}{12}$°) SST data are then considered as the low-resolution input and the corresponding high-resolution target in the 4× downscaling experiments.

In order to distinguish the land and sea points, the SST values over the land points are first set to 0. Then, since all SST values in the dataset are less than 35, the original SST values are normalized to SST′ within the data range [−1, 1), according to Equation (1). This normalization preprocessing is conducted to ensure fairness in the comparison between the proposed DIFFDS method and other algorithms in the experimental results:

$$SST' = 2 \times (\frac{SST}{35}) - 1 \quad (1)$$

The period of the experiment dataset covers from 1990 to 2021. Data from 1990 to 2019 are used as the training set, data within 2020 are the validation set, and the data from March 2021 to February 2022 serve as the test set.

2.2. Diffusion Model

Generative models, such as GANs and variational autoencoders (VAEs), are commonly used to produce highly realistic samples but inherently come with limitations. GANs, for instance, can encounter challenges in training stability and sampling diversity unless carefully designed optimization strategies are employed. In addition, these GAN methods may easily suffer from mode collapse [27]. This phenomenon often occurs in the training process of producing similar or identical samples, which fails to capture the full diversity of the data distribution. It typically happens when the model converges to a limited set of data patterns, ignoring the variety of the actual data, which will decrease the generalization ability of the model and affect its practical application effect. The results generated by VAE may lack detailed information, often leading to blurred results [28].

In contrast, recent advancements in diffusion models (DMs) have demonstrated that employing principled probabilistic diffusion modeling can yield high-quality mapping from randomly sampled Gaussian noise to complex target distribution, without suffering from mode collapse or instabilities. The foundations of the diffusion model can be traced back to the pioneering work in 2015 [29], which was inspired by nonequilibrium thermodynamics. This concept has been further developed and popularized in subsequent works, such as denoising diffusion probabilistic models (DDPM) [30], improved denoising diffusion probabilistic models (IDDPM) [31] and denoising diffusion implicit models (DDIM) [32].

Taking DDPM as an example, DMs typically encompass two processes: the forward diffusion process and the reverse diffusion process, as shown in Figure 2, both of which are characterized by a T-step Markov chain.

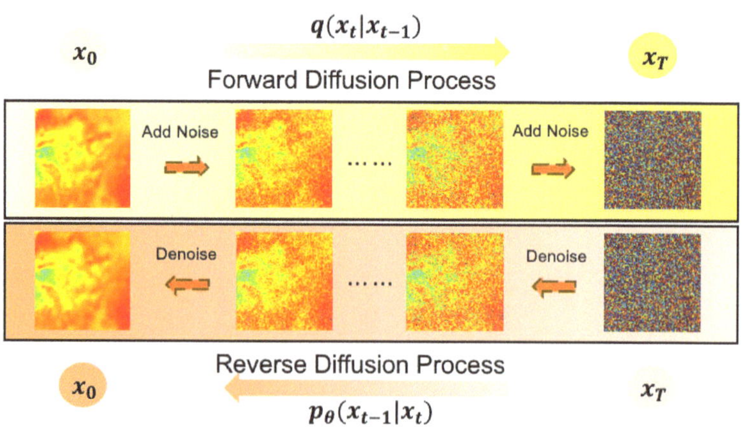

Figure 2. The forward and reverse diffusion processes of diffusion model, where $q(x_t|x_{t-1})$ means the forward process that transforms distribution $q(x_{t-1})$ to $q(x_t)$ and $p_\theta(x_{t-1}|x_t)$ represents the reverse process that transforms distribution $p_\theta(x_t)$ to $p_\theta(x_{t-1})$.

The forward diffusion process transforms the start data distribution into a final Gaussian distribution. Firstly, the initial data x_0 are defined, and then Gaussian noise is progressively added in each timestep until it reaches pure Gaussian noise $x_T \sim \mathcal{N}(\mathbf{0}, \mathbf{I})$ at timestep T by T iterations. During each mid-timestep t, noisy data x_t are generated with the same shape as x_0. The noise incorporated in the diffusion process is specified by a predefined sequence of $\beta_{1:T} \in (0,1]^T$. Denote $\alpha_t = 1 - \beta_t$, $\overline{\alpha_t} = \prod_{n=1}^{T} \alpha_n$, and $\beta_1 < \beta_2 < \ldots < \beta_t (t \in [1, T])$, each iteration of the forward diffusion process can be described as follows:

$$q(x_t | x_{t-1}) = \mathcal{N}(x_t; \sqrt{\alpha_t} x_{t-1}, \beta_t \mathbf{I}) \tag{2}$$

where the process that transforms distribution $q(x_0)$ to $q(x_t)$ can be condensed into one single step:

$$q(x_t | x_0) = \mathcal{N}(x_t; \sqrt{\overline{\alpha_t}} x_0, (1 - \overline{\alpha_t}) \mathbf{I}) \tag{3}$$

Thus, x_t can be directly sampled as follows:

$$x_t = \sqrt{\overline{\alpha_t}} x_0 + \sqrt{(1 - \overline{\alpha_t})} \xi, \xi \in \mathcal{N}(\mathbf{0}, \mathbf{I}) \tag{4}$$

Meanwhile, the reverse process focuses on learning the inverse of the forward diffusion process and generating a distribution that resembles real data distribution. The reverse diffusion process first samples a random noise $x_T \in \mathcal{N}(\mathbf{0}, \mathbf{I})$ and then gradually denoises it until it reaches a high-quality output x_0. Define $p_\theta(x_t)$ as the data distribution at timestep t in the reverse process, and a neural network $\epsilon_\theta(x_t, t)$ is involved in predicting the uncertain variables, where θ represents the network parameters. Each iteration of the reverse diffusion process can be described as follows:

$$p_\theta(x_{t-1} | x_t) = \mathcal{N}(x_{t-1}; \mu_\theta(x_t, t), \sigma_t^2 \mathbf{I}) \tag{5}$$

where $\mu_\theta(x_t, t)$ is the mean value computed using Equation (6):

$$\mu_\theta(x_t, t) = \frac{1}{\sqrt{\alpha_t}} (x_t - \epsilon_\theta \frac{1 - \alpha_t}{\sqrt{1 - \overline{\alpha_t}}}) \tag{6}$$

and σ_t^2 is the variance value calculated using Equation (7):

$$\sigma_t^2 = \frac{1 - \overline{\alpha_{t-1}}}{1 - \overline{\alpha_t}} \beta_t \tag{7}$$

In addition, there is only one unknown variable that should be learned in the reverse process. DMs use a neural network $\epsilon_\theta(x_t, t)$ to estimate it. To train the model of $\epsilon_\theta(x_t, t)$, a timestep t and a noise $\epsilon \in \mathcal{N}(\mathbf{0}, \mathbf{I})$ are randomly sampled to generate noisy data x_t with the given real data x_0, according to Equation (2). Then, the entire network parameters are optimized by the following loss function:

$$Loss = \mathbb{E}_{x_0, \epsilon, t} \left[\| \epsilon - \epsilon_\theta(x_t, t) \|_2^2 \right] \tag{8}$$

Substituting x_t, i.e., Equation (4) into Equation (8), yields the following:

$$Loss = \mathbb{E}_{x_0, \epsilon, t} \left[\| \epsilon - \epsilon_\theta(\sqrt{\overline{\alpha_t}} x_0 + \sqrt{(1 - \overline{\alpha_t})} \xi, t) \|_2^2 \right] \tag{9}$$

In the DMs designed for single-image SR, some models directly generate high-resolution images in the image domain [20], carrying out the forward and reverse diffusion process on the input pixel space. These models demand excessive iteration steps (about 100–1000 steps) on large-scale denoising models to precisely capture data details, which consumes massive computational resources [23]. Furthermore, directly conducting the diffusion process in a high-dimensional data space requires a large amount of GPU memory and training times.

Other models conduct diffusion processes in the latent space. DMs are utilized to generate low-dimensional latent variables. These latent variables are then employed for SR reconstruction, like [33] and DIFFIR. These models offer efficiency advantages by circumventing diffusion processes in high-dimensional data spaces. For the sake of training resources and efficiency, we selected DIFFIR as the base architecture model.

2.3. DIFFDS Method

Computer vision SR tasks focus mainly on the diversity of visual perception. However, when using DMs for SST spatial downscaling, the focus is more on accurately generating meso- or small-scale dynamic SST processes than on their diversity. This is crucial because the downscaled high-resolution SST data may later be used for other downstream tasks like forecasting. In such cases, inaccurate processes can negatively impact the validity of task results.

However, applying the original DIFFIR to SST spatial downscaling can lead to inauthentic details in the downscaled results, resulting in poor subjective and objective evaluations. To ensure that the downscaled results are more consistent with the destination high-resolution SST data and to enhance the accuracy of the reconstructed data, the DIFFDS method is proposed.

DIFFDS consists of three stages: the first stage, training a dynamic IRformer (DIRformer) and a compact prior extraction network (CPEN); the second stage, training the denoising network and finally, the third stage, also the inference stage, where the trained networks are used for the downscaling task. The workflow (illustrated in Figure 3) of DIFFDS employs DDPM to generate a guidance vector called compact IR prior representation (IPR) and then utilizes IPR to direct the DIRformer in the downscaling process. During the training phase, CPEN and DIRformer are first trained. The trained CPEN is integrated into the forward diffusion process of DDPM to train the denoising network. Once the denoising network is trained, the IPR can be predicted through the reverse sampling process of DDPM. Subsequently, guided by the predicted IPR, the pre-trained DIRformer performs the downscaling process.

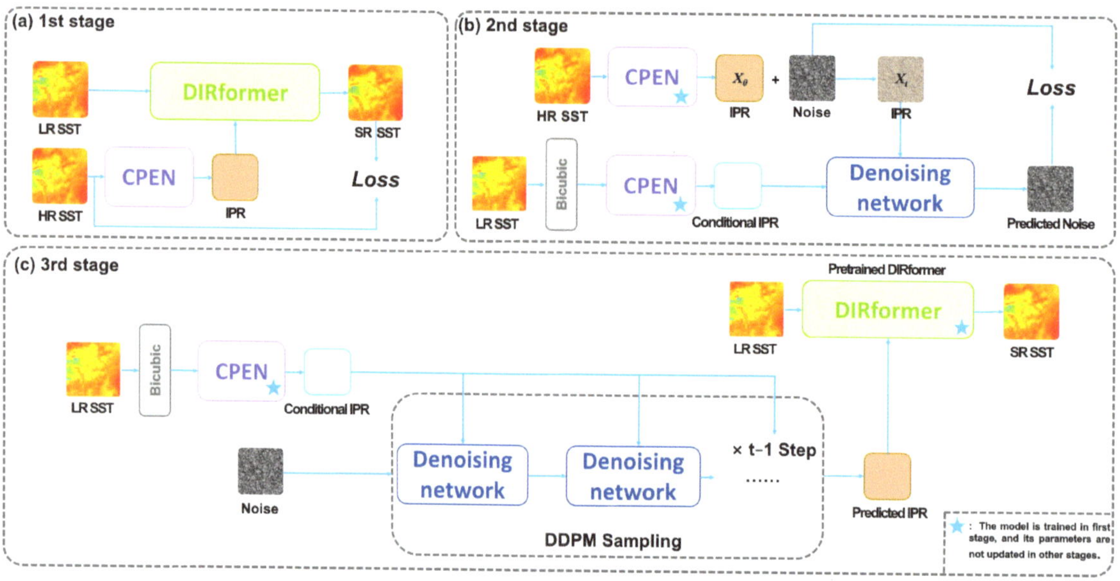

Figure 3. (**a**) illustrates the first training stage, detailing the training processes of CPEN and DIRformer. (**b**) depicts the second training stage, which is also the forward process of DDPM. The 3rd stage (**c**) shows the inference process of DIFFDS.

In DIFFDS, we redesign the structure of the transformer block in DIRformer by incorporating cross-attention and channel-attention mechanisms. Channel attention focuses on increasing the weight of certain significant feature channels [34] while suppressing others within the data. This selective enhancement allows DIFFDS to maintain some SST texture anomalies. Additionally, cross-attention enables DIFFDS to learn the underlying distribution structures in low-resolution SSTs. This ensures that the produced high-resolution SST closely aligns with the original data, minimizing deviations. These adjustments have markedly improved the performance of the model. Specifically, DIFFDS has primarily modified the transformer block, while keeping the rest of the network framework consistent with DIFFIR. Figure 4 depicts the entire architecture.

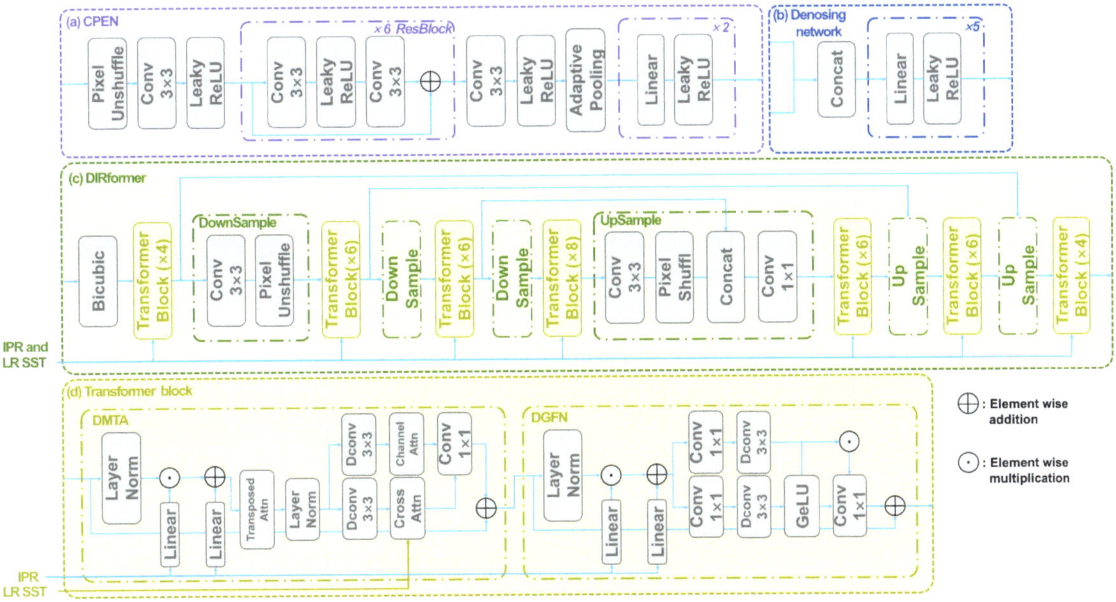

Figure 4. Architecture details of DIFFDS. (**a**) CPEN, (**b**) Denoising network, (**c**) DIRformer, (**d**) Transformer block.

2.3.1. CPEN

With the input of high-resolution SSTs, CPEN learns to generate a low-dimensional IPR, which can guide the DIRformer in the SST downscaling process.

As described in Figure 4a, the CPEN is constructed by a series of residual blocks, convolution layers, and linear layers. First, the original high-resolution SST training data are processed by a pixel unshuffle layer to facilitate training speed. Then, a 3×3 convolution layer and a Leaky ReLU are applied to extract the feature map. After that, multiple residual blocks are utilized to calculate and refine feature representations from the current feature map. Finally, IPR is produced through the transformation of an adaptive pooling layer and several linear and Leaky ReLU layers. The CPEN process can be described as follows.

$$IPR = CPEN(SST_{HR}) \tag{10}$$

2.3.2. DIRformer

Under the guidance of IPR, DIRformer accepts the input low-resolution SST and then generates the corresponding high-resolution versions (as shown in Figure 3a). The DIRformer is constructed by stacking transformer blocks in a U-Net structure. Each of these transformer blocks comprises a modified dynamic multi-head transposed attention (DMTA) part and a dynamic gated feed-forward network (DGFN) module (Figure 4d).

After the CPEN process, IPR information is obtained, denoted as X_0. As provided for dynamic modulation parameters, X_0 is designed into the DMTA of DIRformer to integrate the current feature map with the high-resolution SST feature information and guide the downscaling process of the DIRformer. As shown in Equation (11), F is the input feature map of DMTA:

$$F' = W_l^1 X_0 \odot Norm(F) + W_l^2 X_0 \tag{11}$$

where $\odot$ indicates element-wise multiplication, *Norm* denotes layer normalization, W_l represents linear layer, and F' are the output feature map, respectively.

Next, the original DIRformer employs a transposed multi-head attention mechanism (a kind of efficient muti-head attention [35]) to process F', so the DMTA process of DIFFIR is as follows:

$$\hat{F} = TransposedAttn(F') + F \tag{12}$$

To improve its performance, a 3×3 depth-wise convolution layer is introduced to capture more intricate and detailed spatial features. The extracted feature map is then separately fed into a channel-attention layer and a cross-attention layer. In practice, the channel attention layer is utilized to enhance the representational power of neural networks by emphasizing informative features in vital channels while suppressing incorrect or noisy information channels, allowing DMTA to facilitate more effective utilization of the guidance information generated by the DDPM. Meanwhile, the cross-attention layer is added to fuse low-resolution SST context information and the current feature map. This enhancement can make the results more consistent with the original SST structure, and avoid generating abnormal dynamical processes. Finally, the results are integrated using a 1×1 convolution layer. The DMTA process of DIFFDS can be described as follows:

$$F'' = TransposedAttn(F') \tag{13}$$

$$\hat{F} = W_c(CrossAttn(W_d^1(F'')) + ChannelAttn(W_d^2(F''))) + F \tag{14}$$

where W_c and W_d are the convolution layer and the depth-wise convolution layer.

In the DGFN process, the same feature map F' can be obtained by Equation (11), which integrates IPR information with the DGFN's input feature map F. Then, a 1×1 convolution unit is exploited to aggregate information from different channels. Next, a 3×3 depth-wise convolution unit is added to aggregate information from spatially neighboring pixels. Besides, the gating mechanism is adopted to enhance information encoding. The overall process of DGFN is defined in Equation (15), $\hat{F}$ is the output of DGFN.

$$\hat{F} = GELU(W_d^1 W_c^1 F') \odot W_d^2 W_c^2 F' + F \tag{15}$$

2.3.3. Denoising Network

The structure of the denoising network consists of multiple stacked linear layers and Leaky ReLU layers (Figure 4b). It concatenates the noisy IPR and the reference conditional IPR in the channel dimension as inputs and then produces the previous timestep noise as output.

2.3.4. Training and Inference

The training of DIFFDS incorporates two phases: the first phase trains CPEN and DIRformer, and the second phase is the forward diffusion process of DDPM, which trains the denoising network.

In the first stage, as listed in Algorithm 1, CPEN and DIRformer are trained together, which can make CPEN learn to transform high-resolution SST data into IPR and force DIRformer to reconstruct high-resolution SST guided by IPR. For each pair of low- and high-resolution SST data, the high-resolution SST is sent to CPEN to obtain IPR, after which

IPR and low-resolution SST are sent to the DIRformer to generate downscaled SST. The training loss function in phase 1 is defined as:

$$Loss1 = \left\| SST_{HR} - SST_{SR} \right\|_1 + L_{adv}(SST_{HR}, SST_{SR}) \tag{16}$$

where SST_{HR} and SST_{SR} are the ground-truth and downscaled high-resolution data, respectively. $\| \cdot \|_1$ denotes the L1 norm, and L_{adv} is the adversarial loss used in Real-ESRGAN [36].

Algorithm 1 Training CPEN and DIRformer

Input: low-resolution SST SST_{LR} and high-resolution SST SST_{HR}
1: **for** SST_{LR}, SST_{HR} **do**
2: $X_0 = CPEN(SST_{HR})$
3: $SST_{SR} = DIRformer(X_0, SST_{LR})$
4: Calculate and optimize $Loss1$
5: **end for** if $Loss1$ converges
Output: Trained CPEN and DIRformer

Phase 2 (as shown in Algorithm 2) is the forward diffusion process of DDPM. In this stage, the denoising network is trained to predict noise. After preparing parameters like $\overline{\alpha_t}$ and α_t, for each pair of low- and high-resolution SST data, the pre-trained CPEN extracts IPR X_0 from high-resolution SSTs. Then, X_t is sampled by the forward diffusion equation, according to Equation (4). After substituting variables X_0 and X_t into Equation (4), the formula is obtained as follows:

$$X_t = \sqrt{\overline{\alpha_t}} X_0 + \sqrt{(1 - \overline{\alpha_t})} \xi, \xi \in \mathcal{N}(\mathbf{0}, \mathbf{I}) \tag{17}$$

where the $\overline{\alpha_t}$ is the same parameter in Equation (4), and ξ is a Gaussian noise sampled from $\mathcal{N}(\mathbf{0}, \mathbf{I})$.

Algorithm 2 Training DDPM

Input: Trained CPEN, $\beta_{1:T} \in (0, 1]^T$, low-resolution SST SST_{LR} and high-resolution SST SST_{HR}
1: Init: $\alpha_t = 1 - \beta_t, \overline{\alpha_t} = \prod_{n=1}^{t} \alpha_n$
2: **for** SST_{LR}, SST_{HR} **do**
3: $X_0 = CPEN(SST_{HR})$
4: Sample a $t \in [1, T]$
5: Sample X_t
6: $X_c = CPEN(Bicubic(SST_{LR}))$
7: Calculate and optimize $Loss2$
8: **end for** if $Loss2$ converges
Output: Trained Denoising network

Using bicubic interpolation, the SST_{LR} data are upsampled to the same resolution as SST_{HR}, after which it is sent to CPEN to earn condition IPR X_c. This IPR X_c functions as conditional information, facilitating the denoising network's capability to accurately forecast noise patterns. Finally, the parameters of the denoising network can be updated according to the DDPM loss function. In Loss Equation (18), ϵ_θ is the denoising network, ϵ is the Gaussian noise sampled from $\mathcal{N}(\mathbf{0}, \mathbf{I})$, and t is the timestep, *Concat* is the concatenation step:

$$Loss2 = \left\| \epsilon - \epsilon_\theta(Concat(X_c, X_t), t) \right\|_2^2 \tag{18}$$

For the inference process as illustrated in Algorithm 3, a random noise sample is initialized. Eventually, the reverse diffusion process Equation (19) is deprived of the predicted IPR X_0, which contains the corresponding high-resolution SST information.

$$X_{t-1} = \frac{1}{\sqrt{\alpha_t}}(X_t - \epsilon_\theta(Concat(X_c, X_t), t)\frac{1 - \alpha_t}{\sqrt{1 - \overline{\alpha_t}}}) + \xi\sqrt{\frac{1 - \overline{\alpha_{t-1}}}{1 - \overline{\alpha_t}}\beta_t} \tag{19}$$

where β_t, α_t, $\overline{\alpha_t}$ are the same parameters in Equations (6) and (7), t is the sampled timestep, and ϵ_θ is the denoising network.

Subsequently, X_0 and the low-resolution SST input are passed to the pre-trained DIRformer for spatial downscaling, and then the DIRformer outputs the spatially downscaled high-resolution SST data.

Algorithm 3 Inference

Input: Trained CPEN, DIRformer, $\epsilon_\theta, \beta_{1:T} \in (0,1]^T$, low-resolution SST SST_{LR}
Init: $\alpha_t = 1 - \beta_t, \overline{\alpha_t} = \prod_{n=1}^{t} \alpha_n$
for SST_{LR} **do**
  $X_c = CPEN(Bicubic(SST_{LR}))$
  Sample $X_t \in \mathcal{N}(\mathbf{0}, \mathbf{I})$
  for $t = T, \ldots, 1$ **do**
    Sample $\xi \in \mathcal{N}(\mathbf{0}, \mathbf{I})$ if $t > 1$ else $\xi = 0$
    Update X_{t-1} by Equation (19)
  end for
  $SST_{SR} = DIRformer(X_0, SST_{LR})$
end for
Output: Downscaled results SST_{SR}

2.4. Evaluation Metrics

To thoroughly evaluate the quality and accuracy of the proposed DIFFDS method as well as other baseline models, we have selected a comprehensive set of objective evaluation metrics, including root mean square error (RMSE), mean absolute error (MAE), Bias, peak signal-to-noise ratio (PSNR) and temporal correlation coefficient (TCC). These metrics will measure the differences between spatial resolution improved SST values and actual high-resolution SST values.

In this paper, after obtaining the downscaling results, the data are first denormalized from the range $(-1, 1)$ back to the normal range. RMSE, Bias, MAE, and TCC are calculated directly using these denormalized data. To calculate the PSNR, the data are first normalized to (0, 1) by dividing by 35.

3. Experiments and Results

3.1. Experiments Design

Based on the dataset in Section 2.1.2, we conduct $4\times$ scale downscaling experiments. For comparison, the selected baseline models include Lasso regression, Bicubic, RCAN, ESRGAN and DIFFIR. All models were trained on the aforementioned OSTIA dataset. Since the SST high-resolution results obtained by Bicubic interpolation do not completely align with the edges of the ground truth, we calculate the metrics related to Bicubic using only the overlapping portion of the Bicubic results and the ground truth.

Regarding DIFFDS, in training stage 1, we set the number of transformer blocks per layer in DIRformer to [4, 6, 6, 8], and the number of attention heads per layer to [1, 2, 4, 8]. The number of resblocks in CPEN is set to 6. In training stage 2, the reverse sampling steps of DDPM are set to 200, with a beta scheduler configured to linear, and beta start and beta end values (the β_1 and β_T in Section 2.2) set to 0.0001 and 0.02, respectively. The timestep spacing algorithm is set to leading [37]. The learning rate for both stage 1 and stage 2 starts at 5×10^{-5}, with a cosine scheduler for the learning rate, and the batch size is 16 for both phases.

For the training of DIFFDS, we utilized an NVIDIA RTX 4070Ti graphics (Lenovo, Beijing, China) card equipped with 12 GB of memory. Under these specific hardware conditions, the training process necessitated a total of 400 epochs. Each epoch required approximately 4 min to complete. Upon the completion of the training phase, the model's inference time for performing downscaling operations was recorded to be around 1.8 s. While this setup was sufficient to accomplish training objectives, we observed that training speeds could be further optimized. Based on our practical experience, we recommend employing a graphics card with at least 16 GB of memory. This would likely result in more efficient training processes and a reduced overall training time.

3.2. Results

3.2.1. Metrics Evaluation

As listed in Table 1, the proposed DIFFDS method outperforms other models across various objective evaluation metrics. It achieves an average RMSE of 0.1074 °C, an average Bias of −0.0043 °C, an average MAE of 0.0654 °C, an average PSNR of 50.48 dB and a TCC of 0.9610, demonstrating high precision and effectiveness.

With respect to RMSE, DIFFDS excels in all aspects, including average, maximum, and minimum values. RCAN, ESRGAN, and DIFFIR followed with a 0.02 °C gap. Bicubic performs worse than the above models. The Lasso regression method shows poor performance with a mean of 0.3347 °C, which shows a noticeable gap compared to other models. This highlights the effectiveness of deep learning models in addressing downscaling problems.

For MAE, DIFFDS again takes the lead in average, maximum, and minimum values, followed by RCAN. The differences among the remaining deep learning models are relatively insignificant. Lasso regression and Bicubic show a significant disparity compared to deep learning models on this metric. Regarding Bias, DIFFIR achieves the best average value of −0.0003 °C, ESRGAN performs the best in maximum values at 0.0008 °C, and RCAN performs the best in minimum values at −0.0023 °C, respectively.

As for PSNR, DIFFDS consistently delivers the best results in average, maximum, and minimum values, demonstrating that the downscaling results have very low noise. It is followed by ESRGAN, RCAN, and DIFFIR, while Lasso regression and Bicubic lag behind.

In terms of TCC, DIFFIR achieves the highest value of 0.9634, indicating that the downscaled results exhibit good temporal consistency compared to the true values. DIFFDS, ESRGAN, RCAN and Lasso have lower TCC values than DIFFIR, at 0.9610, 0.9536, 0.9444 and 0.9615, respectively. Bicubic displays the lowest TCC, around 0.88. Although DIFFDS shows a slightly reduced TCC compared to DIFFIR, it still outperforms other benchmark models. An obvious fact is that Lasso regression's TCC is comparable to most deep learning models, indicating that Lasso has lower accuracy regarding reconstruction error and reconstruction quality, but has stronger temporal correlation and can capture the temporal sequence characteristics of SST. Deep learning models, on the other hand, have higher accuracy in reconstructing SSTs in the spatial dimension but have weaker temporal correlation, which may be because deep learning models focus more on capturing spatial features and, to some extent, neglect the information in the temporal dimension.

In general, the improved DIFFDS presents further performance improvements in most metrics, highlighting its superiority over DIFFIR and other baseline models.

Figure 5a shows that DIFFDS surpasses other neural networks in terms of the maximum value, min value, median and upper and lower quartiles. This is consistent with the results presented in Table 1.

The evaluation in Figure 5b,c is largely similar to that in Figure 5a, showing the consistency of DIFFDS across different metrics.

For the Bias box plot in Figure 5d, DIFFIR and DIFFDS outperform other models in terms of the median, and the upper and lower quartiles. ESRGAN performs best in maximum values, while RCAN excels in minimum values. Regarding the data distribution between the maximum and minimum values, and the interquartile, Lasso, DIFFDS, and DIFFIR take the lead. RCAN and ESRAGN have relatively poor performance, whereas Bicubic displays the largest fluctuations. As for DIFFDS, with its modified network architecture, although it is slightly inferior to DIFFIR, it still outperforms most of the comparison models. In the TCC plot, DIFFIR surpasses the other models in maximum, minimum, and median values. DIFFDS and Lasso follow closely, showing only minor differences from DIFFIR. This indicates these models' results have good temporal consistency with the ground truth and are capable of accurately capturing SST change trends over time.

ESRGAN has relatively poor TCC distributions, while RCAN and Bicubic perform the worst in terms of minimum and median values.

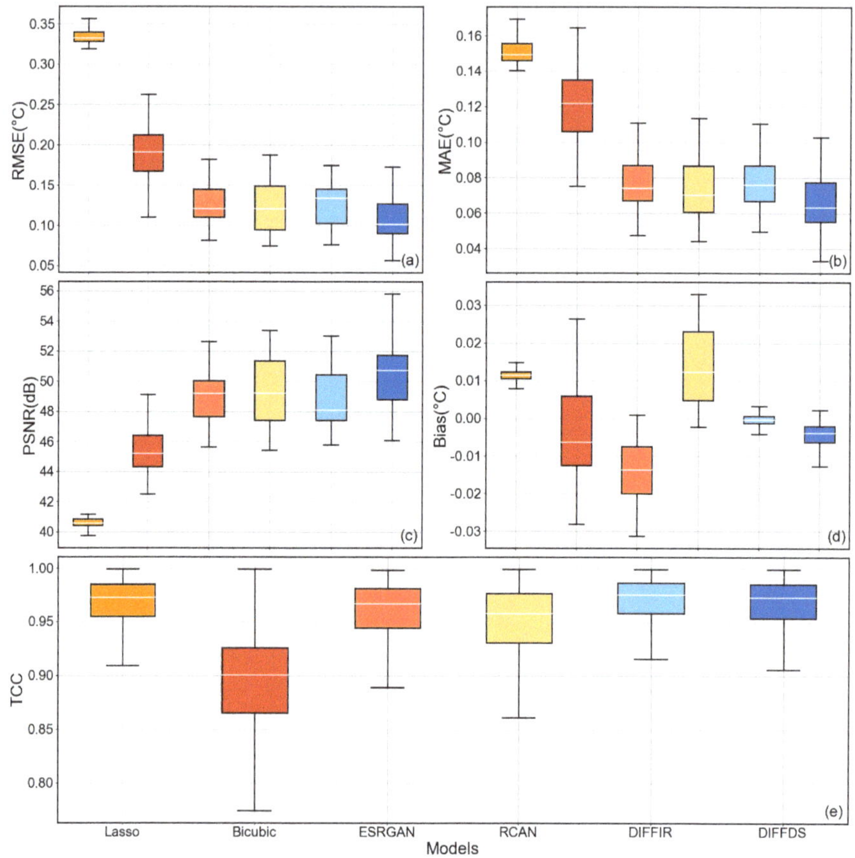

Figure 5. The maximum, minimum, median, both upper and lower quartiles of each metric, (**a**) RMSE, (**b**) MAE, (**c**) PSNR, (**d**) Bias, for each method. (**e**) is the TCC plot for each point in the experiment sea area.

Table 1. The average/maximum/minimum value of RMSE, MAE, Bias, PSNR, and the value of TCC.

Model	RMSE (°C)	MAE (°C)	Bias (°C)	PSNR (dB)	TCC
Lasso	0.3347/0.3644/0.3196	0.1506/0.1692/0.1405	0.0113/0.0155/0.0056	40.60/41.19/39.69	0.9615
Bicubic	0.1891/0.1750/0.0762	0.1206/0.1645/0.0753	−0.0039/0.0264/−0.0282	45.44/50.06/42.50	0.8858
ESRGAN	0.1259/0.1824/0.0819	0.0763/0.1109/0.0474	−0.0138/0.0008/−0.0312	48.99/52.61/45.67	0.9536
RCAN	0.1224/0.1875/0.0747	0.0735/0.1136/0.0442	0.0139/0.0329/−0.0023	49.39/53.41/45.42	0.9444
DIFFIR	0.1269/0.1750/0.0761	0.0770/0.1105/0.0495	−0.0003/0.0040/−0.0066	48.77/53.04/45.82	0.9634
DIFFDS	0.1074/0.1734/0.0567	0.0654/0.1027/0.0331	−0.0043/0.0023/−0.0145	50.48/55.87/46.10	0.9610

3.2.2. Analysis of Temporal Trends

The temporal variations in the metrics reveal similar general trends across deep learning models. In particular, neural network models tend to perform relatively poorly during spring and summer, while their performance improves in autumn and winter. This seasonal pattern may be attributed to the influence of the East Asian monsoon in the study area. During the spring and summer, the ocean–atmosphere interaction intensifies and is more active than in autumn and winter. Factors such as heat flux, evaporation, and advection processes on the sea surface can cause rapid changes in SST. Additionally, warm seasons are often accompanied by stronger solar short-wave radiation and convective activities, resulting in larger SST fluctuation. These fluctuations pose a challenge for model representations, contributing to the relatively poor performance of neural network models during this period.

Unlike spring and summer, SST changes in autumn and winter are relatively stable and exhibit simpler textures. This makes it easier for deep learning models to capture accurate SST representations and features, resulting in better performance during these seasons.

The Lasso regression method exhibits a relatively stable trend without showing significant seasonal variations, compared with other methods. This stability may be attributed to its regularization property, which reduces model complexity and results in more consistent performance across different seasons.

The results for DIFFIR are not ideal due to the lack of specialized adjustments for SST spatial downscaling tasks (Figure 6a). Specifically, DIFFIR performs worse during the summer period compared to other models, as SST fluctuations are more significant during this time, making it more challenging for DIFFIR to learn. In such cases, the unmodified network architecture is more prone to generate some erroneous content, resulting in larger RMSE values. For ESRGAN, its RMSE is slightly smaller than that of DIFFIR. This is because ESRGAN has a relatively simple generator architecture, and incorporates adversarial and perceptual losses. These factors compel GAN to balance different losses during the training process. Consequently, the generated content may contain more detailed textures, but the RMSE remains relatively high. RCAN, which utilizes a channel-attention mechanism, performs better than ESRGAN and DIFFIR in terms of RMSE. However, its relatively simple network structure limits its performance.

The performance of DIFFDS stands out, as it achieves the lowest average, maximum, and minimum RMSE values. The overall RMSE curve reveals that DIFFDS consistently achieves the lowest RMSE on most dates. However, during the latter part of autumn and winter (November 2021–March 2022), RCAN tends to perform better. This suggests that during periods of relatively stable SST variations and simple SST distribution structures, the downscaling performance of DIFFDS is not as pronounced. Other models can also achieve similar results.

DIFFDS generally exhibits lower error than DIFFIR, suggesting that the improved network structure effectively corrects the downscaling results. For ESRGAN, its RMSE gap during the autumn and winter seasons is relatively higher than that of other models, differing from the performance of RCAN. This discrepancy may be attributed to its inherent characteristics, making it less sensitive to data variations in these seasons.

The results for MAE (Figure 6b,c) are similar to RMSE: DIFFDS achieves the best results on most dates. However, in autumn and winter, the differences between DIFFDS and other models tend to be small or even reverse, suggesting that the performance gap narrows during these seasons. The Lasso regression method consistently performs the worst on these metrics.

In terms of Bias variations (Figure 6d), DIFFIR shows the most stable trend, closely followed by DIFFDS and Lasso. This indicates that the predictions of these methods are statistically close to the true values, without systematic overestimation or underestimation of SST over long-term averages. The errors are balanced in both positive and negative directions, resulting in a very small overall error trend. In contrast, RCAN, ESRGAN and Bicubic exhibit greater fluctuations in Bias variation, and their numerical values for Bias are comparatively poorer.

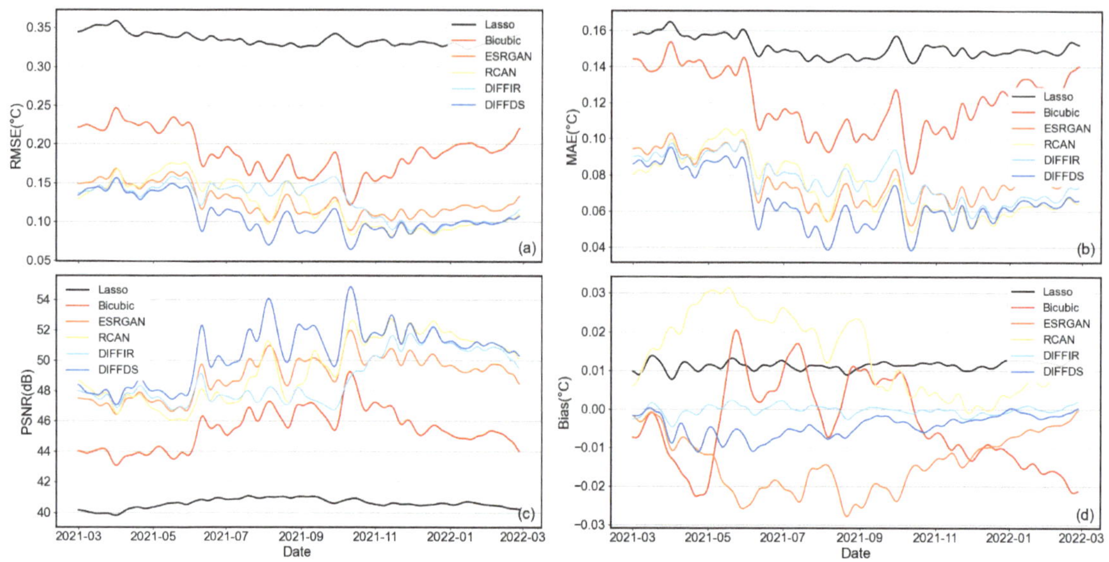

Figure 6. The time series variations for each metric, (**a**) RMSE, (**b**) MAE, (**c**) PSNR, (**d**) Bias, from March 2021 to February 2022. The text highlighted in blue marks the date of the dotted line. The results of these two dates will be used in the discussion section.

3.2.3. Correlation Analysis

Analysis of the correlation distribution across all data points revealed that the results from these neural network models closely approximate the true values (Figure 7). The fitting curves of baseline models almost coincide with the 1:1 line, except for Lasso, which exhibits a relatively noticeable deviation. This suggests that the vast majority of data point predictions from the aforementioned models are accurate, and the deep learning spatial downscaling methods can effectively correct the deviation between low- and high-resolution SST points. Notably, DIFFDS achieves a correlation coefficient of 0.9981, which is the closest to 1, reflecting the highest degree of consistency between its results and the ground truth SST values. The other baselines follow closely, with correlation coefficients very similar to that of DIFFDS, indicating their excellent performance as well.

For Bicubic and Lasso regression methods, the distribution of data points (especially between 22 °C and 32 °C) shows a more pronounced deviation compared to other methods. This is because these methods do not adequately consider the spatial distribution relationship of each data point with its surrounding data points during the downscaling process, resulting in larger errors. For other deep learning models, the distribution of data points between 28 °C and 30.5 °C shows a noticeable deviation, while those above 30.5 °C align well. This could be because SST data points above 30.5 °C make up a smaller proportion, specifically 4 percent of the total data points according to our computation. Additionally, SST data points above 30.5 °C tend to occur in more homogeneous regions (e.g., consistently warm waters), which are easier for the neural network to learn. In contrast, data points in the 28–30.5 °C range constitute a larger portion, accounting for 64 percent of the total data points. These SST points are often significantly impacted by monsoons, exhibiting distinct fluctuations during different periods, making reconstruction quite challenging. Consequently, the number of points where the downscaling results do not match the true values is relatively higher, leading to greater discrepancies in the scatter plot.

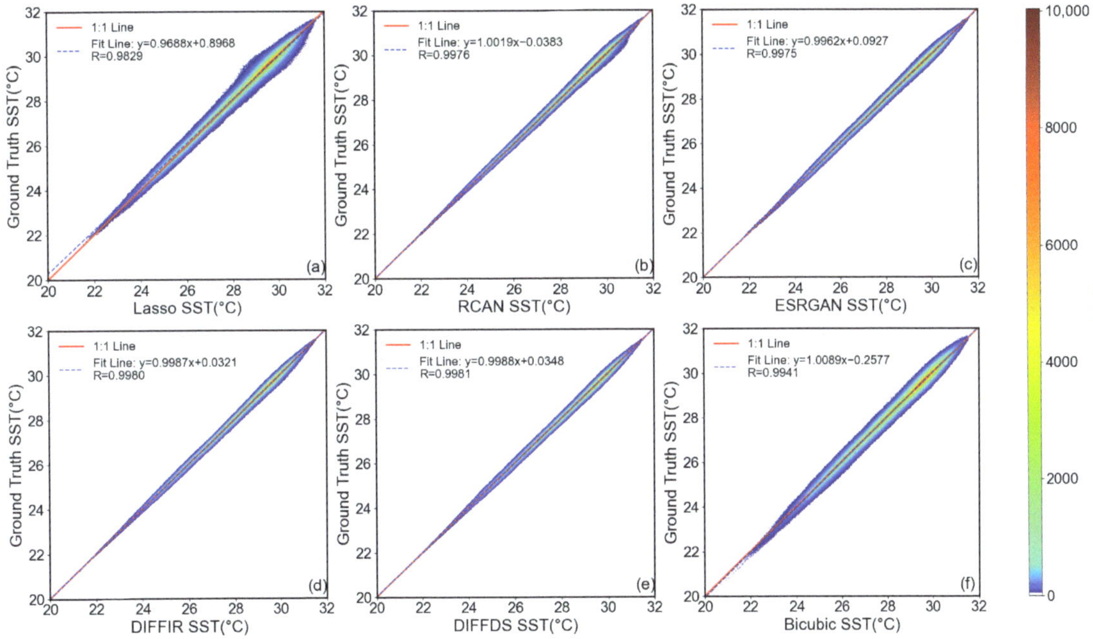

Figure 7. (**a–f**) The density scatter plots of each model.

Furthermore, we randomly selected a 100-day sample subset from the test set to calculate the correlation coefficient, repeating this process 100 times. As shown in the following Table 2, the correlations are similar to those discussed above. Although the correlation coefficient differences between our DIFFDS method and other algorithms are relatively small, it obviously demonstrates that these methods can reserve the large-scale structure from the LR inputs. The reconstruction quality of small-scale structures can evaluate the effectiveness of different algorithms in terms of RMSE, MAE, and PSNR metrics, as shown in Table 1.

Table 2. The correlation coefficient test results.

Model	Mean Correlation	Standard Deviation
Lasso	0.9821	0.0269
Bicubic	0.9942	0.0019
ESRGAN	0.9973	0.0012
RCAN	0.9974	0.0010
DIFFIR	0.9978	0.0011
DIFFDS	0.9981	0.0008

4. Discussion

4.1. Specific Samples Examination

In this section, several samples are selected to analyze the downscaling results and to compare the differences among those models. The first sample is on 29 June 2021, belonging to the summer season (Figure 8). The SST distribution during this period is relatively complex, with many dynamic processes. As shown in the previous time series variations in metrics, summer samples tend to accentuate the differences between models.

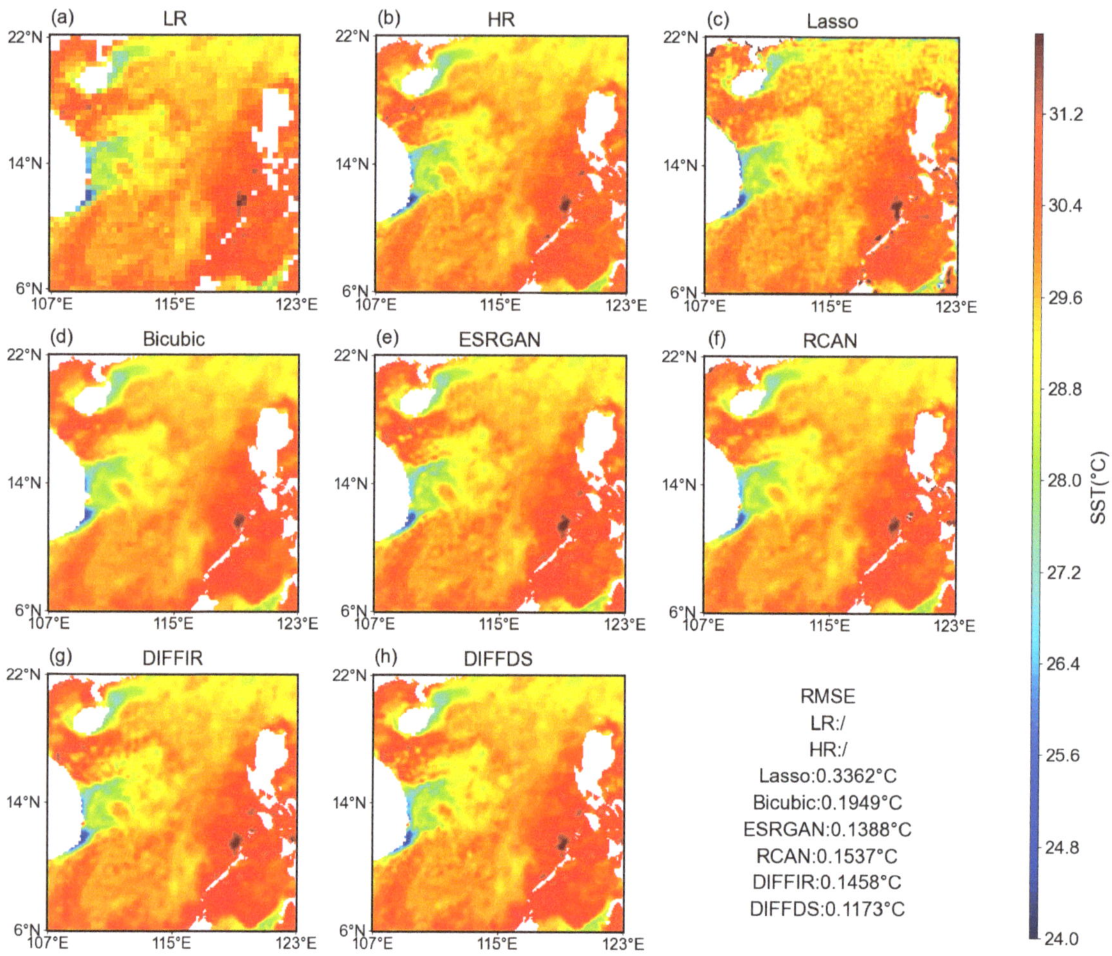

Figure 8. The SST distribution on 29 June 2021 of each model, these eight subplots individually display the high-resolution SST, low-resolution SST, and the downscaled results of each model along with their RMSE (°C).

Firstly, in terms of RMSE, DIFFDS and ESRGAN emerge as the top two models, with the others trailing behind. From Figure 8, we can find that all methods can reconstruct the basic patterns of high-resolution SSTs. However, upon closer inspection, RCAN exhibits relatively blurry and smooth SST structures, similar to Bicubic, whereas ESRGAN, DIFFIR, and DIFFDS reveal more complex reconstructed SST structures. This disparity can be attributed to the fact that models using single L1 loss as a loss function may perform well in metrics such as RMSE and PSNR. However, L1 loss tends to erase high-frequency information [38] during training, making the final downscaled high-resolution SST appear like the smoothed version of ground truth.

In contrast, ESRGAN uses additional adversarial and perceptual losses during training, which enables the generation of richer information while suppressing RMSE and other metrics. The results of the Lasso regression also exhibit relatively complex patterns, but they are not consistent with the true values because the Lasso method does not adequately consider the spatial distribution relationship of each data point with its surrounding data points during the downscaling process, resulting in a higher RMSE.

Generative diffusion models such as DIFFIR, and DIFFDS utilize their distribution fitting capabilities to produce realistic high-resolution SST samples. These models are easier to train than GANs. As a result, they perform better across various metrics in the final downscaled results, showing their potential to generate high-quality SST samples.

Based on the Bias map (Figure 9), the relatively low errors observed in all deep learning models across most marine regions underscore their ability to tackle spatial downscaling problems. However, an acute bias area emerges near the land, specifically between 107°E–114°E and 10°N–17°N, where the models' performance is compromised. As seen in Figure 8, the SST variation in this region is large, indicating a higher level of downscaling complexity compared to other areas. This leads to large deviations in the downscaled results. For Lasso regression, because it did not reconstruct the high-resolution SST features well, its bias regions are relatively larger compared to those of the deep learning model. Compared to Bicubic, deep learning models demonstrate greater accuracy across most areas due to their complex structure and powerful learning capabilities, resulting in lower bias.

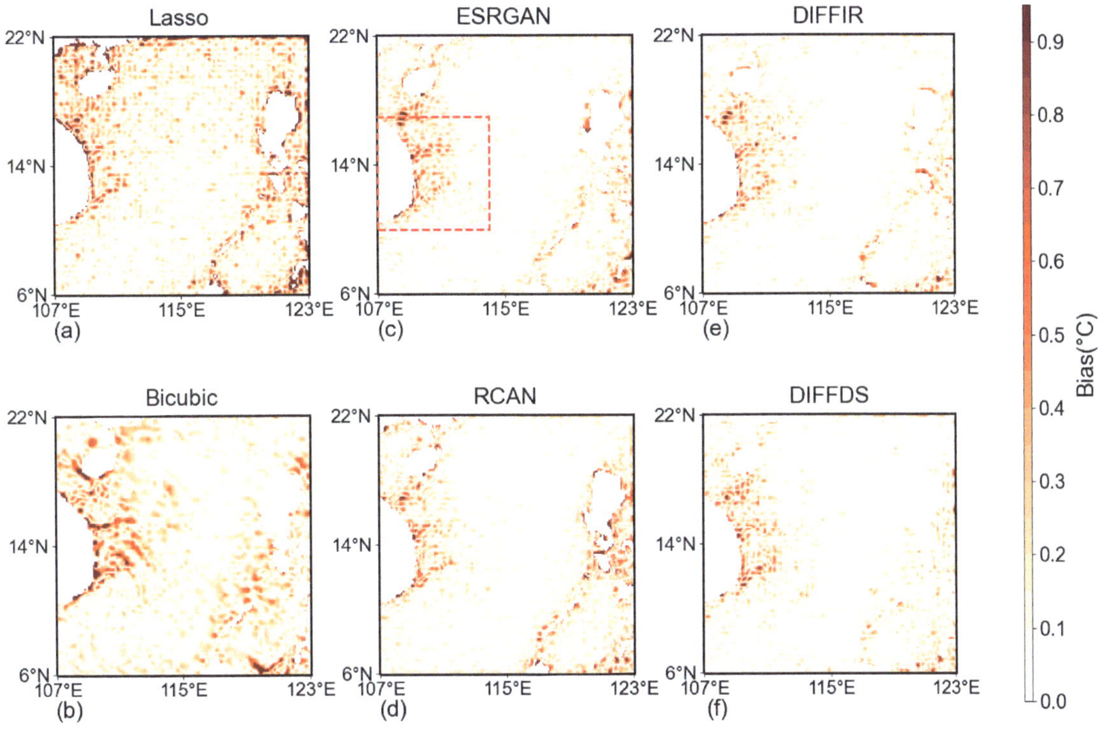

Figure 9. (**a–f**) The absolute Bias map between ground truth and each deep learning model on 29 June 2021. The red box displays the intense Bias area.

As shown in Figure 10, the region from 10°N to 18°N and 107°E to 115°E is selected to examine at a basin scale. Although some discrepancies exist, the DIFFDS-generated results more accurately capture most of the dynamic processes, yielding superior downscaled results. This improvement can be attributed to the adjusted network architecture, which leverages the guidance information provided by the diffusion model while also considering the overall SST distribution structure inherent in low-resolution SSTs.

The original DIFFIR without cross-attention primarily focuses on the guidance information from DDPM, somewhat neglecting the overall SST structure. This oversight results in poor SST structures. For the DIFFIR results (Figure 10), erroneous SST content is shown in the red-boxed area of Figure 10g. These textures cannot be found in the red-boxed area of the input low-resolution SSTs (Figure 10a). Conversely, the corresponding area in DIFFDS is corrected with no obvious anomalous textures. This enhanced structural effectiveness can be directly attributed to the introduction of channel-attention and cross-attention. For ESRGAN, although it generates rich details in downscaled high-resolution SSTs, it still fails to recover certain features, such as the high-temperature area in the block box of subplot Figure 10e.

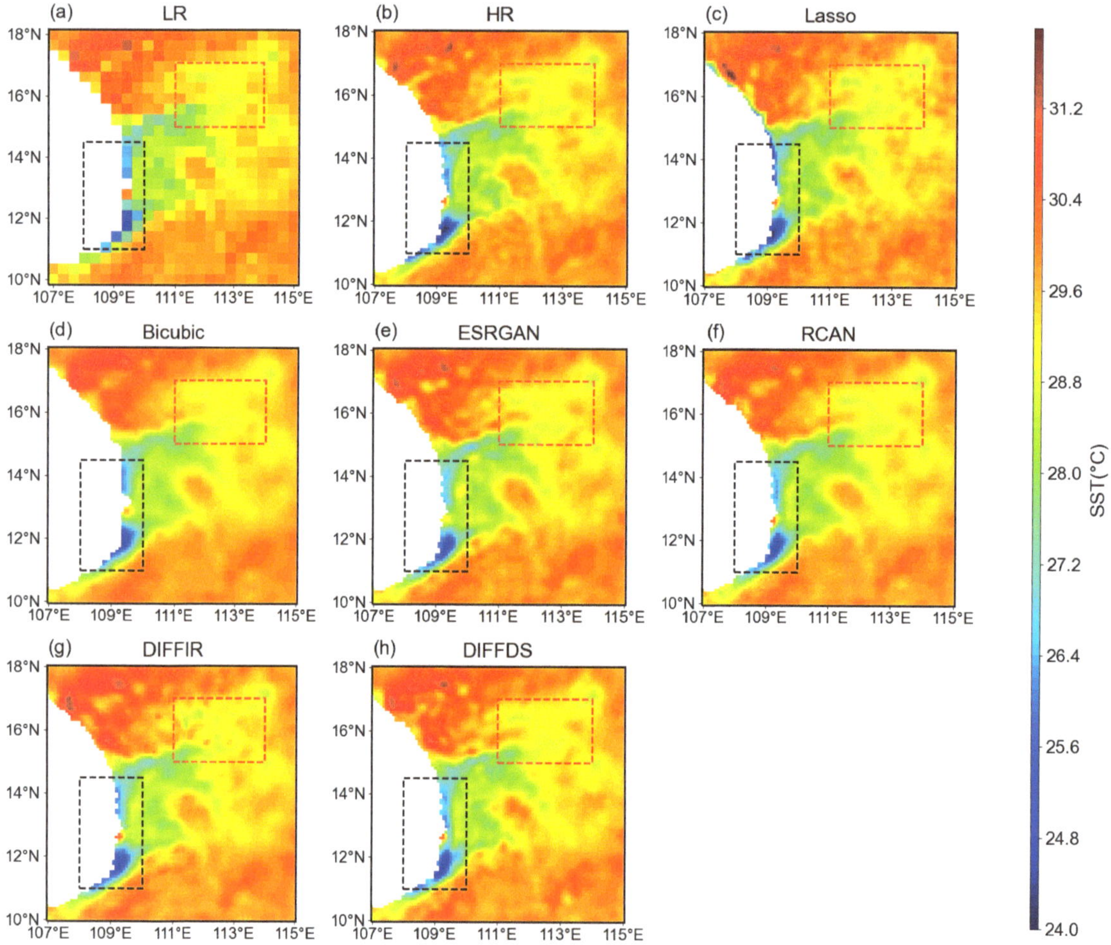

Figure 10. The results on 29 June 2021 zoomed from 9°N–17°N and 108°E–116°E. The red box and black box areas display erroneous SST contents.

Next, we analyze another sample on 1 August 2021. The overall SST pattern on this day is relatively simpler, so the reconstruction task for all deep learning models is less challenging. However, the downscaled results of the Lasso regression show a considerable discrepancy compared to the deep learning methods, with an RMSE of 0.3284 °C, which is higher than that of neural networks. As illustrated in Figure 11, the performance differences among neural networks are further reduced. Their downscaling results are quite similar, effectively restoring SST patterns in this region. This similarity in performance can also be found in their Bias maps (Figure 12). Except for Lasso regression, the bias distribution of these models in this sea area is relatively uniform, with no highly concentrated error regions. This indicates that deep learning models can recover most of the mesoscale dynamic processes in the SST downscaling task, but they still have some difficulty capturing certain small-scale processes.

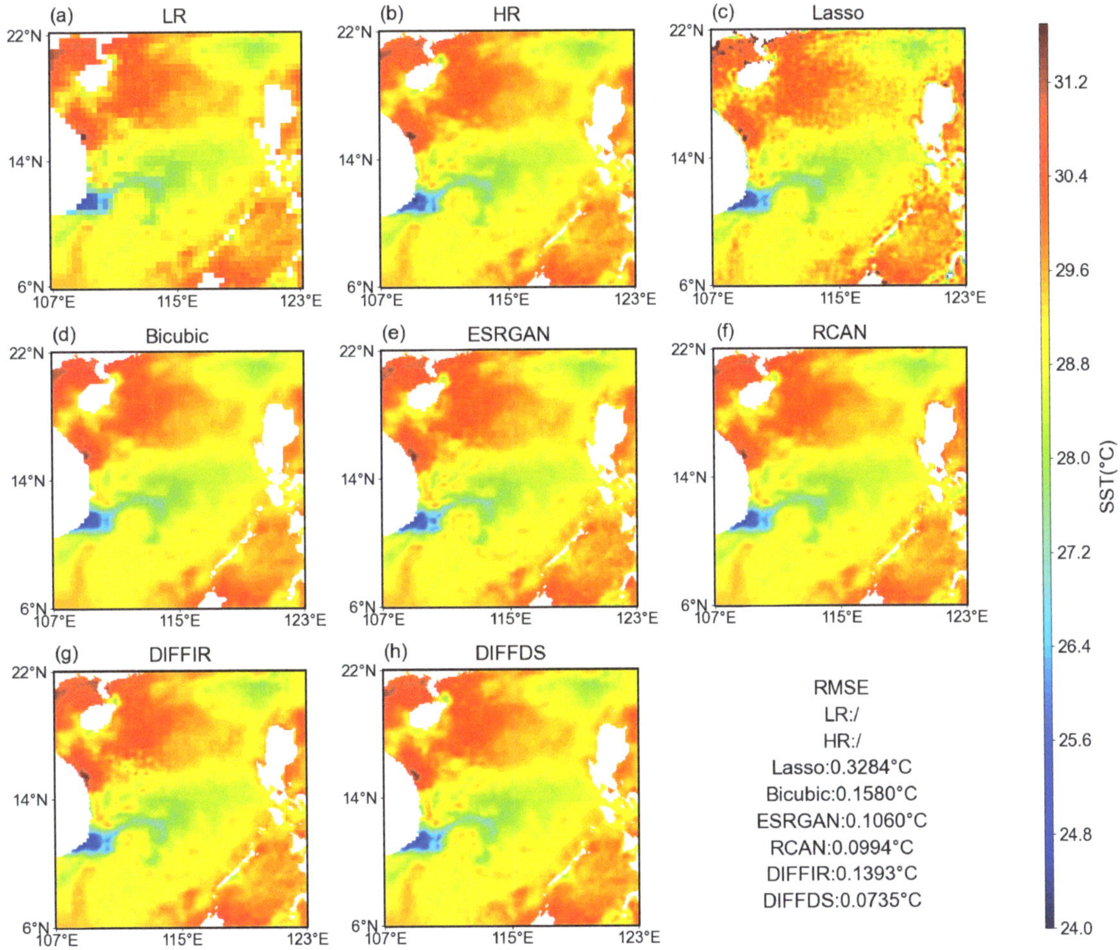

Figure 11. The SST distribution on 1 August 2021 of each model, these eight subplots individually display the high-resolution SSTs, low-resolution SSTs, and the downscaled results of each model along with their RMSE (°C).

The absence of previously significant bias in the 107°E–114°E area, 10°N–17°N area (red box in Figure 9) is likely due to the relatively stable fluctuations in SST in that region. Regarding RMSE, all deep learning models show a noticeable decrease compared to the sample on 29 June 2021, indicating further improvements in the plain SST structure environments. In this context, DIFFDS achieves the lowest RMSE at 0.0735 °C, reflecting the precision of the reconstructed high-resolution SSTs compared to other baselines. The RMSEs of DIFFIR and RCAN are close to those of DIFFDS. As shown in Figure 13, the results from deep learning models show visually minor differences (in the black box area) but are closer to the true SST distribution compared to the Bicubic and Lasso regression methods.

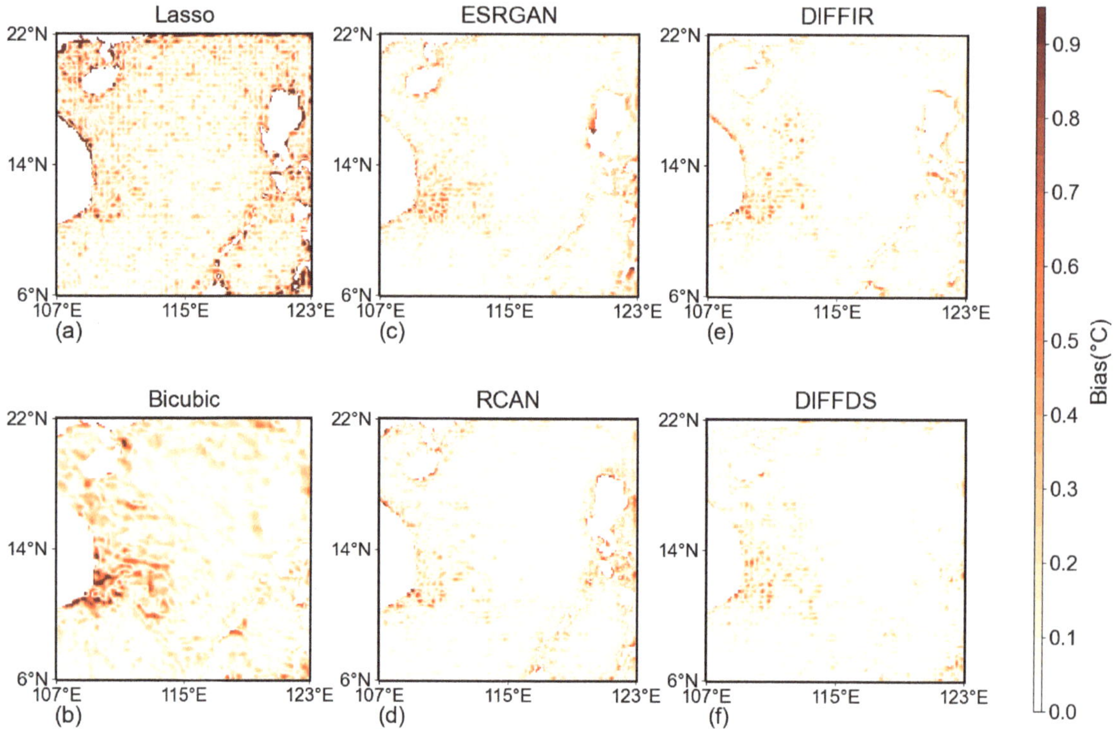

Figure 12. (**a–f**) The absolute Bias map between ground truth and each deep learning model on 1 August 2021.

4.2. Further Comparison of DIFFDS and DIFFIR

In the preceding sections, we analyzed the improvements in DIFFDS over the original DIFFIR using various metrics and specific downscaling results. In this section, to further illustrate the performance differences between these two models, we selected a case study from 27 June to 1 July 2021, in regions where coastal upwelling in Vietnam can be found.

In Figure 14, the coastal upwelling phenomenon in the HR exhibits a continuous variation process, which can be observed from the shape of the red 27 °C isotherms. However, this is not evident in the low-resolution SST data. As seen in the figures, the coastal upwelling morphology (the shape of red isotherms) in the low-resolution SST data over these five days shows almost no differences and fails to reflect a continuous morphological change. For DIFFIR, the downscaled results can not accurately represent this variability across those days. Its upwelling results are similar to those from Bicubic, being merely a simple magnification of the low-resolution data. If there are no significant changes in the upwelling morphology in the LR data, then the results also show no significant changes.

Based on the results of DIFFIR on 27 June, the upwelling morphology is relatively close to the true SST values. However, its upwelling morphology showed little change in the following days, remaining essentially the same as the previous day. It cannot exhibit a noticeable variation, as observed in the ground truth. This contributes to the suboptimal performance of DIFFIR's downscaling results.

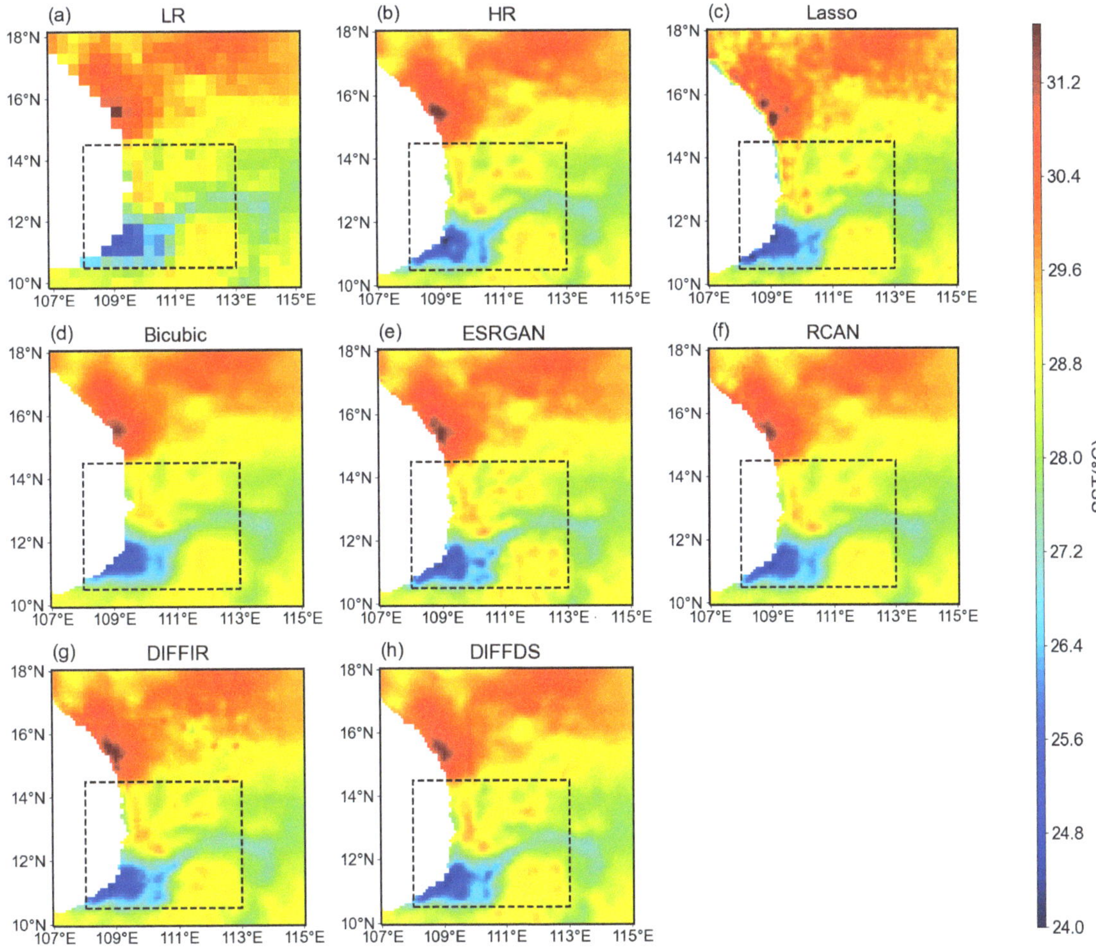

Figure 13. The results on 1 August 2021 zoomed from 10°N–18°N to 107°E–115°E. The black box area displays complex SST contents.

In contrast, the downscaled coastal upwelling from the DIFFDS displays a more pronounced variation over these five days, closely mirroring the ground truth. Additionally, the daily upwelling patterns produced by the DIFFDS are more aligned with the ground truth. This indicates that an improved network structure is more effective at learning the inherent SST distribution patterns in low-resolution SSTs by incorporating channel attention and cross-attention. It captures and reflects the subtle variations within the low-resolution SST data, and then translates them into high-resolution SSTs accurately. Consequently, the final downscaled results are more consistent with the ground truth, making the DIFFDS more suitable for SST spatial downscaling tasks.

In Figure 15, we present an analysis of the spatial distribution RMSE for both DIFFIR and DIFFDS on the test set. The figure clearly illustrates that DIFFDS consistently exhibits a lower RMSE across a majority of the examined regions when compared to DIFFIR. This trend is particularly pronounced within the area highlighted by the red box. In this specific region, DIFFIR demonstrates a significantly higher RMSE, indicating a poorer performance. However, the corresponding region in DIFFDS shows a markedly improved RMSE, highlighting the correction and enhancement achieved by our method.

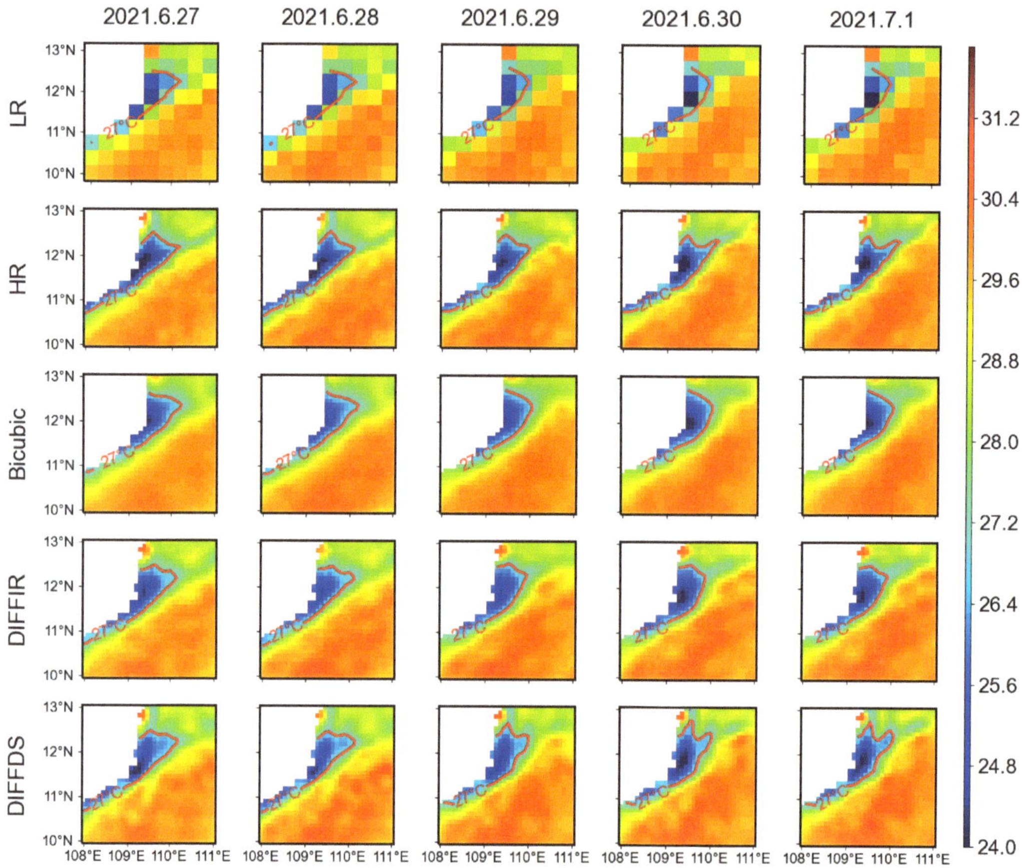

Figure 14. Variations in the low, ground truth, DIFFIR, and DIFFDS SST data over a continuous five-day period. Red isotherms are used to highlight the boundary of the upwelling currents.

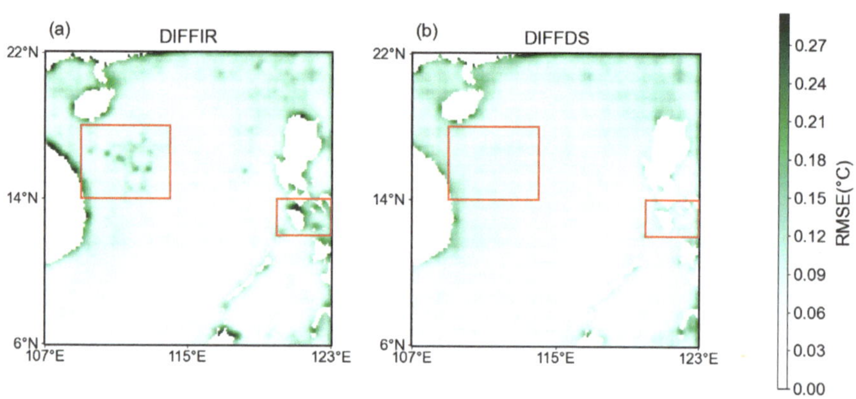

Figure 15. The spatial distribution of RMSE for DIFFIR and DIFFDS on the test set. The red boxed area highlights the significant difference between DIFFIR and DIFFDS.

The improvements observed underline the advantages of the enhanced DIFFDS method over the original DIFFIR approach. The reduction in RMSE across various spatial zones suggests that DIFFDS not only improves overall accuracy but also corrects specific areas of high error, making it a more reliable and robust solution.

4.3. Challenges with High Variability

In Section 3.2.2, we found that the intense fluctuations in SSTs during spring and summer pose a significant challenge to the model's downscaling capabilities. This can ultimately affect the accuracy of the downscaled results. Despite utilizing nearly three decades of data for training, DIFFDS has yet to effectively capture underlying SST patterns under conditions of high variability. We attribute this limitation to several factors. Firstly, recent years have witnessed abnormal changes in weather systems that may not be fully represented in the historical data used for training. Past experiences may not always apply to current or future situations, thereby restricting the model's performance in such cases. Additionally, the inherent limitations of our model's architecture and the complexity of the ocean system may also contribute to this shortcoming. Moreover, relying solely on SSTs as an input variable has its limitations.

When the region is more homogeneous, it is economically feasible to obtain the downscaling reconstruction using simple interpolation (usually Bicubic interpolation). However, the real-world system is very complicated, for instance, extreme ocean events and large changes in ocean elements (high variability situations). Therefore, there is an urgent need for efficient and effective downscaling algorithms that can respond extremely well to real-world systems, which is also the goal of our future work. All of the downscaling methods, including Bicubic interpolation, can greatly preserve the large-scale structure from the low-resolution inputs in terms of correlation coefficient differences. Additionally, the visual results and the corresponding zoomed-in regions, as well as several quantitative metrics (as shown in the figures and tables above), mainly evaluate the reconstruction quality of small-scale processes among different algorithms. On dates with high variability, or during periods when SST variability is relatively stable, our DIFFDS model surpasses the baseline models in most of these metrics, demonstrating that DIFFDS can effectively reconstruct more high-resolution activities under various conditions.

To improve downscaling results during periods of high variability, incorporating other oceanic factors, such as salinity, and atmospheric factors, like surface wind speed, into the neural network is essential. This approach could enable the model to capture a more comprehensive representation of the underlying dynamics and enhance its downscaling capabilities.

5. Conclusions

This study proposes a novel spatial downscaling model (DIFFDS) and applies it to SST spatial downscaling to explore its potential possibility and relieve the growing demand for high-resolution oceanic data. The results of our experiment demonstrate its effectiveness for spatial downscaling of SSTs. The proposed DIFFDS can restore some meso-scale processes that disappeared in low-resolution SSTs, generating accurate results.

To enhance the consistency and accuracy of the downscaled results, we designed a modified DMTA layer by introducing channel-attention and cross-attention mechanisms. The redesign enables the model to extract precise and effective information from the guidance IPR provided by the diffusion model, while also considering the overall SST distribution structure inherent in low-resolution SST data. This approach allows the model to suppress abnormal SST textures in the downscaled results, resulting in more realistic and accurate outputs.

Furthermore, we compared the performance of DIFFDS with commonly used CNN, GAN, and regression methods, to highlight its superiority over other models. Experimental results indicate that DIFFDS achieves an average RMSE of 0.1074 °C and PSNR of 50.48 dB in the $4\times$ scale downscaling task. Meanwhile, the generated content is comparable to the raw high-resolution SST data, such as the coastal upwelling along Vietnam.

Future work will focus on refining this method by incorporating other oceanic or atmospheric elements or integrating physical mechanisms into the model. This will further improve the interpretability and effectiveness of the DIFFDS, enabling more accurate and reliable SST spatial downscaling.

Author Contributions: Methodology, S.W.; experiments, S.W.; writing–original draft, S.W.; data curation, X.Z.; project administration, X.Z.; writing–review, X.Z.; mathematical analysis, X.L.; writing–review and editing, X.L.; resources, S.G. and J.L.; software, S.G. and J.L. All authors have read and agreed to the published version of the manuscript.

Funding: This research was funded in part by the project of Southern Marine Science and Engineering Guangdong Laboratory (Zhuhai) (No. SML2023SP202 and No. SML2023SP219), and the Zhuhai Basic and Applied Basic Research Foundation (No. 2320004002806).

Data Availability Statement: The OSTIA dataset in this paper is available to download in Copernicus at the following addresses: https://data.marine.copernicus.eu/product/SST_GLO_SST_L4_REP_OBSERVATIONS_010_011/description, accessed on 1 August 2024.

Acknowledgments: We acknowledge Natural Earth @naturalearthdata.com for providing the region image in Figure 1.

Conflicts of Interest: The authors declare no conflicts of interest.

References

1. Pastor, F. Sea Surface Temperature: From Observation to Applications. *J. Mar. Sci. Eng.* **2021**, *9*, 1284. [CrossRef]
2. Huang, X.; Rhoades, A.M.; Ullrich, P.A.; Zarzycki, C.M. An evaluation of the variable-resolution CESM for modeling California's climate. *J. Adv. Model. Earth Syst.* **2016**, *8*, 345–369. [CrossRef]
3. Shen, Z.; Shi, C.; Shen, R.; Tie, R.; Ge, L. Spatial Downscaling of Near-Surface Air Temperature Based on Deep Learning Cross-Attention Mechanism. *Remote Sens.* **2023**, *15*, 5084. [CrossRef]
4. Perez, J.; Menendez, M.; Camus, P.; Mendez, F.J.; Losada, I.J. Statistical multi-model climate projections of surface ocean waves in Europe. *Ocean Model.* **2015**, *96*, 161–170. [CrossRef]
5. Dong, C.; Loy, C.C.; He, K.; Tang, X. Image Super-Resolution Using Deep Convolutional Networks. *IEEE Trans. Pattern Anal. Mach. Intell.* **2016**, *38*, 295–307. [CrossRef] [PubMed]
6. Kim, J.; Lee, J.K.; Lee, K.M. Accurate Image Super-Resolution Using Very Deep Convolutional Networks. In Proceedings of the 2016 IEEE Conference on Computer Vision and Pattern Recognition (CVPR), Las Vegas, NV, USA, 27–30 June 2016.
7. Shi, W.; Caballero, J.; Huszar, F.; Totz, J.; Aitken, A.P.; Bishop, R.; Rueckert, D.; Wang, Z. Real-Time Single Image and Video Super-Resolution Using an Efficient Sub-Pixel Convolutional Neural Network. In Proceedings of the 2016 IEEE Conference on Computer Vision and Pattern Recognition (CVPR), Las Vegas, NV, USA, 27–30 June 2016.
8. Tong, T.; Li, G.; Liu, X.; Gao, Q. Image Super-Resolution Using Dense Skip Connections. In Proceedings of the 2017 IEEE International Conference on Computer Vision (ICCV), Venice, Italy, 22–29 October 2017.
9. Dong, X.; Xi, Z.; Sun, X.; Gao, L. Transferred Multi-Perception Attention Networks for Remote Sensing Image Super-Resolution. *Remote Sens.* **2019**, *11*, 2857. [CrossRef]
10. Salvetti, F.; Mazzia, V.; Khaliq, A.; Chiaberge, M. Multi-Image Super Resolution of Remotely Sensed Images Using Residual Attention Deep Neural Networks. *Remote Sens.* **2020**, *12*, 2207. [CrossRef]
11. Wang, X.; Yu, K.; Wu, S.; Gu, J.; Liu, Y.; Dong, C.; Qiao, Y.; Change Loy, C. Esrgan: Enhanced super-resolution generative adversarial networks. In Proceedings of the European Conference on Computer Vision (ECCV) Workshops, Munich, Germany, 8–14 September 2018.
12. Lu, Z.; Li, J.; Liu, H.; Huang, C.; Zhang, L.; Zeng, T. Transformer for Single Image Super-Resolution. In Proceedings of the 2022 IEEE/CVF Conference on Computer Vision and Pattern Recognition Workshops (CVPRW), New Orleans, LA, USA, 18–24 June 2022; pp. 456–465. [CrossRef]
13. Conde, M.V.; Choi, U.J.; Burchi, M.; Timofte, R.; Swin2SR: SwinV2 Transformer for Compressed Image Super-Resolution and Restoration. In *Computer Vision—ECCV 2022 Workshops*; Springer: Berlin/Heidelberg, Germany, 2023; pp. 669–687. [CrossRef]
14. Ducournau, A.; Fablet, R. Deep learning for ocean remote sensing: An application of convolutional neural networks for super-resolution on satellite-derived SST data. In Proceedings of the 2016 9th IAPR Workshop on Pattern Recogniton in Remote Sensing (PRRS), Cancun, Mexico, 4 December 2016. [CrossRef]
15. Khoo, J.J.D.; Lim, K.H.; Pang, P.K. Deep Learning Super Resolution of Sea Surface Temperature on South China Sea. In Proceedings of the 2022 International Conference on Green Energy, Computing and Sustainable Technology (GECOST), Miri, Sarawak, Malaysia, 26–28 October 2022. [CrossRef]
16. Izumi, T.; Amagasaki, M.; Ishida, K.; Kiyama, M. Super-resolution of sea surface temperature with convolutional neural network- and generative adversarial network-based methods. *J. Water Clim. Chang.* **2022**, *13*, 1673–1683. [CrossRef]
17. Zou, R.; Wei, L.; Guan, L. Super Resolution of Satellite-Derived Sea Surface Temperature Using a Transformer-Based Model. *Remote Sens.* **2023**, *15*, 5376. [CrossRef]

18. Saharia, C.; Chan, W.; Saxena, S.; Lit, L.; Whang, J.; Denton, E.; Ghasemipour, S.K.S.; Ayan, B.K.; Mahdavi, S.S.; Gontijo-Lopes, R.; et al. Photorealistic text-to-image diffusion models with deep language understanding. In Proceedings of the 36th International Conference on Neural Information Processing Systems (NeurIPS), New Orleans, LA, USA, 28 November–9 December 2022; pp. 36479–36494.
19. Ramesh, A.; Dhariwal, P.; Nichol, A.; Chu, C.; Chen, M. Hierarchical Text-Conditional Image Generation with CLIP Latents. *arXiv* **2022**, arXiv:2204.06125.
20. Saharia, C.; Ho, J.; Chan, W.; Salimans, T.; Fleet, D.J.; Norouzi, M. Image Super-Resolution Via Iterative Refinement. *IEEE Trans. Pattern Anal. Mach. Intell.* **2023**, *45*, 4713–4726. [CrossRef] [PubMed]
21. Li, H.; Yang, Y.; Chang, M.; Chen, S.; Feng, H.; Xu, Z.; Li, Q.; Chen, Y. SRDiff: Single image super-resolution with diffusion probabilistic models. *Neurocomputing* **2022**, *479*, 47–59. [CrossRef]
22. Shang, S.; Shan, Z.; Liu, G.; Wang, L.; Wang, X.; Zhang, Z.; Zhang, J. Resdiff: Combining cnn and diffusion model for image super-resolution. In Proceedings of the 38th Annual AAAI Conference on Artificial Intelligence, Vancouver, BC, Canada, 20–27 February 2024.
23. Xia, B.; Zhang, Y.; Wang, S.; Wang, Y.; Wu, X.; Tian, Y.; Yang, W.; Van Gool, L. DiffIR: Efficient Diffusion Model for Image Restoration. In Proceedings of the 2023 IEEE/CVF International Conference on Computer Vision (ICCV), Paris, France, 1–6 October 2023.
24. Stark, J.D.; Donlon, C.J.; Martin, M.J.; McCulloch, M.E. OSTIA: An operational, high resolution, real time, global sea surface temperature analysis system. In Proceedings of the OCEANS 2007-Europe, Aberdeen, Scotland, 18–21 June 2007. [CrossRef]
25. Donlon, C.J.; Martin, M.; Stark, J.; Roberts-Jones, J.; Fiedler, E.; Wimmer, W. The Operational Sea Surface Temperature and Sea Ice Analysis (OSTIA) system. *Remote Sens. Environ.* **2012**, *116*, 140–158. [CrossRef]
26. Good, S.; Fiedler, E.; Mao, C.; Martin, M.J.; Maycock, A.; Reid, R.; Roberts-Jones, J.; Searle, T.; Waters, J.; While, J.; et al. The Current Configuration of the OSTIA System for Operational Production of Foundation Sea Surface Temperature and Ice Concentration Analyses. *Remote Sens.* **2020**, *12*, 720. [CrossRef]
27. Liu, K.; Qiu, G.; Tang, W.; Zhou, F. Spectral Regularization for Combating Mode Collapse in GANs. In Proceedings of the 2019 IEEE/CVF International Conference on Computer Vision (ICCV), Seoul, Republic of Korea, 27 October–2 November 2019. [CrossRef]
28. Huang, H.; Li, Z.; He, R.; Sun, Z.; Tan, T. IntroVAE: Introspective variational autoencoders for photographic image synthesis. In Proceedings of the 32nd International Conference on Neural Information Processing Systems (NeurIPS), Montréal, QC, Canada, 3–8 December 2018.
29. Sohl-Dickstein, J.; Weiss, E.; Maheswaranathan, N.; Ganguli, S. Deep unsupervised learning using nonequilibrium thermodynamics. In Proceedings of the 32nd International Conference on Machine Learning (ICML), Lille, France, 6–11 July 2015; pp. 2256–2265.
30. Ho, J.; Jain, A.; Abbeel, P. Denoising diffusion probabilistic models. In Proceedings of the 34th International Conference on Neural Information Processing Systems (NeurIPS), Vancouver, BC, Canada, 6–12 December 2020; pp. 6840–6851.
31. Nichol, A.Q.; Dhariwal, P. Improved denoising diffusion probabilistic models. In Proceedings of the 38th International Conference on Machine Learning (ICML), Virtual, 18–24 July 2021; pp. 8162–8171.
32. Song, J.; Meng, C.; Ermon, S. Denoising Diffusion Implicit Models. In Proceedings of the 8th International Conference on Learning Representations (ICLR), Addis Ababa, Ethiopia, 26–30 April 2020.
33. Rombach, R.; Blattmann, A.; Lorenz, D.; Esser, P.; Ommer, B. High-Resolution Image Synthesis with Latent Diffusion Models. In Proceedings of the 2022 IEEE/CVF Conference on Computer Vision and Pattern Recognition (CVPR), New Orleans, LA, USA, 18–24 June 2022.
34. Hu, J.; Shen, L.; Sun, G. Squeeze-and-Excitation Networks. In Proceedings of the 2018 IEEE/CVF Conference on Computer Vision and Pattern Recognition (CVPR), Salt Lake City, UT, USA, 18–22 June 2018. [CrossRef]
35. Zamir, S.W.; Arora, A.; Khan, S.; Hayat, M.; Khan, F.S.; Yang, M.H. Restormer: Efficient transformer for high-resolution image restoration. In Proceedings of the IEEE/CVF Conference on Computer Vision and Pattern Recognition (CVPR), New Orleans, LA, USA, 18–24 June 2022; pp. 5728–5739.
36. Wang, X.; Xie, L.; Dong, C.; Shan, Y. Real-esrgan: Training real-world blind super-resolution with pure synthetic data. In Proceedings of the 2021 IEEE/CVF International Conference on Computer Vision (ICCV), Montréal, QC, Canada, 10–17 October 2021; pp. 1905–1914.
37. Lin, S.; Liu, B.; Li, J.; Yang, X. Common Diffusion Noise Schedules and Sample Steps are Flawed. In Proceedings of the 2024 IEEE/CVF Winter Conference on Applications of Computer Vision (WACV), Waikoloa, HI, USA, 3–8 January 2024. [CrossRef]
38. Ledig, C.; Theis, L.; Huszar, F.; Caballero, J.; Cunningham, A.; Acosta, A.; Aitken, A.; Tejani, A.; Totz, J.; Wang, Z.; et al. Photo-Realistic Single Image Super-Resolution Using a Generative Adversarial Network. In Proceedings of the 2017 IEEE Conference on Computer Vision and Pattern Recognition (CVPR), Honolulu, HI, USA, 21–26 July 2017.

remote sensing

MDPI

Article

Enhanced Fishing Monitoring in the Central-Eastern North Pacific Using Deep Learning with Nightly Remote Sensing

Jiajun Li [1,2], Jinyou Li [3], Kui Zhang [1,2], Xi Li [4] and Zuozhi Chen [1,2,*]

1. South China Sea Fisheries Research Institute, Chinese Academy of Fishery Sciences, Guangzhou 510300, China; lijiajun@scsfri.ac.cn (J.L.); zhangkui@scsfri.ac.cn (K.Z.)
2. Key Laboratory for Sustainable Utilization of Open-Sea Fishery, Ministry of Agriculture and Rural Affairs, Guangzhou 510300, China
3. Goergen Institute for Data Science, University of Rochester, Rochester, NY 14627, USA
4. State Key Laboratory of Information Engineering in Surveying, Mapping and Remote Sensing, Wuhan University, Wuhan 430072, China; lixi@whu.edu.cn

* Correspondence: chenzuozhi@scsfri.ac.cn

Abstract: The timely and accurate monitoring of high-seas fisheries is essential for effective management. However, efforts to monitor industry fishing vessels in the central-eastern North Pacific have been hampered by frequent cloud cover and solar illumination interference. In this study, enhanced fishing extraction algorithms based on computer vision were developed and tested. The results showed that YOLO-based computer vision models effectively detected dense small fishing targets, with original YOLOv8 achieving a precision (P) of 89% and a recall (R) of 79%, while refined versions improved these metrics to 93% and 99%, respectively. Compared with traditional threshold methods, the YOLO-based enhanced models showed significantly higher accuracy. While the threshold method could identify similar trend changes, it lacked precision in detecting individual targets, especially in blurry scenarios. Using our trained computer vision model, we established a dataset of dynamic changes in fishing vessels over the past decade. This research provides an accurate and reproducible process for precise monitoring of lit fisheries in the North Pacific, leveraging the operational and near-real-time capabilities of Google Earth Engine and computer vision. The approach can also be applied to dynamic monitoring of industrial lit fishing vessels in other regions.

Keywords: nighttime lights; fishing monitoring; deep learning; VIIRS DNB

Citation: Li, J.; Li, J.; Zhang, K.; Li, X.; Chen, Z. Enhanced Fishing Monitoring in the Central-Eastern North Pacific Using Deep Learning with Nightly Remote Sensing. *Remote Sens.* **2024**, *16*, 4312. https://doi.org/10.3390/rs16224312

Academic Editors: Benoit Vozel and Wenfang Lu

Received: 10 September 2024
Revised: 1 November 2024
Accepted: 11 November 2024
Published: 19 November 2024

1. Introduction

Due to the convergence of the Kuroshio and Oyashio currents (Figure 1), the North Pacific has become a significant fishing ground, especially renowned for its long-standing tradition of light-based fisheries [1,2]. Despite a noted decline in recent years, the lit fishing industry continues to make a substantial contribution to the distant-water fishery [3]. The lit fisheries of the North Pacific are primarily concentrated in two regions: the west stock and the central-east stock (Figure 1). Approximately 600 fishing vessels using lights operate in the North Pacific. These vessels can be categorized into three main types: around 280 using stick-held dip nets to catch Pacific saury, 260 employing jiggers to catch squid, and 60 purse seiners primarily targeting chub mackerel, spotted mackerel, and Japanese sardines. In the western stock, all three types of gear are utilized, whereas fishing in the central-east stock is predominantly conducted by squid vessels from China and Japan. These vessels operate from mid-May to early August, with peak activity occurring in June and July, during which nearly all vessels are actively engaged in fishing [3–5]. Although the fishing industry in the North Pacific has generally shown a downward trend, the central-east stock has experienced growth [4]. Comprehensive and efficient monitoring of these vessels is instrumental for the effective management of distant-water fisheries. Additionally, with the recent drastic climate changes, the precise dynamics of fishing movements combined with ecological environments could also aid in understanding the factors influencing the dynamics of fishing grounds, thereby providing a scientific basis for forecasting distant-water fishery conditions [6–8]. However, monitoring efforts for such lit fisheries in the high seas have been hindered by the lack of effective vessel monitoring data. Precise monitoring methods and long-term surveillance data for such fisheries remain insufficient.

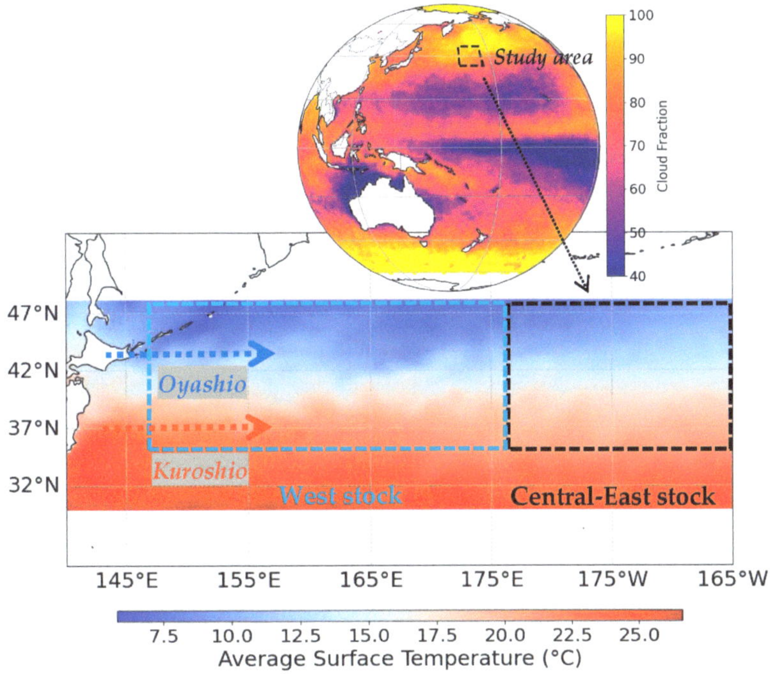

Figure 1. Study region of the lit fishing grounds in the North Pacific (central-east stock, 176E–165W, 35N–48N), overlaid with the average sea surface temperature from the summer fishing season of 2020. The top global map shows the location of the study area, overlaid with the average cloud fraction from the summer fishing season of 2020. The cloud fraction data are sourced from GlobColour (http://globcolour.info), which has been developed, validated, and distributed by ACRI-ST, France. The sea surface temperature data come from the Copernicus Marine Service (https://data.marine.copernicus.eu), which is the marine component of the Copernicus Programme of the European Union.

In recent years, nighttime remote sensing technology has been widely used for monitoring fishing vessels using lights and analyzing dynamic changes [9–12]. Among the most prominent sources of nighttime remote sensing imagery are the Defense Meteorological Satellite Program's Operational Linescan System (DMSP-OLS) and the Visible Infrared Imaging Radiometer Suite's Day/Night Band (VIIRS-DNB), which are valued for their global coverage and long time series [13,14]. Lit fishing vessels typically appear as small bright spikes in nighttime imagery, and traditional detection methods have primarily relied on visual interpretation, followed by the application of various threshold-based methods [15–17]. Elvidge et al. [9] analyzed the spike features in VIIRS DNB data collected during moonless and cloudless nights and developed an algorithm that detected vessels by evaluating the sharpness of the light spikes. They also established a comprehensive process for filtering out land-based and gas flare sources. The results of the detection process were validated through vessel position data, confirming the algorithm's effectiveness. Building on this workflow, the Earth Observation Group (EOG) has further developed and refined the automated fishing vessel extraction process. They have designed an adaptive threshold method to accommodate varying weather conditions and released the corresponding dataset: VIIRS Boat Detection (VBD) [https://eogdata.mines.edu/products/vbd/]. However, the improved algorithm's effectiveness in extracting lit fishing vessels diminishes significantly under complex weather conditions [18,19]. During the central-east stock fishing season, which overlaps with summer, the VIIRS DNB nighttime imagery is significantly impacted by strong solar illumination and cloud cover (Figure 1). This affects the reliability of VBD products in the region, necessitating further comparison and validation. To achieve more accurate monitoring and advance related research, there is an urgent need not only to develop more suitable extraction algorithms but also to supplement long-term monitoring data. In recent years, advances in computer technology have enabled computer vision algorithms to excel in detecting small, obscured, or blurred targets, demonstrating great potential for vessel detection in nightly imagery impacted by clouds [20–23]. These methods automatically learn target features and have been successfully applied across various scenarios. Therefore, leveraging computer vision technology could significantly enhance the ability to extract and monitor lit fishing vessels in complex and cloudy weather conditions, which is crucial for advancing relevant research.

In this study, we initially established a dataset of nighttime remote sensing images of fishing vessels operating in the central-east stock, spanning various meteorological conditions from 2012 to 2022, along with corresponding labeled vessel targets. By leveraging this image database and the You Only Look Once (YOLO) architecture, we refined the network's sensitivity to small objects and developed improved methods for extracting fishing vessels under varying cloud conditions [20–24]. The effectiveness of these methods was evaluated by comparing visual inspection, VBD threshold extraction, and YOLO-based models using data from the 2020 fishing season. We also extracted dynamic information on fishing vessels in the study area over the past decade, focusing on peak activity in June and July. Finally, a discussion is provided to explore the advantages, limitations, necessity, and future development directions of nighttime remote sensing-based fishing vessel extraction and monitoring methods.

2. Data and Method

2.1. Data

2.1.1. Nighttime Light Imagery and Cloud Condition Data

The VIIRS provides comprehensive global data across both the visible and the infrared spectra [25]. Specifically, the VIIRS Day/Night Band (DNB) offers daily global measurements of nighttime illumination in these spectra, making it particularly well suited for tracking fishing vessels that utilize lights. In addition to the DNB, VIIRS includes long-wave infrared channels, which are crucial for assessing sea surface temperature and distinguishing cloud formations. In our study, we employed a Level 3 VIIRS product from the Suomi NPP (VNP46A1), a daily radiance product that captured nighttime brightness at the sensor level, referred to as the VIIRS/NPP Daily Gridded Day Night Band 15 arc-second Linear Lat/Lon Grid Night [26]. This product, available since January 2012, was derived from VIIRS sensors and was processed daily within 3–5 h post-acquisition, supporting both near-real-time applications and long-term monitoring. For our research, we utilized Google Earth Engine (GEE) to filter and obtain VNP46A1 data for the North Pacific central-east stock during the fishing seasons (June–July) from 2012 to 2023 [27]. In addition to using DNB radiance as nighttime light imagery for detecting fishing vessels using lights, we also incorporated the M16 band brightness temperature to account for cloud reflections in our analysis.

2.1.2. VIIRS Boat Detection Data

The VIIRS Boat Detection (VBD) data product captures illuminated pixels of fishing boats in a VIIRS DNB image for a single night. The basic VBD algorithms were described by Elvidge et al. [10]. The VBD data files provide detailed information, including the geo-location of illuminated pixels, radiance values, satellite overpass times, satellite viewing zenith angles, and various threshold parameters used to differentiate potential light sources from fishing vessels and other sources [28]. In this study, nightly VBD data from the 2020 fishing seasons in the study area were utilized to compare and evaluate the performance of deep learning methods for detecting lit fishing activities (https://eogdata.mines.edu/products/vbd/). We extracted only VBD points with quality flags 1 (strong boat detection), 2 (weak boat detection), 3 (blurry boat detection), and 10 (weak and blurry lights), which corresponded to radiance spikes more likely to have originated from marine vessels. Additionally, for VBD points in overlapping satellite images, only a selected portion corresponding to GEE filtered parts were used, and pixels with radiance values below 10 were excluded from the analysis [29].

2.1.3. Satellite AIS Data

AIS data for the study area during the 2020 fishing season were obtained from the satellite service provider ORBCOMM (https://www.orbcomm.com/en/solutions/maritime/ais-data) to evaluate the performance of deep learning methods for detecting illuminated fishing activities [30]. We filtered information from AIS message types 1 (scheduled position report), 2 (assigned scheduled position report), 3 (special position report responding to an interrogation), 18 (standard Class B equipment position report), 19 (extended Class B equipment position report), and 27 (long-distance message), all of which included a timestamp, longitude and latitude, MMSI number, speed, and course. Squid fishing vessels authorized to operate in the central to east Pacific were listed by the North Pacific Fisheries Commission (NPFC). This list was downloaded from the NPFC (https://www.npfc.int/compliance/vessels) and used to filter and extract the relevant vessels from the AIS data based on their unique MMSI numbers.

2.2. Method

2.2.1. Selection and Annotation of Nighttime Light Imagery

To develop a more accurate algorithm for detecting illuminated fishing vessels in nighttime imagery, we first compiled a dataset of labeled images containing these vessels. The nighttime images from the study area, obtained through GEE, were preprocessed by being logarithmically transformed and normalized between 0 and 1. These images were then resized to a uniform dimension of 640 $\times$ 640 pixels, and those without any vessels were discarded. The remaining images were categorized into three classes: clearly visible and distinguishable (Figure 2A); not clearly visible but locatable and distinguishable (Figure 2B); and not clearly visible, locatable but indistinguishable (Figure 2C). For the first two categories, accurate quantification and location determination were feasible, whereas the third category did not allow for precise numerical differentiation but permitted the identification of the operational location. The images from the first two categories were annotated using LabelImg (1.8.2). Ultimately, we obtained a dataset comprising 541 images with 10,837 vessels. This dataset has been uploaded to an online repository, and it is available to the scientific community on demand.

2.2.2. Development and Test of Computer Vision Models

Detecting small and dense objects presents inherent challenges. Over time, numerous improvements have been introduced to object detection frameworks to enhance the accuracy of detecting [23]. In our research, we developed a computer vision model using YOLOv8 (Figure 3), which consisted of a backbone for feature extraction, a neck for combining multi-scale features, and a head for object detection. YOLOv8 improved detection performance with its efficient architecture, combining CSPNet and PANet for better feature propagation and localization accuracy. This allowed it to perform real-time detection while maintaining high precision across varying object scales. To address the challenge of detecting dense clusters of small fishing vessels, we employed two key optimization modifications: (1) We integrated multi-scale feature fusion to combine low-level feature maps, which provide detailed positional information, with high-level feature maps, rich in semantic context. This integration was key to improving the detection performance, particularly for small targets like dense clusters of fishing vessels (Figure 3, YOLO-P2). (2) We replaced strided convolution and/or pooling layers, which often lead to fine-grained information loss, with a CNN building block called SPD-Conv to create a more effective model [31] (Figure 3, YOLO-SPD).

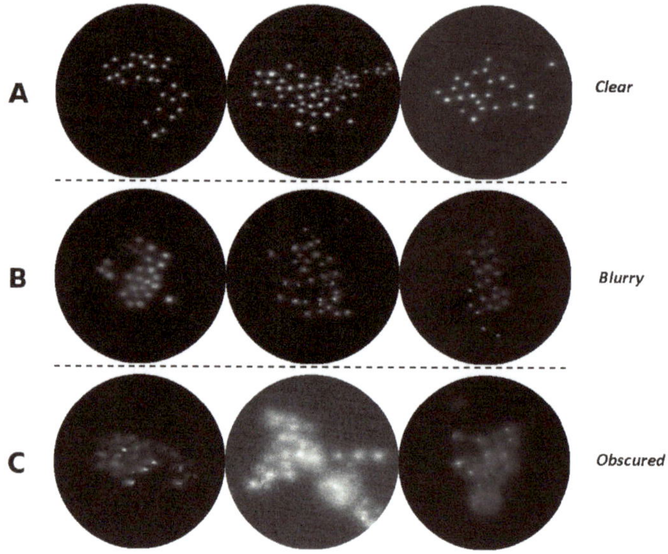

Figure 2. Typical nighttime remote sensing images observed by VIIRS-DNB with squid jiggers: (**A**) Clearly identifiable and distinguishable lit fishing vessels. (**B**) Blurry but distinguishable and locatable lit fishing vessels. (**C**) Obscured lit fishing vessels, only locatable.

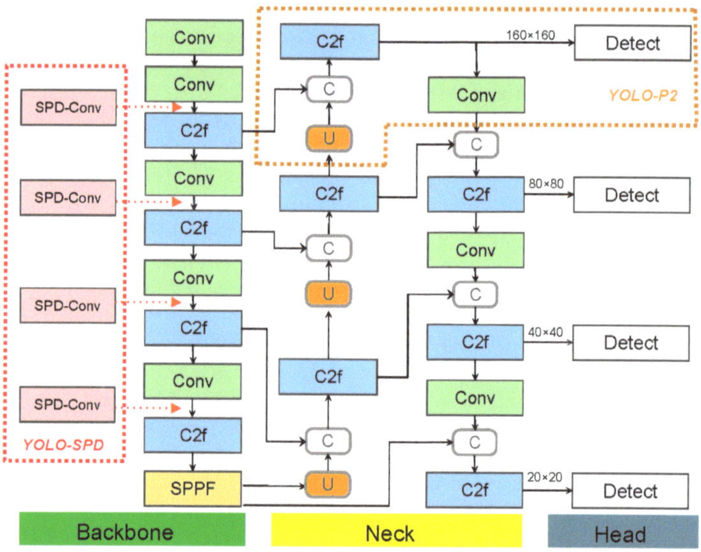

Figure 3. The overall framework of the original YOLOv8, YOLO-P2, and YOLO-SPD networks. In the YOLO-SPD model, SPD-Conv is integrated into the YOLOv8 backbone to enhance performance. Meanwhile, YOLO-P2 retains the YOLOv8 backbone but adds a P2 detection head to improve small target detection.

Labeled images from the year 2020 were selected for comparative testing, while the remaining images were randomly allocated, with 90% used for training and 10% for validation. Due to the nature of dense small targets, the criterion for a true positive was adjusted to an intersection over union threshold of 10%. The training parameters were configured as follows: a batch size of 16, a total of 500 iterations, and an input image resolution of 640 × 640 pixels. All other parameters were kept at their default settings. The experimental setup is outlined in Table 1.

Table 1. Setup of the experiment environment.

Item	Value
CPU	Intel(R) Xeon(R) E5-2686 v4 @ 2.30 GHz
RAM	60 GB
GPU	NVIDIA RTX A4000
Operating system	Ubuntu 20.04
Cuda	CUDA 11.3
Data processing	Python 3.9
Deep learning framework	Pytorch 1.12.1

In the experiment, computer vision models were evaluated using several metrics, including precision, recall, accuracy, F1 score, mean average precision (mAP), frames per second (FPS), and giga floating point operations per second (GFLOPS). The formulas for these key metrics are as follows:

$$\text{Precision (P)}: \ P = \frac{TP}{TP+FP}$$
$$\text{Recall (R)}: \ R = \frac{TP}{TP+FN}$$
$$\text{F1 Score}: \ F1 = 2 \times \frac{P \times R}{P+R}$$

The definitions of the parameters in the formulas are as follows: true positive (TP) represents the samples correctly identified as positive, false positive (FP) represents the samples incorrectly identified as positive, false negative (FN) represents the samples incorrectly identified as negative, and true negative (TN) represents the samples correctly identified as negative. Additionally, mAP (mean average precision) measures the average area under the precision–recall curve for each target across all images, providing a comprehensive assessment of the precision and recall performance. FPS is the inverse of the total time required to process a single frame, encompassing both the inference time and the non-maximum suppression time. GFLOPS (giga floating point operations per second) represents the computational cost of performing the forward pass (inference) in billions of floating point operations per second.

3. Results

3.1. Fishing Detection with YOLO-Based Computer Vision Methods

Three computer vision-based models were trained for the extraction and detection of central-east stock lit fishing vessels. For the validation set, YOLOV8 achieved its highest F1 score at a very low confidence (Conf) threshold, with a value of Conf = 0.004, resulting in an F1 score of 0.947. The two improved YOLO models exhibited a parabolic variation in their F1 scores. Specifically, the model designated as YOLO-SPD reached its peak F1 score at a confidence threshold of Conf = 0.182, while YOLO-P2 achieved its maximum at Conf = 0.247, with respective F1 scores of 0.977 and 0.964.

Labeled images from 2020 served as the test set for evaluating the performance of these three models. The confidence thresholds for the test data were derived from the validation results at the point where the F1 scores peaked. Table 2 illustrates the performance parameters of the three detection models. The model validation outcomes demonstrated that both YOLOv8 and its enhanced counterparts were adept at efficiently filtering and extracting results under complex meteorological conditions, with high accuracy rates. YOLOv8 itself achieved an accuracy of 89%, while the precision of both enhanced YOLOv8 models improved by approximately 5%, reaching around 93%. Representative nighttime imagery confirmed that all models were capable of fully and effectively detecting lit fishing vessels in the North Pacific under clear skies (Figures 4, A1 and A2).

Table 2. The overall test results for YOLO and tiny targets improved models.

Model	P	R	F1	Confidence
YOLOV8	0.886	0.79	0.947	0.031
YOLO-SPD	0.932	0.987	0.977	0.182
YOLO-P2	0.928	0.989	0.964	0.247

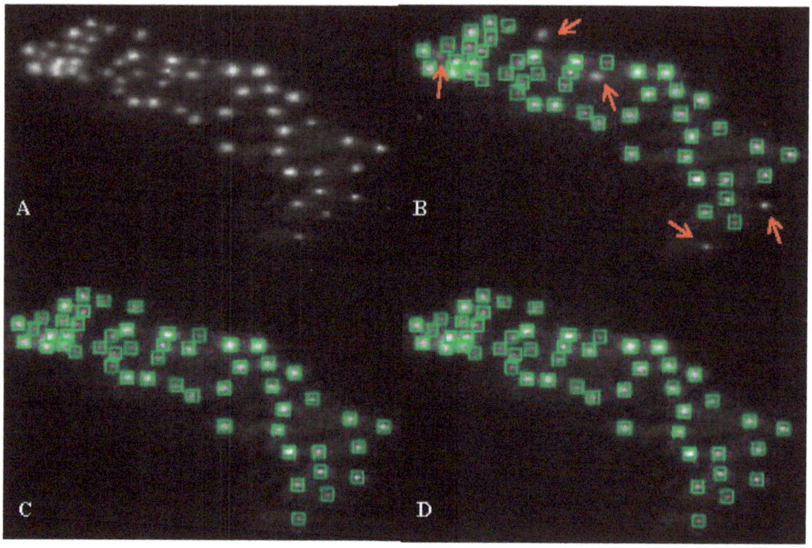

Figure 4. Typical nighttime imagery and detection results from different methods: (**A**) Original nighttime image. (**B**) Detection results by YOLO, where missed targets are denoted by red arrows. (**C**) Detection results by YOLO-P2. (**D**) Detection results by YOLO-SPD.

Comparatively, the original YOLOv8 model's performance in filtering is marginally less effective than its improved counterparts, particularly evident in a lower recall rate, leading to a higher likelihood of missed detections (Figures 4 and A1). In the test set, the original YOLOv8 architecture had a 21% omission rate for dense fishing boat targets. However, the integration of a small target detection head (YOLO-P2) and the Space-to-Depth (YOLO-SPD) model both significantly improved their performance in detecting fishing vessels, resulting in an effective increase in monitoring precision. The recall rate saw a notable increase, with both models improving by about 18%. Both enhancements effectively compensated for the original YOLOv8's tendency to miss detections. Regarding computational speed, the original YOLO architecture boasted the FPS. While the FPS of the improved models increased, the addition of the SPD model led to a more substantial enhancement compared with the addition of the detection head. Computational requirements followed a similar pattern, with YOLO having the lowest demand and YOLO-SPD significantly exceeding that of YOLO-P2 (Table 3).

Table 3. Parameters, mAP, FPS, and GFLOPS of improved YOLO models at different scales.

Model	Size	Parameters (MB)	mAP	FPS (s)	GFLOPS
YOLO-P2	n	2.92	0.970	138.89	12.2
	s	10.63	0.971	107.53	36.6
	m	25.03	0.965	49.75	97.9
	l	42.82	0.968	45.05	204.7
	x	66.56	0.969	30.67	316.1
YOLO-SPD	n	2.86	0.973	73.53	45.1
	s	10.54	0.957	39.84	164.7
	m	24.85	0.949	21.55	423.9
	l	42.29	0.953	13.11	843.3
	x	66.07	0.956	8.32	1317

Beyond the n scale version, we also evaluated models of varying sizes for the improved YOLO models (Table 3). Contrary to expectations, the performance accuracy from the training results of the s, m, l, and x versions did not improve with the increase in model size. Instead, there was a notable escalation in the number of parameters and computational demands, accompanied by a significant drop in FPS. Hence, for the central-east stock of the North Pacific, there was no necessity to opt for larger models in training. The n-version model, characterized by minimal parameters and swift computation, was sufficient to yield excellent outcomes.

3.2. Comparision of Different Fishing Detection Methods

In this section, we conducted a comparative analysis of fishing monitoring using visual examination, the VBD algorithm, and computer vision techniques. During the peak fishing season (June and July) of 2020, visual inspection successfully identified operational fishing vessels over a span of 60 days. Only one day (4th June) yielded unrecognizable images (Figure 5, type 3). On 29 days, the imagery allowed for the identification and precise location of most fishing vessels (Figure 5, type 1). However, on the remaining 31 days, although fishing vessels were visually confirmed, their exact positions could not be determined, or some vessels were entirely obscured (Figure 5, type 2, and Figure A3).

We then analyzed the results obtained from both the VBD threshold algorithm and computer vision models. Even though these light-based high-seas fishing vessels operated daily throughout the fishing season, a consistent pattern of variation emerged across the VBD algorithm and the three computer vision models (Figure 5). The highest vessel counts detected by all three methods corresponded with clear visual images, while lower counts were associated with blurred images. These high counts were also comparable to the AIS-derived ground truth, with maximum vessel numbers reaching approximately 80 (Figure 5). Utilizing the M16 parameter as an indicator of meteorological conditions, the findings further demonstrated that fluctuations in the number of detected fishing vessels corresponded with changes in M16 (Figure 5). In terms of overall counts, the VBD algorithm detected approximately 2000 vessels throughout the fishing season, while the YOLO-based and related computer vision algorithms observed a 25% increase, identifying around 2500 vessels.

In cases where location and identification were largely feasible, we conducted a comparative analysis of various methods against the results of visual inspection (Figure 6). The validation findings revealed that the VBD algorithm produced the least accurate results compared with the visually derived ground truth (R^2 = 0.64). In contrast, the regression consistency between the YOLO methods and visual inspection was stronger than that of the VBD algorithm. YOLOv8 and its improved versions showed significant correlation, with YOLOv8 demonstrating slightly lower precision (R^2 = 0.82), while the two refined models achieved comparable and superior results (R^2 = 0.97). For clearly visible targets, the results of the VBD algorithm closely aligned with those of the computer vision models (Figure A1). However, in instances where the targets were indistinct but still locatable, computer vision outperformed the VBD threshold algorithm, effectively extracting these more ambiguous instances (Figure A2). Further detailed analysis showed that both the YOLO-P2 and YOLO models produced more false detections from the background compared with YOLO-SPD, with SPD demonstrating overall better performance.

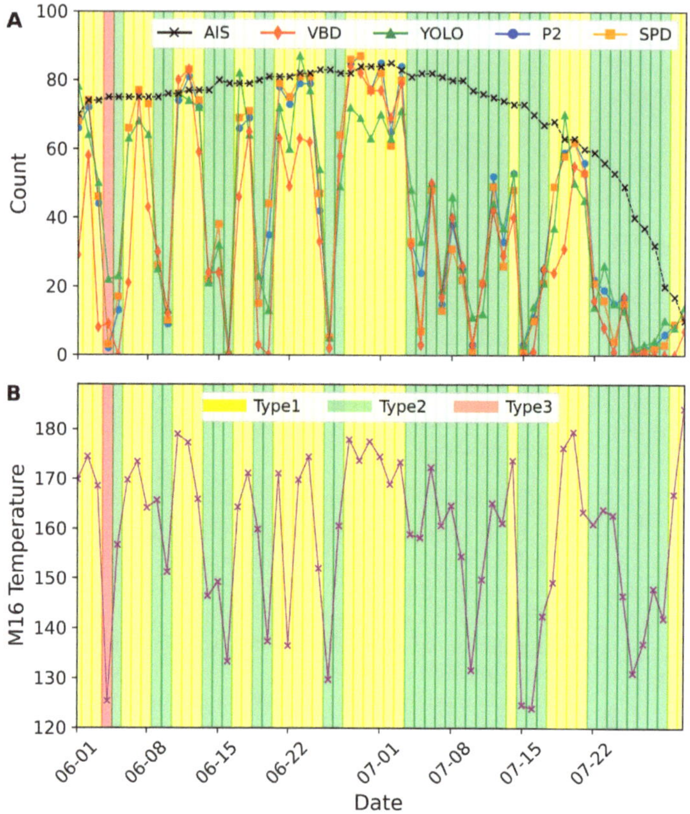

Figure 5. Changes in the number of vessels during the summer fishing season in the North Pacific central-east stock obtained using AIS, VBD, YOLO, and improved YOLO algorithms (with P2 representing YOLO-P2 and SPD representing YOLO-SPD) (**A**) and changes in M16 brightness temperature during the fishing season (**B**). The background colors in both subplots represent visual classification categories: yellow (type 1) for cases where most light fishing vessels can be located and distinguished, green (type 2) for cases where a significant portion of the vessels are blurred or partially obscured, and red (type 3) for cases where all vessels are completely obscured.

Although the VBD and computer vision methods differed in their precision of extraction, the monthly distribution results revealed a consistent trend across both methods (Figure 7). Despite variations in the number of detections, with computer vision identifying a greater quantity than VBD, the distribution patterns from the different methods showed a notable correlation, indicating a shared pattern in the data collected by each technique. Additionally, the false detections produced by the YOLO-P2 and YOLOv8 models were observable in the horizontal distribution, particularly in scattered operational points that were distant from the dense fishing hot spots (Figure 7).

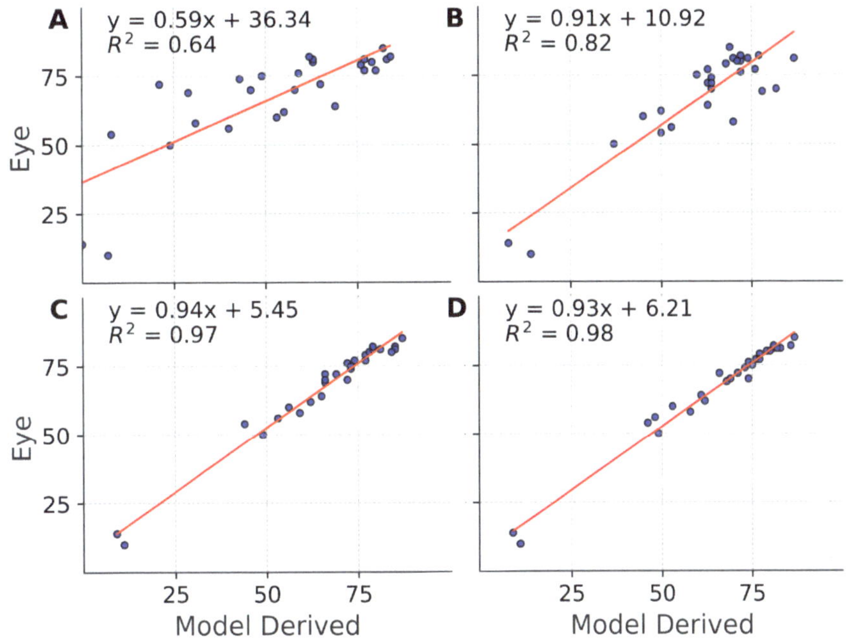

Figure 6. Regression fit between visual observations (Eye) and results from VBD, YOLO, and improved YOLO models under conditions where most lit fishing vessels can be visually distinguished. (**A**) Visual observations vs. VBD. (**B**) Visual observations vs. YOLO. (**C**) Visual observations vs. YOLO-P2. (**D**) Visual observations vs. YOLO-SPD.

3.3. Decadal Variation of Fishing Vessels in the Central-East Stock

Based on the optimal computer vision model (YOLO-SPD) we developed, fishing vessels using lights during the fishing season (June to July) in the study area from 2012 to 2023 were extracted. The results indicated that the area with the highest concentration of fishing activity, which showed considerable variation in distribution and was located between 180°W and 168°W longitude and 39°N and 43°N latitude, maintained a consistent pattern of change, with significant differences between June and July (Figure 8A). In June, the center of activity was generally situated farther south, while in July, it shifted approximately 2 degrees northward, demonstrating a clear seasonal variation in fishing activity concentration.

Regarding the number of vessels, there was a noticeable increase from around 30 fishing vessels in 2012–2016 to a peak of approximately 70 in 2017–2020 (Figure 8B). However, from 2021 to 2023, the number of fishing vessels declined, likely due to the impact of the COVID-19 pandemic. This decline reflected the broader economic and operational challenges faced by the fishing industry during this period.

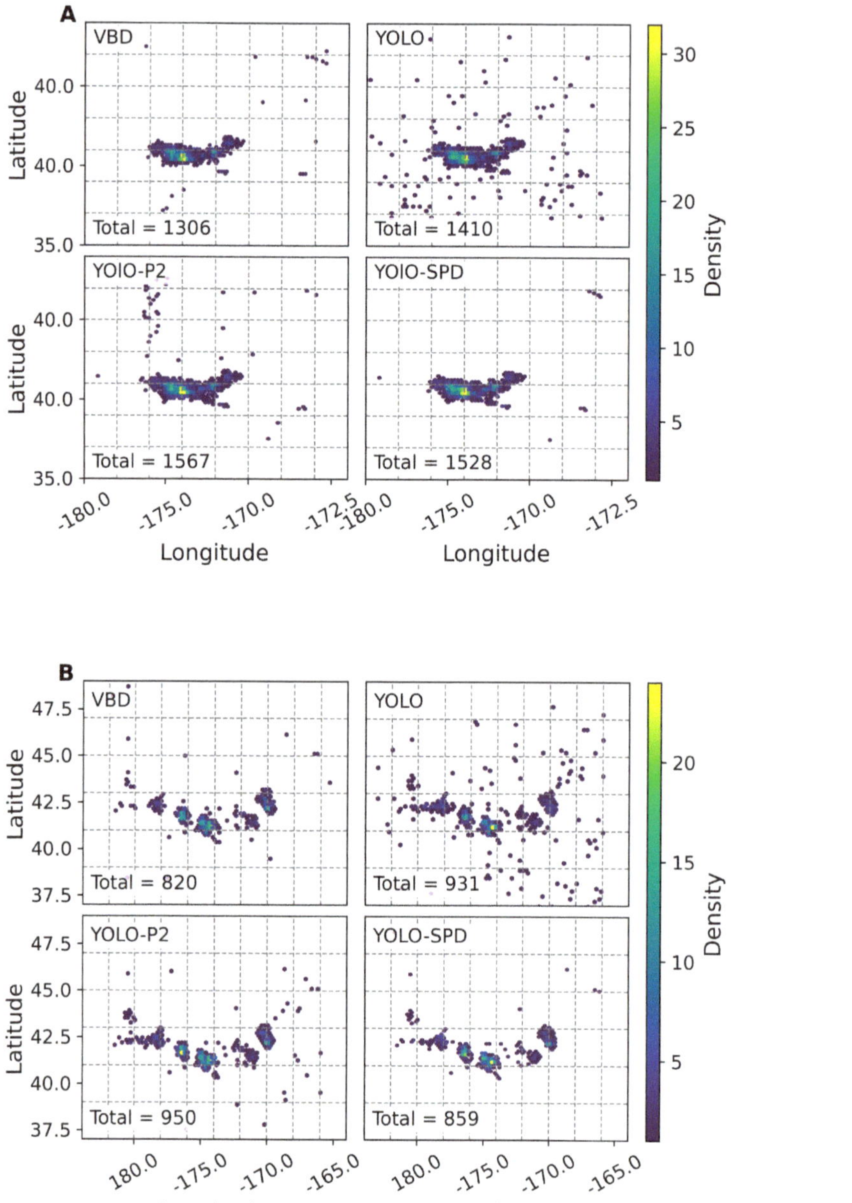

Figure 7. Monthly density scatter plots of lit fishing vessels during the summer fishing season in the North Pacific central-east stock, obtained using VBD, YOLO, and their improved algorithms. (**A**) Distribution in June. (**B**) Distribution in July.

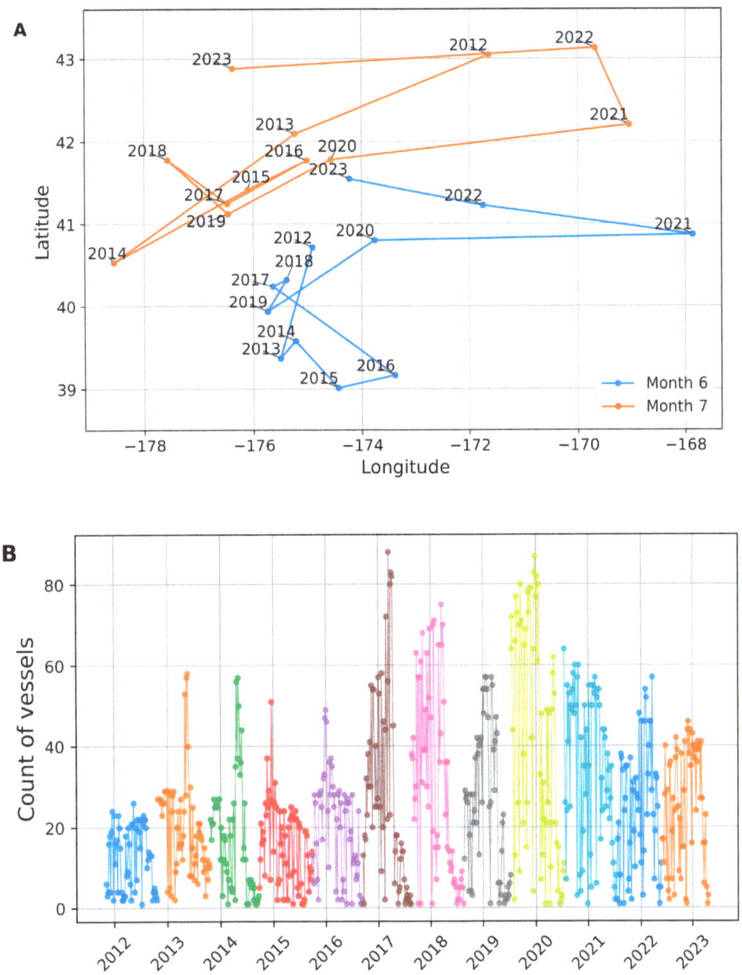

Figure 8. Dynamic changes in the centroid position of fishing vessels (**A**) and the number of operating vessels (**B**) during the summer fishing seasons from 2012 to 2023 in the central-east stock of the North Pacific.

4. Discussion

4.1. The Need for Fishing Vessel Monitoring with Nighttime Remote Sensing

The effective management of fishing vessels hinges on precise monitoring. Compared with coastal regions, obtaining comprehensive, effective, and accurate data on the operating locations of fishing vessels in the open ocean is significantly more challenging [32]. Currently, most commercial fishing companies operating on the high seas are equipped with vessel monitoring systems, AIS (Automatic Identification System), and fishing logbooks. While vessel monitoring systems and AIS can record vessel positions, the frequency of position updates is limited, sometimes providing only one or two locations per day [31–33]. This makes it difficult to accurately pinpoint the operating locations of light fishing vessels based solely on monitoring data, especially during transition of different fishing grounds (Figure A4). Furthermore, illegal, unreported, and unregulated (IUU) fishing vessels often deliberately disable their monitoring systems, resulting in untrackable positions [33]. The overlay of typical nighttime remote sensing imagery with AIS data reveals discrepancies between the monitored positions and the actual operating locations of vessels (Figure A4).

On the other hand, the coverage of vessel monitoring data was sparse in earlier years, with AIS satellite coverage also being low. As equipment coverage increased and satellite constellations were established, data coverage has become more comprehensive in recent years [34,35]. Only after 2018 did the data coverage become sufficient to allow for a good match with nighttime remote sensing imagery. This discrepancy means that data from vessel monitoring systems cannot be used for long-term quantitative studies, as the coverage in earlier years was comparatively low, whereas nightly remote sensing data can be used for long-term continuous comparative studies, as they adhere to a unified standard. Fishing logbooks not only record operational locations but also contain rich information on yields [28]. However, they lack near-real-time capabilities. Moreover, acquiring data from vessel monitoring systems, AIS, and fishing logbooks is challenging, especially in fishing grounds jointly developed by multiple countries and regions, where data sharing between different countries is often difficult.

Nighttime remote sensing offers a publicly accessible and comprehensive coverage technology for monitoring light fishing vessels. To date, research on the dynamic changes in fishing grounds in the study area has primarily relied on the integration of positional data from Japanese commercial fishing vessels' logbooks with ecological and environmental factors [4,6,7,36]. Fishing vessels from different countries and regions often operate in the same areas, and having comprehensive operational information from multiple nations could significantly enhance the ability to forecast fishing ground conditions. Using nighttime light remote sensing, we observed that the number of fishing vessels operating in the central-east North Pacific during 2020 peaked at around 80 vessels, which is closely aligned with the vessel counts recorded by the NPFC. Similarly, AIS data also reported approximately 80 active vessels, further supporting the reliability of the nighttime light-based observations. This strong consistency between remote sensing and AIS-derived ground truth demonstrates the effectiveness of nighttime light remote sensing as a complementary monitoring tool for fishing activities. However, a daily comparison between AIS and nighttime light remote sensing revealed that AIS reported significantly higher vessel counts under cloudy conditions, whereas in moonless conditions, nighttime light remote sensing was able to detect nearly all active vessels. This suggests that when extracting and interpreting fishing vessel data from nighttime light imagery, special attention must be given to the meteorological conditions in specific fishing grounds. Adverse weather conditions may limit the accuracy of nighttime light remote sensing, preventing it from fully reflecting real-world vessel activity. It is also important to note that while nighttime light remote sensing offers an effective supplementary method for monitoring lighted fishing vessels, it is insufficient for identifying the vessels' identities. Nighttime light imagery can confirm the presence of vessels but cannot determine their nationality or whether they are engaging in illegal, unreported, and unregulated (IUU) fishing. To accurately identify these vessels and assess their compliance with regulations, AIS data and the NPFC's list of registered vessels permitted to operate must be integrated.

Additionally, with the launch of the VIIRS satellite series (including NPP, JPSS1, and JPSS2), the frequency of observations further increased, allowing for even more detailed monitoring [37,38]. Therefore, it is of critical importance to continue developing precise and efficient monitoring methods and processes for light fishing vessels based on nighttime remote sensing. This advancement will provide significant support for the accurate and comprehensive monitoring of light fisheries on the high seas, ensuring better management and sustainability of these vital resources.

4.2. Opportunities for High-Seas Lit Fishing Detection with Computer Vision

Maritime lights captured in nightly imagery may be influenced by cloud cover and lunar phases [39]. Particularly when both cloud cover and moonlight exposure are at play, small-power maritime lights can be easily obscured [11]. However, under conditions of thin cloud cover, the bright lights of industry lit fishing vessels can still be effectively located and distinguished (Figure 2). Our visual observations indicate that during the fishing season, these very bright squid fishing vessels can still be effectively located and distinguished under thin cloud conditions (Figure 2). Nevertheless, the visual detection method is both time-consuming and labor-intensive and is susceptible to subjective judgments. Achieving automated accurate identification to a level of interpretability comparable to visual results is the main challenge in extracting and monitoring these industry fishing vessels. Under cloudless and moonless conditions, the threshold method performs exceptionally well in extraction and recognition [10,17]. However, under thin cloud conditions, a fixed threshold may not be effective for extraction and recognition. To extend the applicability of the threshold method, EOG has proposed an adaptive threshold approach to address issues under various meteorological conditions. Yet, to avoid false detections caused by moonlit clouds, the threshold is set relatively high, which may result in some situations that are still observable under visual conditions being unable to be extracted by traditional threshold methods. Our research findings demonstrate the advantages of computer vision technology over traditional methods, effectively enhancing the accurate extraction capability of industry fishing vessels under thin cloud conditions. The robust capabilities and swift progress in computer vision technology present a wealth of opportunities for the precise extraction, monitoring, and study of fishing vessels using lights. Visual interpretation adeptly decodes the majority of lighting images within the research area. Cloudy conditions, despite their challenges, cannot obscure the discernible anomalies in nightly imagery, which still offer a broad evaluation of the operational zones and vessel positions (Figure A3). In many cases, the dynamics and shifts within fishing grounds do not require a count of the fishing vessels; pinpointing the operational areas is itself a crucial piece of information for such studies. Nonetheless, current methodologies fall short in interpreting these images, highlighting a notable challenge for the future of automated extraction and monitoring in designated research fishing zones. Additionally, in the South Atlantic Anomaly region, another high-seas lit fishing "hot spot", the interference of high-energy particles can lead the standard maritime imagery threshold method astray, misidentifying these particles as fishing vessels and thus generating a significant number of false positives [40]. The differentiation between high-energy particles and genuine fishing vessels is beyond the scope of threshold methods alone [41]. Computer vision, leveraging its formidable prowess in classification, extraction, and recognition, offers the potential to significantly bolster the capabilities for extracting and monitoring light fishing vessels in this region.

4.3. Optimized Procedure for Monitoring Localized High-Seas Fishing Vessels

EOG's VBD vessel extraction algorithm offers a comprehensive workflow for extracting and distributing fishing vessel data. This process begins with downloading extensive raw VIIRS-DNB data, followed by vessel extraction using the threshold method, and ends with public distribution (Figure 9). By providing ready-to-use CSV files with vessel location information, researchers are spared the challenges of handling massive raw imagery and can focus directly on monitoring and analyzing the dynamics of lit fishing vessels. However, this approach has limitations. Currently, the VBD data are produced in near-real time, with nightly records dating back to April 2012 for Asia and 2017 for other regions [42]. Reproducing the VBD algorithm to extract long-term time series data for specific fishing grounds outside Asia remains challenging for most researchers. Additionally, our study shows that the high adaptive threshold used in the VBD data product reduces the efficiency of detecting industry lit fishing vessels under cloudy conditions. To address these issues, we propose an improved extraction method tailored for specific "hot spots", using the north-central western Pacific fishing grounds as a case study. This method leverages GEE for near-real-time selection of VNP46A1 data, which outputs DNB imagery reflecting faint light (Figure 9). An enhanced YOLOv8 model then accurately extracts vessels in the targeted areas. Compared with the massive raw nighttime imagery, VNP46A1 has been preprocessed, reducing the data volume to 40 MB per file [26], and GEE allows for precise regional selection, enabling focused analysis.

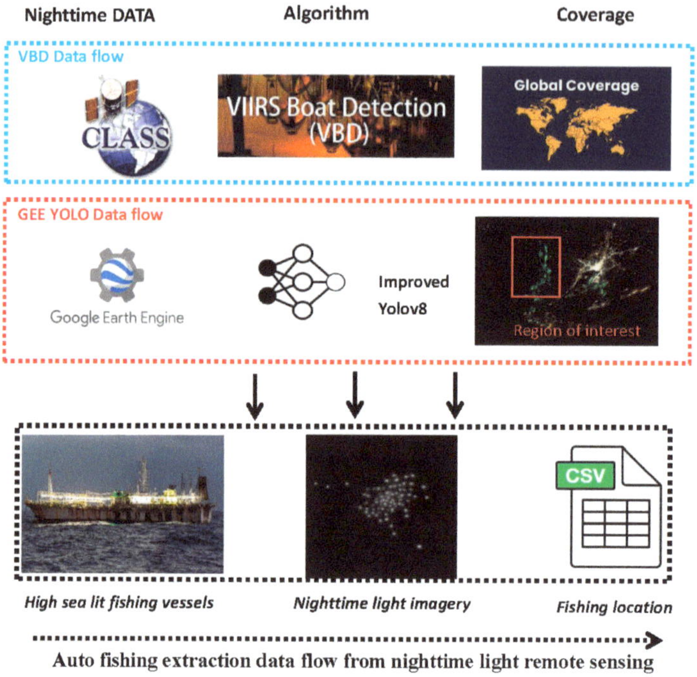

Figure 9. Comparison of VBD and our data flow for high-seas lit fishing monitoring.

Our procedure could also include daily meteorological parameters for specific study area, such as cloud cover extent and intensity, which are crucial for accurate vessel count comparisons. Weather conditions significantly affect the detection of industry lit vessels on the high seas (Figure 5). Since not all daily data are usable, determining the number of analyzable images is vital for comparing different months and years. Our approach first filters data based on weather conditions, then enhances the quantitative accuracy by considering interannual and monthly weather variations. In summary, we propose a more reproducible procedure for extracting and monitoring lighted fishing vessels using nighttime remote sensing. This method simplifies the process for personalized monitoring and research of localized industry lit fishing vessels in the high seas.

5. Conclusions

This study introduced an advanced methodology for monitoring and extracting industry lit fishing vessels in specific fishing grounds, particularly under cloudy conditions, using computer vision techniques and nightly imagery. By leveraging the YOLO model and its enhanced variants, the proposed approach demonstrated a high accuracy in detecting fishing vessels in the central-east stock of the North Pacific, with improvements in precision and recall achieved through the small target detection enhancements model. The machine learning methods outperformed conventional threshold-based algorithms (VBD), particularly in handling blurred images affected by thin cloud cover, with the SPD model proving to be the most effective, closely aligning with visual inspection results and minimizing false detections. This research provides a reproducible framework that enhances the recognition of vessels in challenging conditions and offers valuable insights into the cloud cover characteristics of specific fishing grounds. Additionally, the study contributes a decade-long dataset for the North Pacific central-east stock fishing grounds, offering a valuable resource for further research and fisheries management.

Author Contributions: Conceptualization, J.L. (Jiajun Li) and Z.C.; data curation, J.L. (Jiajun Li), J.L. (Jinyou Li) and K.Z.; formal analysis, J.L. (Jiajun Li); funding acquisition, Z.C.; methodology, J.L. (Jiajun Li), J.L. (Jinyou Li) and K.Z.; software, X.L.; validation, J.L. (Jiajun Li) and J.L. (Jinyou Li); visualization, J.L. (Jiajun Li) and J.L. (Jinyou Li); writing—original draft, J.L. (Jiajun Li) and J.L. (Jinyou Li); writing—review and editing, Z.C. All authors have read and agreed to the published version of the manuscript.

Funding: This work was supported by the National Natural Science Foundation of China (32303008) and the Central Public-interest Scientific Basal Research Fund, CAFS (Nos. 2024RC07 and 2023TD05).

Data Availability Statement: The original contributions presented in the study are included in the article; further inquiries can be directed to the corresponding author.

Acknowledgments: The authors would like to thank the anonymous reviewers for their very competent comments and helpful suggestions.

Conflicts of Interest: The authors declare no conflicts of interest.

Appendix A

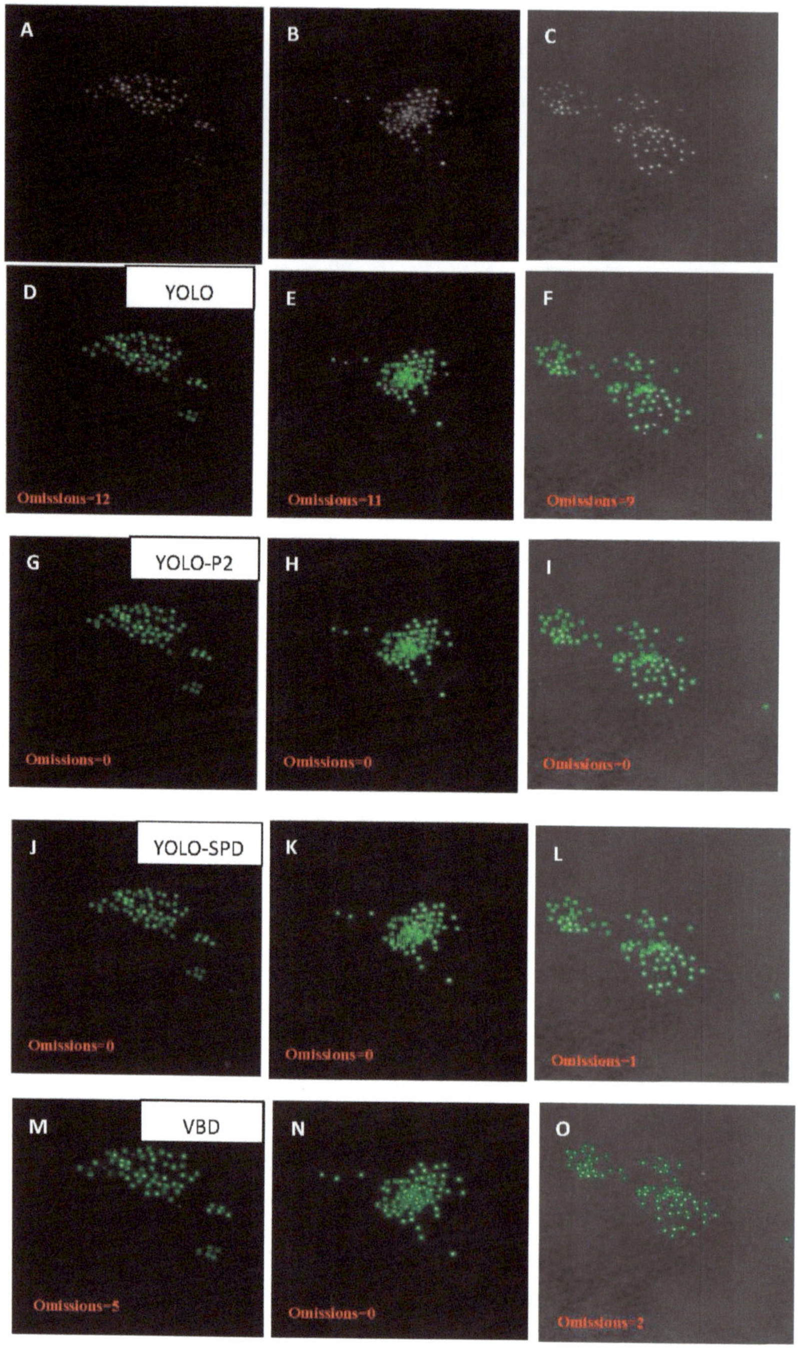

Figure A1. Typical clear nightly images (**A–C**) and detections using various automated fishing extraction methods, (**D–F**) nightly images with detections by YOLO, (**G–I**) nightly images with detections by YOLO-P2, (**J–L**) nightly images with detections by YOLO-SPD, and (**M–O**) nightly images with detections by VBD.

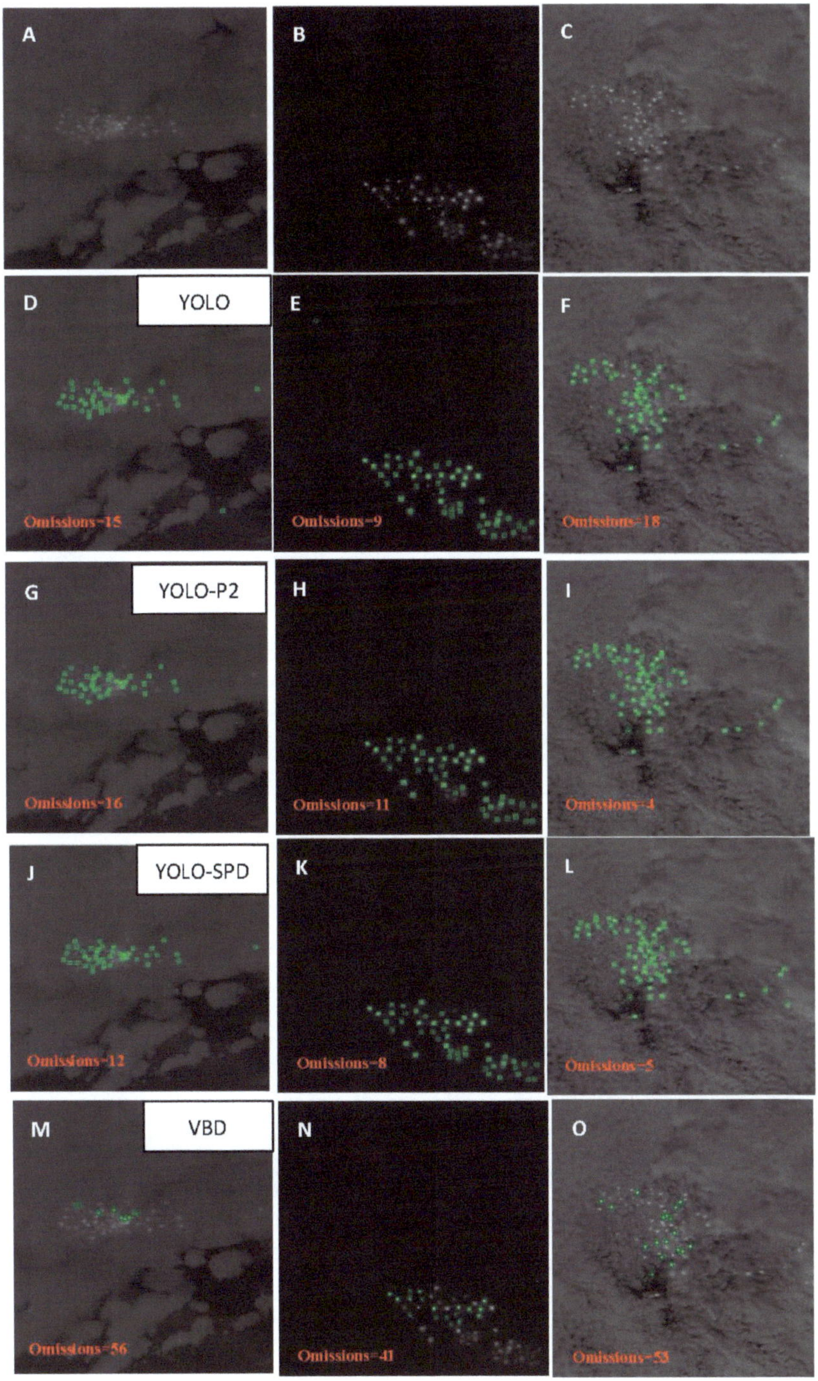

Figure A2. Typical blurry nightly images (**A–C**) and detections using various automated fishing extraction methods, (**D–F**) nightly images with detections by YOLO, (**G–I**) nightly images with detections by YOLO-P2, (**J–L**) nightly images with detections by YOLO-SPD, and (**M–O**) nightly images with detections by VBD.

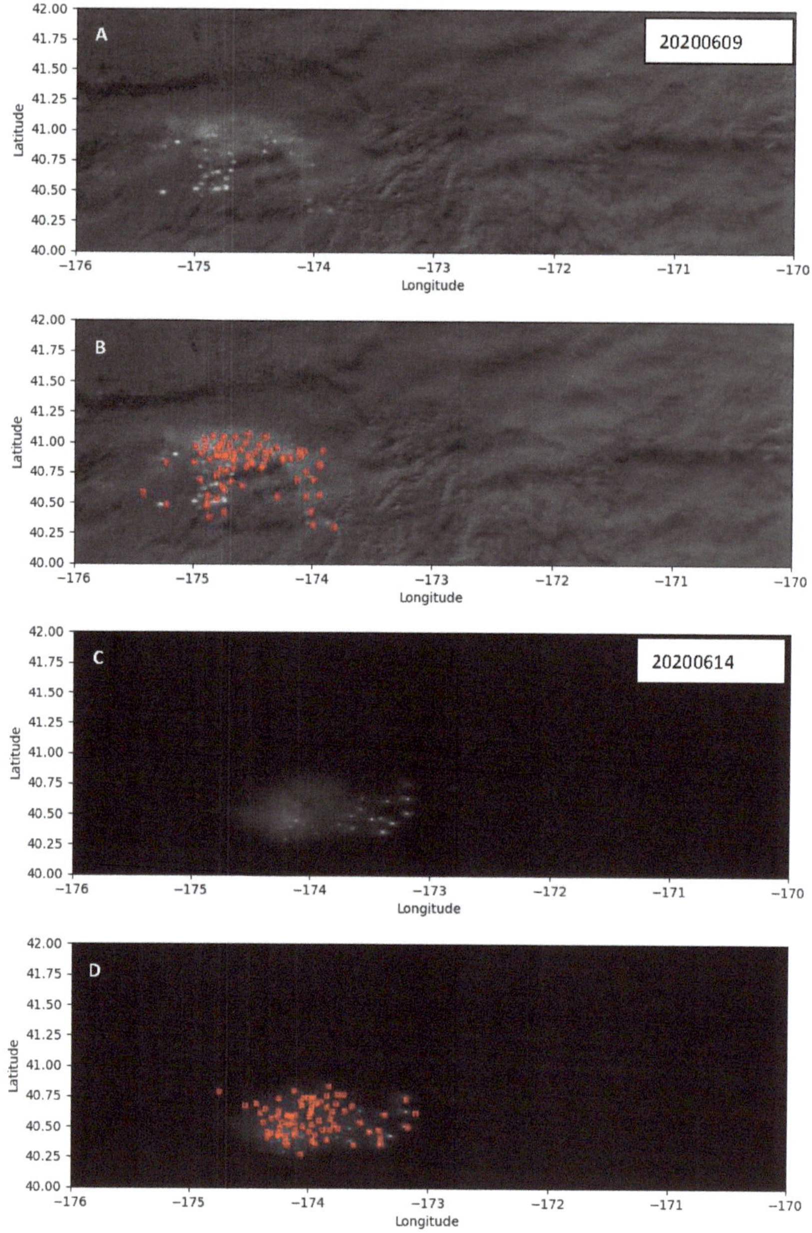

Figure A3. *Cont.*

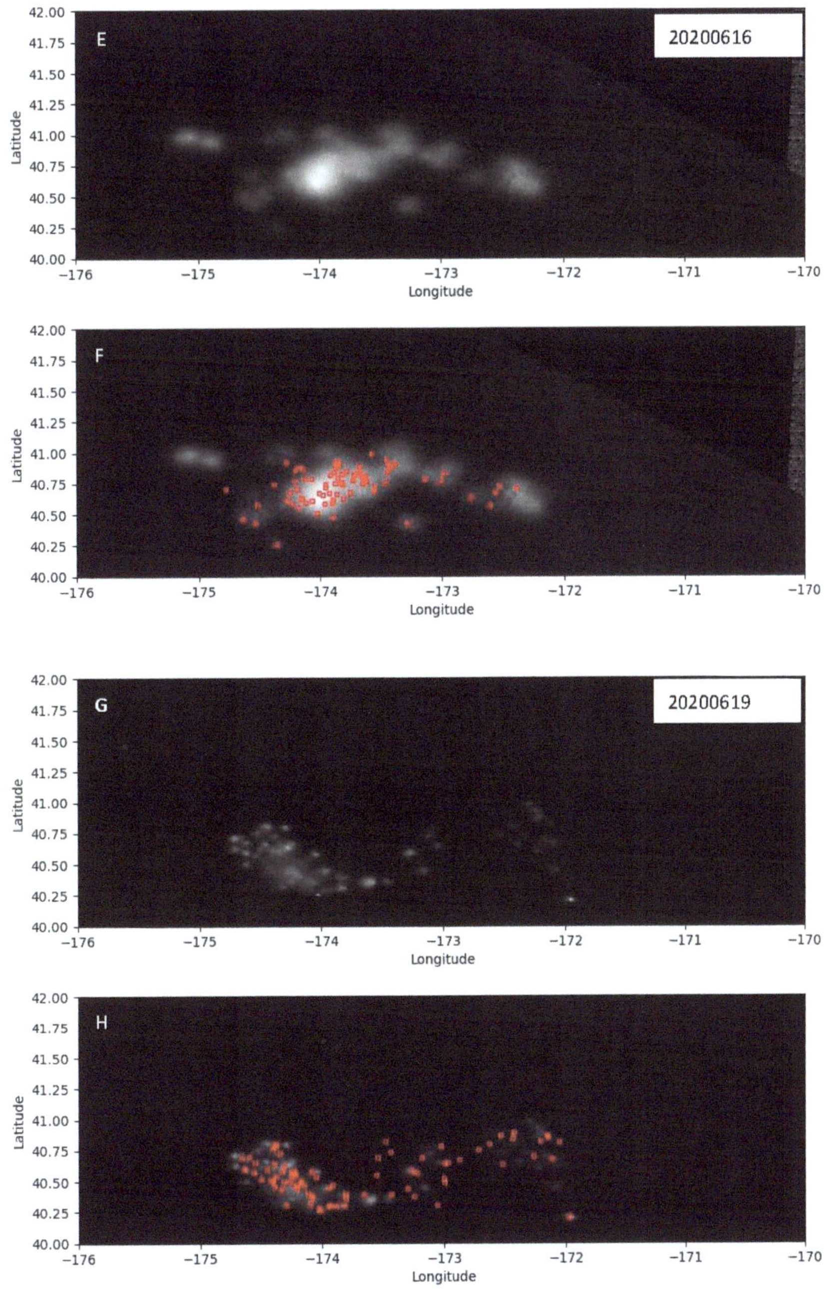

Figure A3. Satellite AIS-derived fishing vessels overlaid with typical obscured nightly images. While fishing locations can be roughly identified, the exact number of vessels cannot be determined. Panels (**A**,**C**,**E**,**G**) display typical obscured nightly remote sensing images, and panels (**B**,**D**,**F**,**H**) show AIS-derived fishing activity locations overlaid on nightly images.

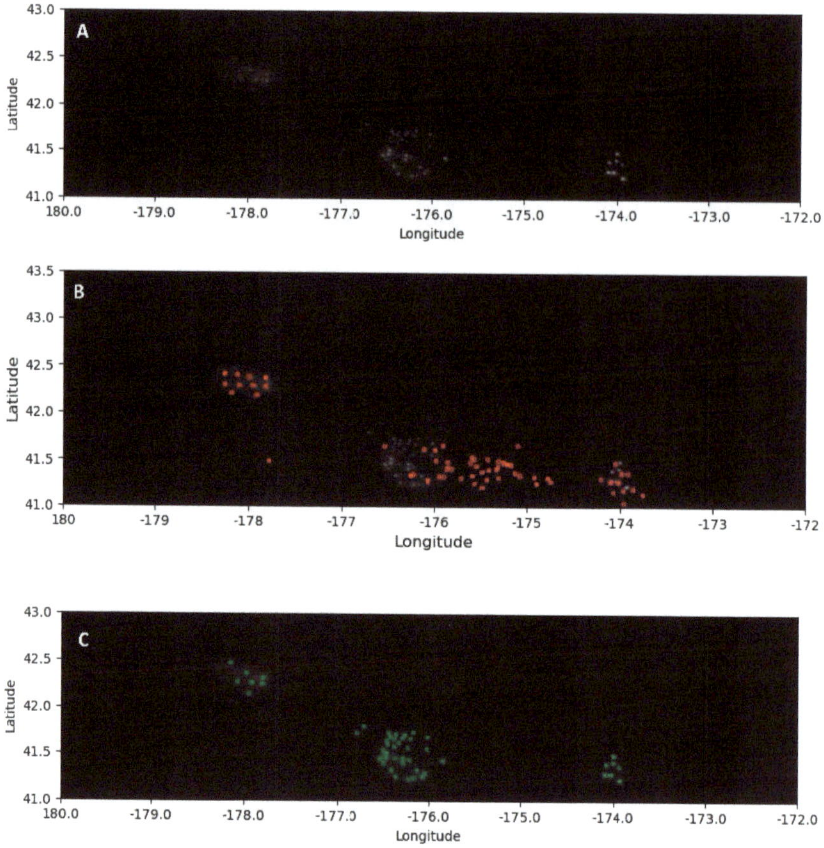

Figure A4. A typical scenario where discrepancies exist between AIS vessel positions and actual fishing locations during fishing ground transitions. (**A**) Nightly remote sensing images, (**B**) AIS vessel positions overlaid on nightly images, and (**C**) vessel positions extracted using machine vision algorithms, also overlaid on nightly images.

References

1. Nguyen, K.Q.; Winger, P.D. Artificial Light in Commercial Industrialized Fishing Applications: A Review. *Rev. Fish. Sci. Aquac.* **2019**, *27*, 106–126. [CrossRef]
2. Schiller, L.; Bailey, M.; Jacquet, J.; Sala, E.; Ñiquen, M.; Sumaila, U.R. High Seas Fisheries Play a Negligible Role in Addressing Global Food Security. *Sci. Adv.* **2018**, *4*, eaat8351. [CrossRef] [PubMed]
3. Chen, X.; Liu, B.; Chen, Y. A Review of the Development of Chinese Distant-Water Squid Jigging Fisheries. *Fish. Res.* **2008**, *89*, 211–221. [CrossRef]
4. Alabia, I.D.; Saitoh, S.I.; Mugo, R.; Seikawa, H. Seasonal Potential Fishing Ground Prediction of Neon Flying Squid (*Ommastrephes bartramii*) in the Western and Central North Pacific. *Fish. Oceanogr.* **2015**, *24*, 190–203. [CrossRef]
5. Arkhipkin, A.I.; Rodhouse, P.G.; Pierce, G.; Sauer, W.; Sakai, M.; Allcock, A.L.; Arguelles, J.; Shcherbich, Z.; González, A.F. World Squid Fisheries. *Rev. Fish. Sci. Aquac.* **2015**, *23*, 92–252. [CrossRef]
6. Alabia, I.D.; Saitoh, S.I.; Hirawake, T.; Seikawa, H. Elucidating the Potential Squid Habitat Responses in the Central North Pacific to the Recent ENSO Flavors. *Hydrobiologia* **2016**, *772*, 215–227. [CrossRef]
7. Alabia, I.D.; Saitoh, S.I.; Igarashi, H.; Hirawake, T.; Mugo, R. Future Projected Impacts of Ocean Warming on Potential Squid Habitat in the Western and Central North Pacific. *ICES J. Mar. Sci.* **2016**, *73*, 1343–1356. [CrossRef]
8. Geronimo, R.C.; Franklin, E.C.; Brainard, R.E.; Asher, J.; Oliver, T.A. Mapping Fishing Activities and Suitable Fishing Grounds Using Nighttime Satellite Images and Maximum Entropy Modelling. *Remote Sens.* **2018**, *10*, 1604. [CrossRef]
9. Waluda, C.M.; Yamashiro, C.; Elvidge, C.D.; Hobson, V.J.; Rodhouse, P.G. Quantifying Light-Fishing for Dosidicus gigas in the Eastern Pacific Using Satellite Remote Sensing. *Remote Sens. Environ.* **2004**, *91*, 129–133. [CrossRef]
10. Elvidge, C.D.; Zhizhin, M.; Baugh, K.E.; Hsu, F.C.; Ghosh, T. Automatic Boat Identification System for VIIRS Low Light Imaging Data. *Remote Sens.* **2015**, *7*, 3020–3036. [CrossRef]

11. Li, J.; Cai, Y.; Zhang, P.; Shi, L.; Wu, Y. Satellite Observation of a Newly Developed Light-Fishing "Hotspot" in the Open South China Sea. *Remote Sens. Environ.* **2021**, *256*, 112312. [CrossRef]
12. Zhao, M.; Zhou, Y.; Li, X.; Cao, W.; Li, D.; Xiao, J. Applications of Satellite Remote Sensing of Nighttime Light Observations: Advances, Challenges, and Perspectives. *Remote Sens.* **2019**, *11*, 1971. [CrossRef]
13. Croft, T.A. Nighttime Images of the Earth from Space. *Sci. Am.* **1978**, *239*, 86–101. [CrossRef]
14. Elvidge, C.D.; Baugh, K.E.; Zhizhin, M.; Hsu, F.C.; Ghosh, T. VIIRS Night-Time Lights. *Int. J. Remote Sens.* **2017**, *38*, 5860–5879. [CrossRef]
15. Waluda, C.M.; Griffiths, H.J.; Rodhouse, P.G. Remotely Sensed Spatial Dynamics of the Illex argentinus Fishery, Southwest Atlantic. *Fish. Res.* **2008**, *91*, 196–202. [CrossRef]
16. Rodhouse, P.G.; Elvidge, C.D.; Trathan, P.N. Remote Sensing of the Global Light-Fishing Fleet: An Analysis of Interactions with Oceanography, Other Fisheries and Predators. In *Advances in Marine Biology*; Academic Press: London, UK, 2001; Volume 39, pp. 261–303.
17. Cozzolino, E.; Lasta, C.A. Use of VIIRS DNB Satellite Images to Detect Jigger Ships Involved in the Illex argentinus Fishery. *Remote Sens. Appl. Soc. Environ.* **2016**, *4*, 167–178. [CrossRef]
18. Li, J.; Qiu, Y.; Cai, Y.; Wu, Y.; Zhang, P. Trend in Fishing Activity in the Open South China Sea Estimated from Remote Sensing of the Lights Used at Night by Fishing Vessels. *ICES J. Mar. Sci.* **2022**, *79*, 230–241. [CrossRef]
19. Kim, E.; Kim, S.W.; Jung, H.C.; Jung, W.Y. Moon Phase-Based Threshold Determination for VIIRS Boat Detection. *Korean J. Remote Sens.* **2021**, *37*, 69–84.
20. Xianbao, C.; Guihua, Q.; Yu, J.; He, J. An Improved Small Object Detection Method Based on YOLO V3. *Pattern Anal. Appl.* **2021**, *24*, 1347–1355. [CrossRef]
21. Li, H.; Zhu, M. A Small Object Detection Algorithm Based on Deep Convolutional Neural Network. *Comput. Eng. Sci.* **2020**, *42*, 649.
22. Bashir, S.M.A.; Wang, Y. Small Object Detection in Remote Sensing Images with Residual Feature Aggregation-Based Super-Resolution and Object Detector Network. *Remote Sens.* **2021**, *13*, 1854. [CrossRef]
23. Shao, J.; Yang, Q.; Luo, C.; Yu, H. Vessel Detection from Nighttime Remote Sensing Imagery Based on Deep Learning. *IEEE J. Sel. Top. Appl. Earth Obs. Remote Sens.* **2021**, *14*, 12536–12544. [CrossRef]
24. Jiang, P.; Ergu, D.; Liu, F.; Cai, Y.; Ma, B. A Review of YOLO Algorithm Developments. *Procedia Comput. Sci.* **2022**, *199*, 1066–1073. [CrossRef]
25. Cao, C.; Xiong, J.; Blonski, S.; Shao, X.; Uprety, S. Suomi NPP VIIRS Sensor Data Record Verification, Validation, and Long-Term Performance Monitoring. *J. Geophys. Res. Atmos.* **2013**, *118*, 11664–11678. [CrossRef]
26. Román, M.O.; Wang, Z.; Sun, Q.; Kalb, V.; Miller, S.D.; Molthan, A.L.; Schultz, L.; Inoue, K.; Hilburn, K.; Golpayegani, N.; et al. NASA's Black Marble Nighttime Lights Product Suite. *Remote Sens. Environ.* **2018**, *210*, 113–143. [CrossRef]
27. Mutanga, O.; Kumar, L. Google Earth Engine Applications. *Remote Sens.* **2019**, *11*, 591. [CrossRef]
28. Li, J.; Zhang, P.; Cai, Y.; Qiu, Y. Performance of VMS and Nightly Satellite in Monitoring Light Fishing Vessels in the Open South China Sea. *Fish. Res.* **2021**, *243*, 106100. [CrossRef]
29. Park, J.; Lee, J.; Seto, K.; Hochberg, T.; Leland, C.; Elvidge, C.D.; Watts, A.C. Illuminating Dark Fishing Fleets in North Korea. *Sci. Adv.* **2020**, *6*, eabb1197. [CrossRef]
30. Kroodsma, D.A.; Mayorga, J.; Hochberg, T.; Miller, N.A.; Boerder, K.; Ferretti, F.; Wilson, A.; Bergman, B.; White, T.D.; Block, B.A.; et al. Tracking the Global Footprint of Fisheries. *Science* **2018**, *359*, 904–908. [CrossRef]
31. Hsu, P.H.; Lee, P.J.; Bui, T.A.; Huang, K.C. YOLO-SPD: Tiny Objects Localization on Remote Sensing Based on You Only Look Once and Space-to-Depth Convolution. In Proceedings of the 2024 IEEE International Conference on Consumer Electronics (ICCE), Las Vegas, NV, USA, 6–8 January 2024; IEEE: New York, NY, USA, 2024; pp. 1–3.
32. Poloczanska, E. Keeping Watch on the Ocean. *Science* **2018**, *359*, 864–865. [CrossRef]
33. Longépé, N.; Hajduch, G.; Ardianto, R.; Guennec, S.; Tournadre, J.; Chapron, B. Completing Fishing Monitoring with Spaceborne Vessel Detection System (VDS) and Automatic Identification System (AIS) to Assess Illegal Fishing in Indonesia. *Mar. Pollut. Bull.* **2018**, *131*, 33–39. [CrossRef] [PubMed]
34. Tickler, D.; Meeuwig, J.J.; Palomares, M.L.; Pauly, D. Far from Home: Distance Patterns of Global Fishing Fleets. *Sci. Adv.* **2018**, *4*, eaar3279. [CrossRef] [PubMed]
35. Welch, H.; Clavelle, T.; White, T.D.; Froehlich, H.E.; Best, B.D. Hot Spots of Unseen Fishing Vessels. *Sci. Adv.* **2022**, *8*, eabq2109. [CrossRef] [PubMed]
36. Alabia, I.D.; Saitoh, S.I.; Igarashi, H.; Hirawake, T.; Yasuda, I. Ensemble Squid Habitat Model Using Three-Dimensional Ocean Data. *ICES J. Mar. Sci.* **2016**, *73*, 1863–1874. [CrossRef]
37. Oudrari, H.; McIntire, J.; Xiong, X.; Weng, F. An Overall Assessment of JPSS-2 VIIRS Radiometric Performance Based on Pre-Launch Testing. *Remote Sens.* **2018**, *10*, 1921. [CrossRef]
38. Oudrari, H.; McIntire, J.; Xiong, X.; Weng, F. JPSS-1 VIIRS Radiometric Characterization and Calibration Based on Pre-Launch Testing. *Remote Sens.* **2016**, *8*, 41. [CrossRef]
39. Li, X.; Ma, R.; Zhang, Q.; Li, D.; Cao, W. Anisotropic Characteristic of Artificial Light at Night—Systematic Investigation with VIIRS DNB Multi-Temporal Observations. *Remote Sens. Environ.* **2019**, *233*, 111342. [CrossRef]

40. Nasuddin, K.A.; Abdullah, M.; Abdul Hamid, N.S. Characterization of the South Atlantic Anomaly. *Nonlinear Process. Geophys.* **2019**, *26*, 25–35. [CrossRef]
41. Seto, K.L.; Miller, N.A.; Kroodsma, D.; White, T.D.; Zhang, C.I.; Watson, R. Fishing through the Cracks: The Unregulated Nature of Global Squid Fisheries. *Sci. Adv.* **2023**, *9*, eadd8125. [CrossRef]
42. Elvidge, C.D.; Tilottama, G.; Namrata, C.; Mikhail, Z. Lights on the Water? Accumulating VIIRS Boat Detection Grids in Southeast Asia spanning 2012–2021. *Fish People* **2023**, *21*, 33–38.

remote sensing

Article

Improving Atmospheric Correction Algorithms for Sea Surface Skin Temperature Retrievals from Moderate-Resolution Imaging Spectroradiometer Using Machine Learning Methods

Bingkun Luo [1,2], Peter J. Minnett [2,*] and Chong Jia [2]

1 Harvard–Smithsonian Center for Astrophysics, 60 Garden Street, Cambridge, MA 02138, USA; bkluo@cfa.harvard.edu
2 Department of Ocean Sciences, Rosenstiel School of Marine, Atmospheric, and Earth Science, University of Miami, 4600 Rickenbacker Causeway, Miami, FL 33149, USA; chong.jia@earth.miami.edu
* Correspondence: pminnett@earth.miami.edu

Abstract: Satellite-retrieved sea-surface skin temperature (SST_{skin}) is essential for many Near-Real-Time studies. This study aimed to assess the potential to improve the accuracy of satellite-based SST_{skin} retrieval in the Caribbean region by using atmospheric correction algorithms based on four readily available machine learning (ML) approaches: eXtreme Gradient Boosting (XGBoost), Support Vector Regression (SVR), Random Forest (RF), and the Artificial Neural Network (ANN). The ML models were trained on an extensive dataset comprising in situ SST measurements and atmospheric state parameters obtained from satellite products, reanalyzed datasets, research cruises, surface moorings, and drifting buoys. The benefits and shortcomings of various ML methods were assessed through comparisons with withheld in situ measurements. The results demonstrate that the ML-based algorithms achieve promising accuracy, with mean biases within 0.07 K when compared with the buoy data and ranging from −0.107 K to 0.179 K relative to the ship-derived SST_{skin} data. Notably, both XGBoost and RF stand out for their superior correlation and efficacy in the statistical results of validation. The improved SST_{skin} derived using the ML-based algorithms could enhance our understanding of vital oceanic and atmospheric characteristics and have the potential to reduce uncertainty in oceanographic, meteorological, and climate research.

Keywords: sea surface skin temperature; atmospheric correction algorithms; machine learning

Citation: Luo, B.; Minnett, P.J.; Jia, C. Improving Atmospheric Correction Algorithms for Sea Surface Skin Temperature Retrievals from Moderate-Resolution Imaging Spectroradiometer Using Machine Learning Methods. *Remote Sens.* **2024**, *16*, 4555. https://doi.org/10.3390/rs16234555

Academic Editors: Xiao-Hai Yan, Hua Su and Wenfang Lu

Received: 2 November 2024
Revised: 25 November 2024
Accepted: 29 November 2024
Published: 4 December 2024

1. Introduction

Infrared imaging radiometers onboard geostationary or polar orbiting satellites have been used to measure sea surface skin temperature (SST_{skin}) for more than 60 years [1]. For over two decades, the Moderate-Resolution Imaging Spectroradiometer (MODIS) sensors on both the Terra and Aqua satellites have been consistently providing data with an unprecedented spectral resolution [2,3]. It is important to improve the accuracy of SST_{skin} retrievals, including in challenging situations, and to quantify its errors and uncertainties [4,5] as this facilitates the appropriate use of fields in all applications, especially in forecast model assimilation schemes.

The areas where the impact of accurate SST_{skin} fields delivered in Near-Real Time (NRT) is the greatest are weather forecasting [6] and operational oceanography [7]. The accurate forecasting of severe storms, especially landfalling Atlantic hurricanes, typhoons and cyclones in other oceans [8], and extra-tropical storms [9], requires the accurate and timely determination of the SST_{skin} around the storm and along the forecast trajectory [10–13]. A well-known example is the rapid intensification of Hurricane Katrina in the Gulf of Mexico in 2006 as it passed over the northern edge of the Loop Current, and then again over a warm eddy in the northern Gulf shortly before landfall. Providing local authorities with timely information on likely hurricane intensification or dissipation is of great value in implementing measures to reduce injuries and the loss of lives. The recent Atlantic hurricanes have caused record property damage, disruptions to communities, and losses of lives [14,15], and these are expected to grow in future years [16]. Recent analyses have shown that the rate of peak intensification of Atlantic hurricanes has increased significantly in the last fifty years as a result of global warming [17], and the global occurrence of tropical cyclones that undergo multiple rapid intensification events has nearly doubled in the last two decades [18]. This study focuses on improving SST_{skin} retrieval in the Caribbean Sea and the surrounding region, which are vulnerable to hurricane formation and intensification, and from which we have an extensive and representative in situ SST dataset.

In this study, we applied several readily available machine learning (ML) methods capable of Near-Real-Time (NRT) application to assess the potential to improve the accuracy of atmospheric correction algorithms for MODIS SST_{skin} retrieval. This work compared the performance of the widely used existing ML algorithms with that of the NLSST (Non-Linear SST) of Walton, et al. [19], which is the basis of the standard atmospheric correction algorithm applied to cloud-free MODIS measurements [20] as a precursor to identifying where research into improvements can be focused. To achieve this goal, we need a sufficient number of factors in the training dataset used in ML to realize a statistically meaningful model. In situ oceanic and atmospheric data measured by ships, surface buoys, and drifters were used in the study to train the model and assess the accuracy of the derived SST_{skin} field. Some of the variables used in the development of the ML algorithms are from reanalysis products.

1.1. Atmospheric Corrections for MODIS SST_{skin} Retrieval

As a prelude to the ML approaches, we first discuss the standard atmospheric correction algorithms for SST_{skin} retrieval. The SST_{skin} can be retrieved from relatively transparent "atmospheric windows" with a minimum effect caused by water vapor and other gases at wavelengths ~4 μm and 10–13 μm. It is worth noting that a shorter wavelength window ~4 μm can only be used during nighttime due to the contamination by reflected sunlight within the atmosphere and at the surface.

According to the MODIS SST_{skin} R2019 Algorithm Theoretical Basis documents (from Goddard Space Flight Center (GSFC [21]); accessed on 21 June 2021), the existing MODIS SST_{skin} retrieval method is based on the NLSST algorithm adapted for MODIS measurements:

$$\begin{aligned} SST_{skin} = a_{ij0} \quad & + a_{ij1} BT_{11\mu m} + a_{ij2}(BT_{11\mu m} - BT_{12\mu m}) \times T_{sfc} + a_{ij3}(\sec(\theta) - 1) \times (BT_{11\mu m} - BT_{12\mu m}) \\ & + a_{ij4} \times M + a_{ij5}(\theta) + a_{ij6}(\theta)^2 \end{aligned}$$

where $BT_{11\mu m}$ and $BT_{12\mu m}$ indicate the brightness temperatures in the MODIS 11 and 12 μm bands. M represents the two sides of the paddle-wheel scan mirror to correct for small differences in spectral reflectivity between the two sides. T_{sfc} is a preliminary estimation of the SST, which is reliant on two sources, SST4 (derived from measurements at λ = 3.95 and 4.05 μm [20]) when available for nighttime data, and otherwise, the Canadian Meteorological Center Global Foundation SST [22]. θ is the satellite zenith angle. a_{ij0-6} are coefficients derived by the regression of match-ups between the in situ and satellite measurements and vary by month in several latitude bands [20] to account for regional and seasonal variations in the properties of the atmosphere.

An example of the limitations of the NLSST algorithm to compensate for anomalous atmospheres, such as Saharan dust aerosols, is given by Luo, et al. [23], who introduced an additional index from other MODIS IR channels to provide correction to nighttime SST_{skin} retrieval in aerosol-contaminated regions, with significant improvements in accuracy. The ability to improve SST_{skin} retrievals through aerosol-containing and dry layers is important as these conditions are known to inhibit the development of Atlantic hurricanes [24]. Accurate SST_{skin} retrieval depends on a good understanding of the uncertainties caused by controlling parameters, such as aerosol distributions [25–27], dry layers [28], and air–sea temperature differences [29]. As it is not straightforward to derive explicit functional forms between satellite brightness temperature measurements and the associated surface and atmospheric factors, we will apply the ML methods to attempt to improve the MODIS SST_{skin} atmospheric correction algorithms. The incorporation of multiple sources of information on atmospheric state in an algorithm optimized by ML may improve SST_{skin} retrieval.

1.2. Machine Learning Applications

An early application of ML to satellite oceanography was developing algorithms to estimate the air–sea exchanges of CO_2 from satellite data. Given that CO_2 in the atmosphere is well mixed, this problem is dominated by the variability in CO_2 fugacity in the upper ocean, which is temperature-dependent. At that time, the best estimation of air–sea exchanges of CO_2, derived by Olsen, et al. [30] from the regression of ship-based measurements, was a linear expression of temperature, latitude, and longitude. However, the geographical variables could be proxies for other factors. Using an ML method, Genetic Algorithm Discovery, on measurements taken from repeated tracks around the Caribbean Sea of the Explorer of the Seas [31], Wickramaratna, et al. [32] derived sets of functions of SST and surface pressure clustered into five regions, in each of which the ML equations were more accurate than those of Olsen, Triñanes and Wanninkhof [30]. Reassuringly, the areas of the clustered algorithms fell into those of major ocean surface currents in the region.

A subsequent study using similar techniques on the global Terra MODIS Match-Up Data Base (MUDB; [20]) "rediscovered" the formulation of the NLSST atmospheric correction algorithm [19]. When the ML technique was tasked with deriving NLSST algorithms without the T_{sfc} term, a set of regionally dependent equations and coefficients was found. The boundaries between the regions were primarily zonal, indicating dependence on the general nature of the water vapor distribution related to large-scale atmospheric circulation [33].

A cloud mask is necessary to identify and remove the cloud-contaminated pixels for MODIS SST_{skin} retrieval. Kilpatrick, et al. [34] developed an improved cloud-screening algorithm using boosted Alternating Decision trees (ADtrees). Compared to the binary tests forming the previous decision tree cloud mask, the ADtrees approach better identifies cloud types and improves the retention of SST gradients; it also provides confidence levels of the clear sky determination for each pixel.

Subsequently, RF and Cubist Decision Trees were used to represent and predict the errors and uncertainties in MODIS SST_{skin} retrieval [5]. Both the methods performed well, but the Cubist method was explored further to gain an insight into the sources of inaccuracies in retrieval, identifying seven Rule Sets with different error distributions. Each Rule Set occupied different, but overlapping volumes in the parameter space.

In addition to these early studies related to SST, different kinds of ML method have been used to derive other remote sensing products; a scalable end-to-end gradient-boosting tree (XGBoost; [35]) approach has been used to process non-linear satellite product estimations such as Chlorophyll-a [36,37], insolation at the sea surface [38], and urban heat storage [39]. The Support Vector Machine (SVM) and Artificial Neural Network (ANN) approaches were used in atmospheric correction for satellite ocean color sensors [36,40], water quality retrieval [41], and oceanic particulate organic carbon concentrations [42]. Overall, these preliminary studies demonstrate the potential of using ML to produce accurate geophysical variables and provide new insight into the errors of satellite-derived fields in different measurement conditions.

Here, our goal is to assess the ability of four ML approaches to improve on the currently used NLSST algorithms for the derivation of MODIS SST_{skin}.

2. Data

This study involves the joint analysis of variables derived from the measurements of satellite sensors, in situ instruments, and reanalysis data to generate a training dataset for ML development. Table 1 lists the relevant variables and data sources for this study.

Table 1. Relevant variables and data sources.

Variable	Satellite or Reanalysis Data	In Situ Source
SST_{skin}	MODIS	M-AERI
IR brightness temperatures of 11 μm and 12 μm bands, solar zenith angle, satellite zenith angle	MODIS	
In situ SST	---	Buoys and drifters
Near surface air-temperature	MERRA-2	M-AERI; Ship
Near-surface humidity	MERRA-2	Ship
Near-surface winds	MERRA-2	Ship
First guess SST	OSTIA, Canadian Meteorological Center's reanalysis of SST	
Latitude, months	---	---

2.1. Satellite Data

The SST_{skin} data are available from MODIS starting from 2000 [20]. The newest version of MODIS SST_{skin} retrieval (R2019) generated at NASA GSFC has some significant improvements, including applying the ADtrees cloud screening algorithm [43]; newly-derived high-latitude coefficients for areas north of 60°N [44]; and nighttime dust aerosol corrections [23]. Our research uses the Level 1B brightness temperatures of the infrared Channels 20 (λ = 3.75 μm), 29 (λ = 8.55 μm), 31 (λ = 11.03 μm), and 32 (λ = 12.02 μm). The satellite data can be downloaded from different sources, primarily the NASA PO.DAAC (Physical Oceanography Distributed Active Archive Center) at the JPL (Jet Propulsion Laboratory) and the Langley Atmospheric Science Data Center.

2.2. In Situ Data Measurements from Ships

This study used in situ and remotely sensed data from research ships and from the Royal Caribbean Group (RCG) cruise ships. All the ships were equipped with infrared radiometers to derive the SST_{skin}. Three ships of the RCG, Celebrity Equinox, Allure of the Seas, and Adventure of the Seas, have become a major source of measurements in the Caribbean area. Figure 1 (top) illustrates the positions of the RCG ships where the SST_{skin} measurements have been included in the MODIS MUDBs.

Self-calibrating radiometers on ships facilitate the more direct evaluation of the accuracy of satellite-derived SST_{skin}. A hyperspectral interferometric ship-board radiometer, the Marine-Atmospheric Emitted Radiance Interferometer (M-AERI), is a dependable and accurate seagoing Fourier-transform infrared spectroradiometer mounted on ships to derive the spectra originating from the marine atmosphere and sea surface to derive the SST_{skin} [45] and near-surface air temperature [46]. M-AERIs include two internal black body cavities with SI-traceable calibration to provide the real-time at-sea calibration of infrared spectra.

Another important factor in SST_{skin} retrieval is the air–sea temperature difference (air–sea δT), which is not explicit in the NLSST algorithms, but which can introduce large errors in SST_{skin} retrieval, as shown by Luo and Minnett [47] for GOES ABI (Geostationary Operational Environmental Satellite Advanced Baseline Imager) and by Luo, et al. [48] for Sentinel-3 SLSTR (Sea and Land Surface Temperature Radiometer). Utilizing the measurements of the emission of CO_2 in the M-AERI spectra produces a highly accurate retrieval of the near-surface air temperature used in the determination of air–sea δT, which can be used in ML model analyses.

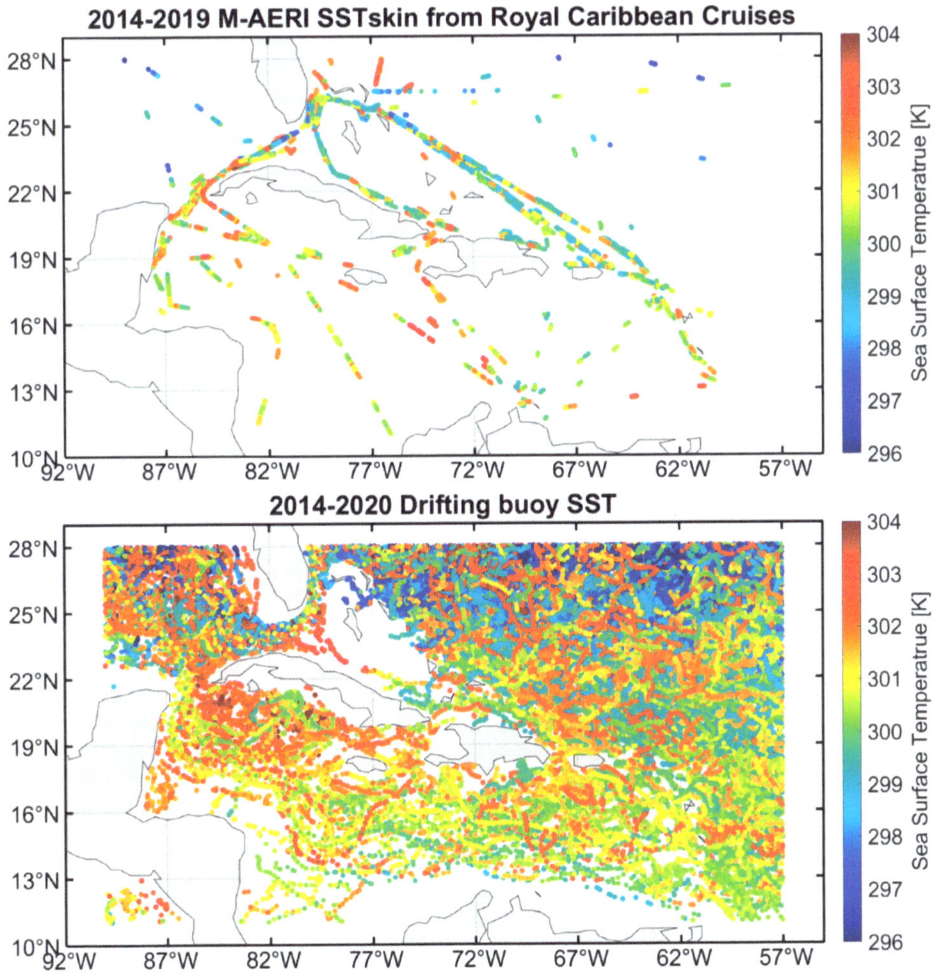

Figure 1. (**Top**): The routes of the RCG vessels equipped with M-AERIs providing the data for this investigation. The colors are the SST_{skin} derived from M-AERI measurements, as shown in the scale on the right in K. Note that the cruise ships repeat the same tracks multiple times, and so there are many more measurements than those that appear here. (**Bottom**): The drifting buoy-measured subsurface SSTs which have been matched to satellite SST_{skin} retrievals in the study region.

2.3. In Situ Data Measurements from Buoys

The National Oceanic and Atmospheric Administration (NOAA) took a significant step in the validation process of satellite-derived SST_{skin} products through the development of the in situ SST Quality Monitor (iQuam), which provides quality-assured data from various sources, including drifters, moored buoys, and marine vessels [49,50]. The drifters in the iQuam have a surface float and a subsurface drogue. At about 20 cm depth, the surface floats are equipped with thermometers [51,52], but it is worth mentioning that the exact thermometer depth can be influenced by surface wave conditions with potential occurrences of the float being momentarily fully immersed. The data are transmitted to land via satellite communication in NRT. The iQuam database is significant in our study because the distribution of drifters is extremely widespread, as shown in Figure 1 (bottom), for multiple years, thereby enhancing data analyses.

2.4. Reanalysis Fields

This study used the atmospheric state vectors from the NASA MERRA-2 (Modern-Era Retrospective Analysis for Research and Applications, Version 2) database [53]; these reanalyzed data guarantee internal consistency and present an array of geolocated and derived geophysical factors, such as wind, atmospheric temperature, and humidity recorded at 72 pressure levels (accessed on 21 June 2021). The role of the MERRA-2 repository is to furnish a detailed depiction of the atmospheric conditions at the time when both the in situ and satellite data were recorded. MERRA-2 has accurate SST values, with average differences of less than 0.1 K [54]. This, in turn, assists in the precise simulation of satellite radiometer measurements and offers inputs to the ML models.

3. Methods

3.1. Overview

Four ML methods were adopted initially to improve the atmospheric correction algorithms for MODIS SST_{skin} retrievals: eXtreme Gradient Boosting (XGBoost), Support Vector Regression (SVR), the Artificial Neural Network (ANN), and Random Forest (RF). Figure 2 shows the overall flow diagram of this study. Introductions of the four ML models are summarized here:

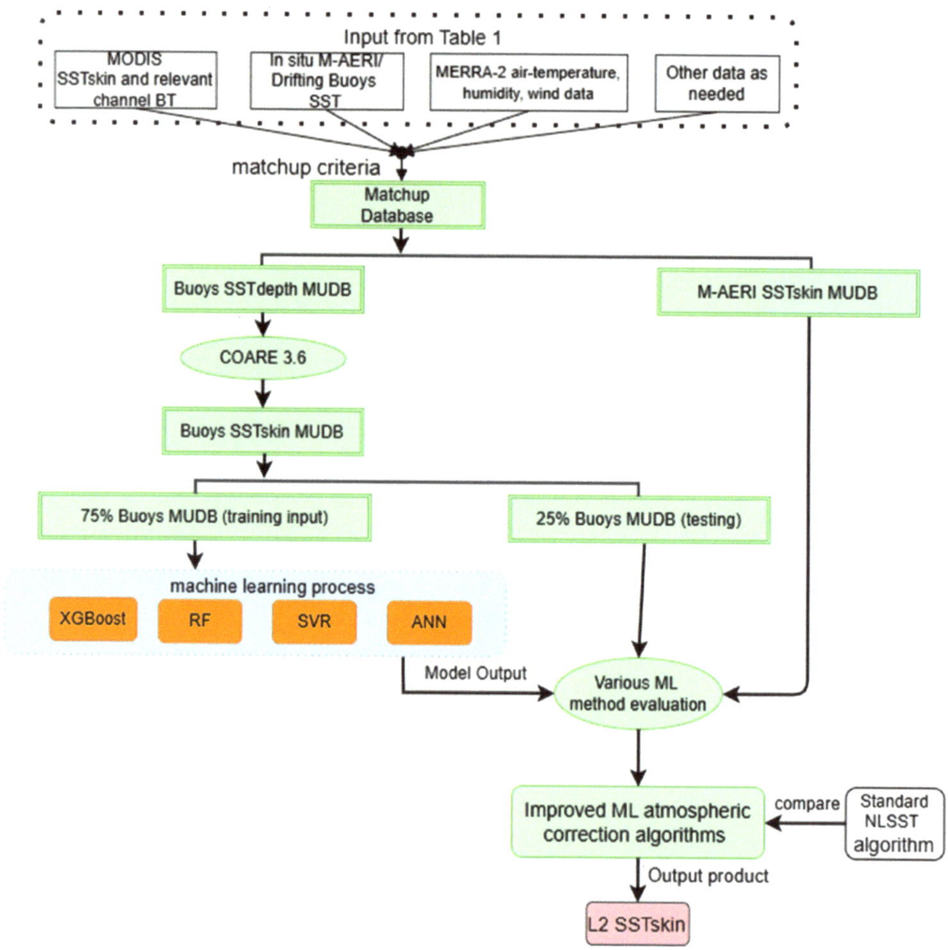

Figure 2. The overall framework of this study.

XGBoost stands as a contemporary variant of the gradient-boosting decision trees proposed by Chen and Guestrin [35]. XGBoost applies the entire training dataset with many regression trees rather than resampling partial samples to construct a strong predictor. The workflow of XGBoost commences with an equal weight allocation to training samples during inaugural iteration to formulate the initial tree. As the process progresses, these weights undergo adjustment, finely tuning them in accordance with the performance demonstrated by alignment with the training dataset. Subsequently, each tree assimilates a designated weight, influenced by observed fitting errors. To deduce the ultimate class allocated to an observation, the collective outputs of all trees are synthesized, with individual tree outputs being proportionally scaled by their respective weights. This methodology leverages gradient optimization to fine-tune cost functions through the least squares technique, thereby minimizing variance and precluding the possibility of overfitting.

RF constructs a forest of standard recursive partitioning trees and optimizes binary splits based on atmospheric and environmental variables by minimizing the mean squared error in the target variable at each split, thereby facilitating efficient data partitioning. Not only does this approach ensure higher predictive accuracy, as introduced by Breiman [55], but it also offers robustness against potential data fluctuations, a trait highlighted in the studies by Roy and Larocque [56]. The cumulative impact of this ensemble technique significantly mitigates the errors likely from single tree prediction, reducing variance, while maintaining a small bias, thereby promising not just an improved accuracy, but also a robust defense against overfitting issues in the data retrieval processes. Thus, the RF approach heralds a potential revolution in the fields of meteorological and marine science research, fostering enhanced accuracy and reliability in MODIS SST_{skin} retrieval accuracy assessment [5], Aerosol Optical Depth (AOD) retrieval [57], land cover [58], etc.

The ANN was devised as a network of interconnected artificial neurons assembled in a layered configuration [59]. Neurons in the input layer represent the various input variables that describe atmospheric conditions, radiative transfer properties, and other relevant parameters. For SST_{skin} retrieval, these neurons include the parameters listed in Table 1. At its core, an ANN encompasses a minimum of three tiers, the input layer, an intermediary or hidden layer, and the output layer, each being a nexus of intricately interconnected neurons. This multi-layered structure facilitates complex computational operations, enabling sophisticated analyses and predictions. Gross, et al. [60] used the ANN method to retrieve chlorophyll pigments from SeaWiFS (Sea-viewing Wide Field-of-view Sensor) ocean color measurements; they showed the advantages of the ANN in performing the bio-optical inversion, non-linear complexity, and noise filtering compared to those of the classical polynomial inverse methods.

SVR functions as a supervised learning algorithm grounded in kernel-based principles, which are used to transform input data into a higher-dimensional space to handle the non-linear relationships and complexities for SST_{skin} retrieval. Initially, the training dataset undergoes a transformation, projected into a higher-dimensional space via a kernel function. The optimization process then focuses on identifying the ideal hyperplane that aligns well with the training dataset, as described by Drucker, et al. [61]. Mountrakis, et al. [62] discussed the applications of SVR in remote sensing, showing it can provide a non-linear fitting ability between input variables and the target variable. Su, et al. [63] studied Subsurface Temperature Anomalies (STAs) in the Indian Ocean, utilizing SVR to analyze a compilation of satellite-derived data, including SST, sea surface elevation, and salinity. This analysis underscored the proficiency of SVR in delineating deeper oceanic thermal configurations and enhancing the precision of STA estimation. Furthermore, SVR has exhibited a remarkable aptitude in data categorization and regression analysis, coupled with an adeptness in pattern recognition, and thus its main contribution here is to identify the regions where atmospheric anomalies introduce large inaccuracies in SST_{skin} retrieval, and consequently require a different set of algorithms.

3.2. Match-Up Process

A comprehensive MUDB has been developed to facilitate the comparison between the MODIS SST_{skin} fields and the iQuam in situ measurements and the M-AERI data. The term 'match-up' in this context refers to a data vector comprising both in situ and satellite-derived variables, including attributes such as brightness temperatures from various MODIS infrared bands, captured as a 5 × 5 pixel array centered on each match-up location. Moreover, it incorporates other supplementary data, including preliminary SST estimates, in situ SST, meteorological variables, latitude and longitude, satellite zenith angle, and time stamps. Each MUDB record comprises 145 variables.

These match-ups are also synchronized between the satellite and in situ data within a 30 min window and within a geographical radius of 10 km [20,64]. In this study, we utilized data records extracted from the SST MUDB for the MODIS onboard NASA's Aqua satellite. We conducted a rigorous filtration process for MODIS Aqua match-up, selectively incorporating the MODIS data that met quality levels of 0, 1, or 2. Quality level 0 data are of the best quality, being confidently cloud-free with a satellite zenith angle of <55°; quality level 1 data are also confidently cloud-free, but have satellite zenith angles >55°. Quality level 3 data are also likely to be good, but with a risk of some degradation; they were included in this study to explore the ability of the ML algorithms to deal with less-than-perfect data. Removing the data from this study with quality level 3 prevented potential inaccuracies arising from significant cloud interference or other underlying factors that could compromise analyses. Despite the broad geographical scope encompassed by these match-ups, our study predominantly focused on the Caribbean area, a decision driven by the abundance of available M-AERI measurements in that region (Figure 1).

3.3. Machine Learning Model Setup

In this research, the configuration and validation of the model were conducted using the Scikit-learn library (v0.24) in Python (v3.9). We used the grid search methodology facilitating the comprehensive exploration of potential parameter values for a chosen estimator. This process, fundamental to the identification of optimal parameters, operates on the principles of a cross-validation system, meticulously examining every possible combination of parameters within this framework. Our strategy used the k-fold method, a robust approach that segments the entire sample pool into k equally sized subsets, commonly referred to as 'folds'. This procedure fosters a learning environment where the predictive function is cultivated utilizing data from k − 1 folds, reserving the remaining fold exclusively for testing purposes. We used k = 4 to balance computational efficiency and analytical accuracy.

Data collected between 2014 and 2020 within the region from 11°N to 28°N latitude and 57°W to 90°W longitude were selected, covering the Caribbean area. Using iQuam as the in situ data source, a total of 123,612 samples were available after filtering. Of these, 75% were used as the training set, and the remaining 25% (30,903 samples) were allocated to the testing set.

To assess the accuracy of each ML model, several parameters were calculated, including the mean differences between the ML-predicted SST and the in situ measured SST, median, standard deviation (STD), robust standard deviation (RSD), and runtime to validate the model's performance. STD serves as an indicator of the model's statistical dispersion, reflecting the deviation between the predicted and actual values. RSD is less sensitive to outliers. Smaller STD and RSD values signify a smaller deviation, indicating the higher capability of the model. The runtime parameter was analyzed to determine the ML models' computational costs.

Aiming to enhance the atmospheric correction algorithms for MODIS SST_{skin} retrievals using ML, the partition of training and test datasets aids in constructing robust ML models trained with a substantial number of samples to recognize complex patterns and correlations.

4. Results

It is widely acknowledged that the SST_{skin} is generally cooler than the water beneath the thermal skin layer, a phenomenon resulting from upward heat flux from the ocean to the atmosphere caused by a combination of factors, including net longwave radiation and sensible and latent heat fluxes at the air–sea interface. The characterization of the upper-ocean vertical temperature structure has been outlined by the Science Team of the Group for High-Resolution Sea Surface Temperature (GHRSST) using schematic profiles [4].

To facilitate the more accurate representation of the SST_{skin}, it becomes necessary to adapt buoy measurements, which are typically obtained approximately 0.2 m beneath the surface [51,52], to align with the SST_{skin}. This represents the temperature of the thermally conductive skin layer, typically ranging between 10 μm and 100 μm in thickness, and which is approximated by the measurements of infrared radiometers. The previous studies have addressed this discrepancy by applying a fixed adjustment of −0.17 K to buoy SST measurements to approximate the SST_{skin} [5,65,66], representing the global mean skin–subsurface temperature difference. However, such static corrections can introduce systematic biases, particularly when training ML models, as they fail to account for the dynamic variability in cool skin and diurnal heating effects due to changes in the environmental conditions, such as wind speed, solar radiation, and surface fluxes. Accurately modeling the cool skin and warm layer (diurnal heating) effects presents a significant challenge, prompting numerous refinements to the existing parameterizations and models [25,29,67–69]. In this study, we use the Coupled Ocean Atmosphere Response Experiment (COARE; Fairall, et al. [70]) model to more accurately account for both the cool skin and diurnal warming effects in buoy SST measurement. The COARE model dynamically corrects these processes by incorporating meteorological and oceanographic variables, such as air–sea fluxes, solar radiation, and wind speed, allowing for the more accurate estimation of the SST_{skin}. By applying these corrections, we generate a more accurate in situ SST_{skin} dataset, which serves as a reliable basis for training and validating ML models.

We compared the ML models' performance against that of the currently implemented NLSST atmospheric correction algorithm to establish any improvements.

4.1. Comparison with iQuam

In Figure 3, the scatter plots illustrate the correlation between the NLSST-retrieved SST_{skin} and the ML predictions and the in situ measured SST from iQuam, converted to SST_{skin} using the COARE model, in which the dashed line represents one-to-one agreement, indicating a perfect match between the predictions and the observations. The outputs from RF show the best correspondence with the in situ SSTs, having a linear regression equation of y = 1.00x + 0.03. In comparison, the slopes for XGBoost and ANN are approximately equal to 0.99, whereas it is 0.95 for the SVR method. The central panels of Figure 3 show histograms of the ML-derived SST_{skin} and the iQuam SST data, providing a visualization of the frequency of occurrence of different temperatures. The ML models were found to surpass the conventional NLSST algorithm in multiple metrics. Specifically, NLSST presented a mean bias of −0.319 K and an STD of 0.547 K. When focusing on mean bias as a performance indicator, the XGBoost model demonstrated superior accuracy, recording a negligible mean bias of 0.0012 K. This was followed by RF (−0.0103 K), SVR (0.0133 K), and the ANN (−0.0762 K) in this respective order. In relation to STD and RSD, a consistent performance was shown among XGBoost, RF, and the ANN. The ANN displayed the lowest STD of 0.243 K, with those of RF and XGBoost at 0.247 K and 0.252 K, respectively. However, SVR was slightly inferior, with the highest ML STD of 0.341 K. The distribution of SST predicted by the ANN best fits that of iQuam SST.

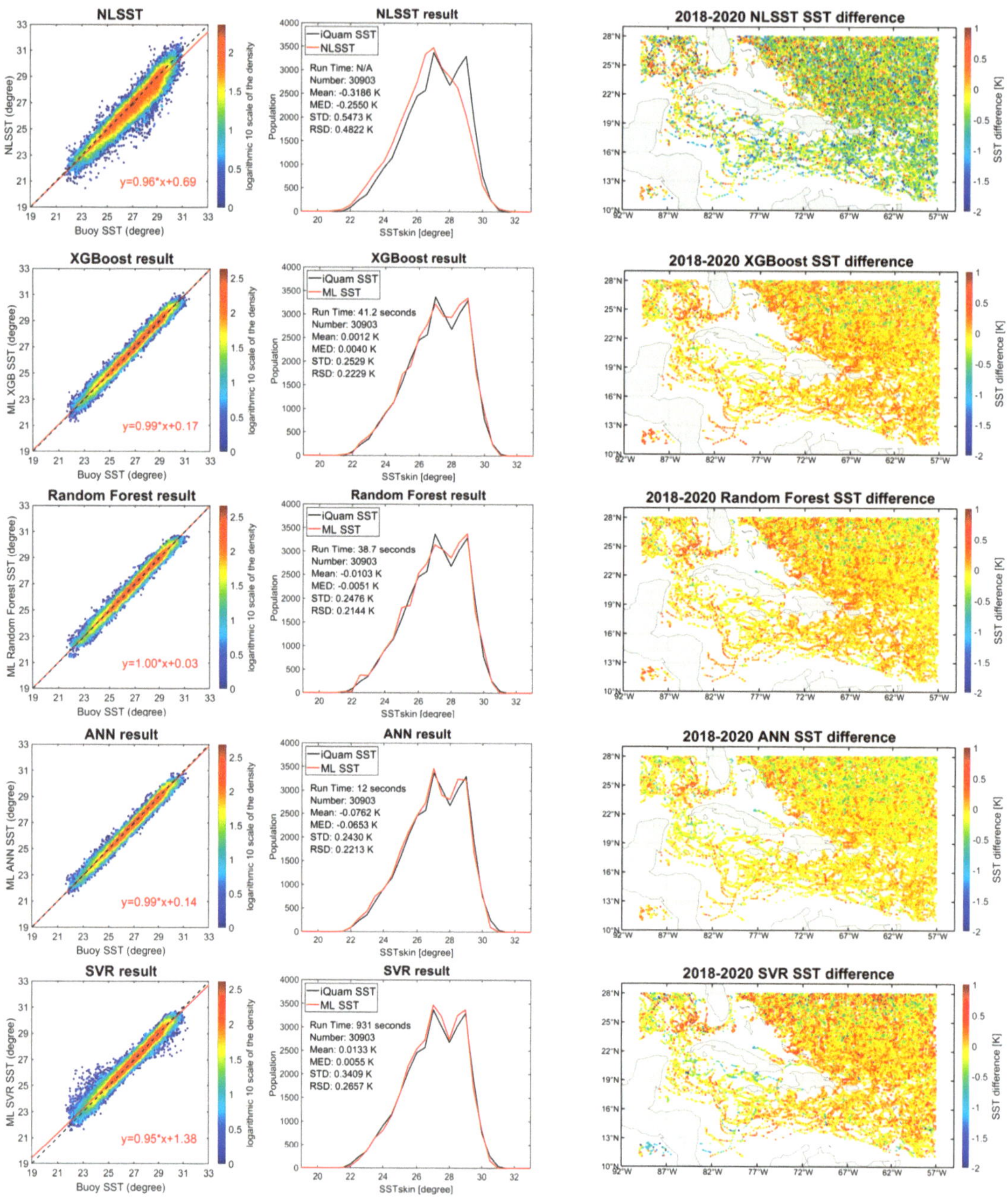

Figure 3. Diagnostic plots illustrating the performance of the standard NLSST atmospheric correction in the study area with the SST_{skin} derived from the ML approaches. The buoy SSTs have been adjusted to the SST_{skin} using the COARE model. **Left** column: scatter plots of the predicted and buoy SST_{skin}; the dashed line is a one-to-one correspondence, while the red line represents the linear least-squares fit. **Center** column: histograms of the predicted SST_{skin} and those from iQuam. **Right** column: maps of the disparities between the predicted SST_{skin} and those from iQuam using a color-coded scheme to show the magnitudes in degrees, given on the right.

An additional aspect of the evaluation was the computational efficiency of these ML models. In the dataset comprising 30,903 match-ups, the ANN delivered results in only 12 s. XGBoost and RF demonstrated comparable computational efficiency, taking around 40 s to process the same dataset. SVR fell behind significantly, taking 931 s due to its high computational complexity when dealing with large datasets. XGBoost, RF, and the ANN can handle high-dimensional feature spaces efficiently through their network topology; geospatial variables, such as longitude, latitude, and various atmospheric parameters, can be easily accommodated. The traditional non-linear relationships between these factors are optimized by using XGBoost's and RF's decision trees and the ANN's activation functions and multiple layers, while SVR requires calculating the distances in this feature space and uses different kernel functions, which can be computationally intensive. Furthermore, XGBoost, RF, and the ANN are parallelized across multiple CPU cores, making them highly efficient for analyzing big datasets. SVR has a computational complexity ranging from $O(N^2)$ to $O(N^3)$, depending on the choice of kernel and other settings, and this complexity makes SVR less efficient for large-scale problems. To sum up, among the ML algorithms, XGBoost and RF appear to offer the most reliable and efficient approaches for improving the accuracy of MODIS SST_{skin} retrieval, outperforming both the ANN and SVR.

The right column of Figure 3 shows maps of the discrepancies between NLSST retrieval and the ML predictions and the in situ SST measurements after correction for the warm layer and cools skin effects; the color bars illustrate the temperature differences in Kelvin. The ML models yielded more-accurate spatial distribution patterns with respect to iQuam SST than those from the NLSST. However, notable discrepancies were observed in the SVR model's predictions. Specifically, SVR tended to underestimate the SST in the southwestern regions, while overestimating in the northern parts of the study area. One of the intricacies of using SVR is its reliance on kernel functions to model the non-linear relationships in data. Selecting and fine-tuning the appropriate kernel for SVR is challenging, especially when dealing with high-dimensional geospatial datasets.

Figure 4 shows the discrepancies between NLSST retrievals and the ML predictions and the buoy-derived SST_{skin} values varying with various environmental factors, including wind speed, 2 m air temperature, geographic latitude, in situ SST_{skin} values, the month, and specific humidity. Each subplot includes error bars to highlight the variability in temperature differences, while the dots signify the mean discrepancy associated with each parameter. Grey histograms depict the data distributions in the parameters.

The vertical gradients of the SST in the near-surface ocean layer are controlled by multiple factors, including solar radiation absorption, the vertical diffusion of heat, and mixing due to wind turbulence [29,67,71]. We used the OSTIA surface temperature, T_{sfc}, as the initial guess of the SST. This was then used as the input for the ML models and in the NLSST algorithm. As shown in Figure 4A, the observational SST data and the ML model-predicted SST showed a good agreement with 10 m wind speeds (U10) ranging from 3 m/s to 8 m/s, where the U10 values were concentrated, having a minimal bias below 0.1 K. At very low wind speeds, both the SVR and NLSST exhibited pronounced inaccuracies. For U10 > 10 m/s, there is a larger bias in the predicted SST_{skin}, especially for the SVR method. Ideally, all ML algorithms should not exhibit a large bias at high wind speeds because the SST_{skin} relationship with subsurface temperature shows very little variation due to wind-driven turbulence mixing. In high-dimensional spaces, especially when the sample size is small, the risk of overfitting is high. Therefore, SVR's performance might be affected more in such conditions without proper regularization.

The effect of surface air temperature on these ML schemes is shown in Figure 4B. The temperature difference between the SST_{skin} and the overlying air influences the effect of the atmosphere on the propagation of the sea surface emission to satellite height [72–74]. Within the air temperature range from 24 to 29 °C, the SST_{skin} simulated by the ML models is close to the observed values. However, the deviations for SVR become distinct at air temperatures below 22 °C or above 30 °C, with an average bias up to 2 K.

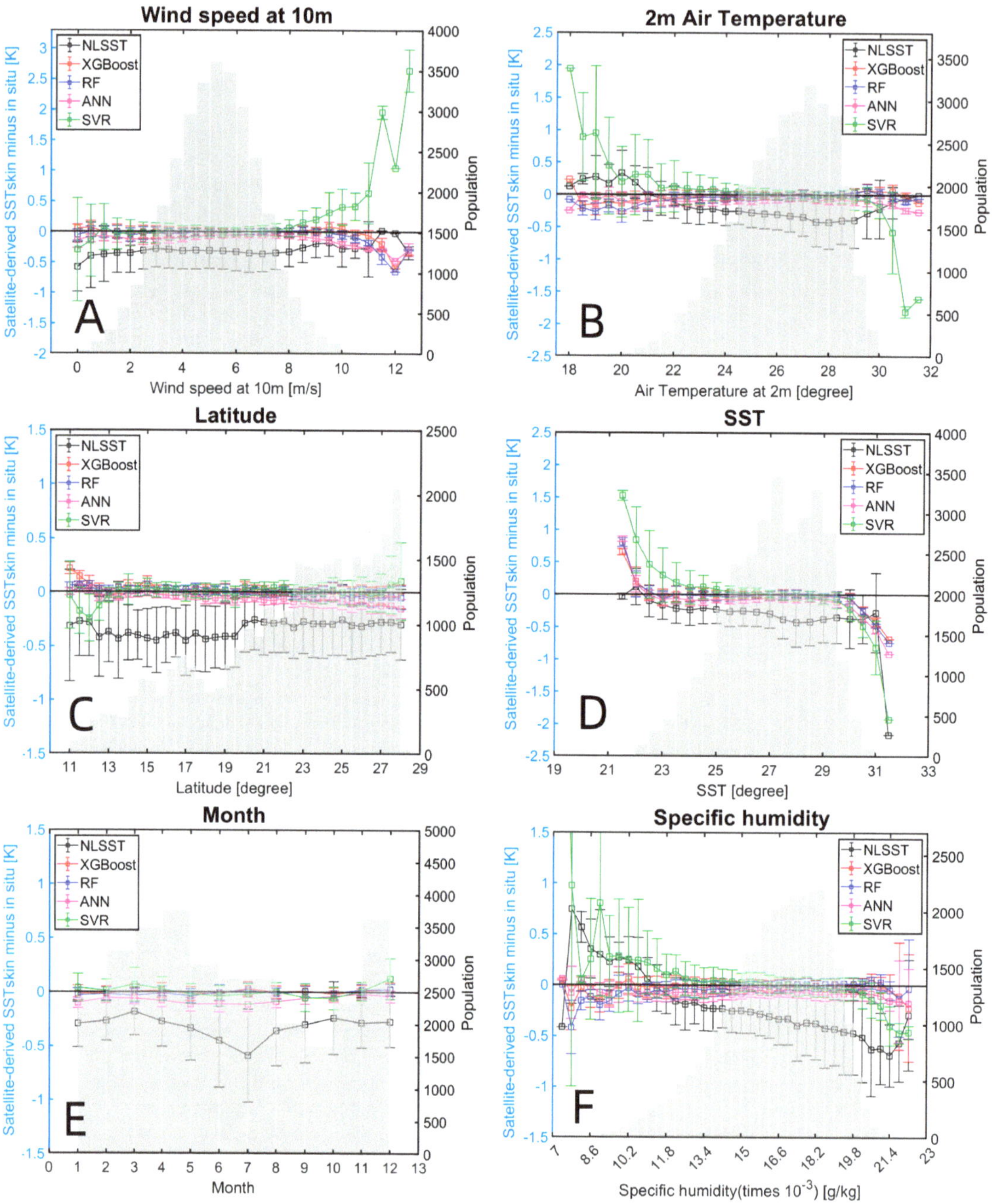

Figure 4. The discrepancies between the ML and NLSST outputs and the buoy-determined SST_{skin} influenced by various environmental parameters: wind speed (**A**), air temperature at 2 m above sea level (**B**), geographic latitude (**C**), SST_{skin} (**D**), the month (**E**), and specific humidity at 2 m (**F**). Each subplot includes error bars, denoting the STD in the temperature differences, with a dot representing the mean discrepancy. The grey histograms illustrate the data distribution of parameters on the *x*-axis.

The coefficients used in the NLSST algorithm are dependent upon both the latitude and the month [21]. The latitude- and month-based variations in the NLSST-derived SST_{skin} discrepancies are apparent in Figure 4C,E. The SVR ML model displays a negative bias reaching up to 0.25 K, while the other ML models exhibit a positive bias for latitudes below 13°N. The scarcity of buoy measurements south of 13°N, as depicted in Figure 1 (bottom), results in inadequate training data for the ML algorithms, contributing to such large biases. No significant seasonal impact was found on the different ML models, but the NLSST shows a pronounced bias during summer. Furthermore, between 11°N and 20°N, the NLSST reveals a significant bias of −0.5 K, which indicates that the coefficients or the formulation of the NLSST algorithm derived at this latitude range for the entire range of longitudes require further refinement and optimization for this region, particularly in summer.

Figure 4D illustrates the biases across different SST values. All the ML models exhibit pronounced positive biases for colder SSTs and negative biases under warmer SSTs, with a paucity of input training the data for these conditions. It is essential to understand that an abundant dataset is important for ML models to reduce the risk of overfitting and to enable the successful identification and depiction of the underlying patterns and tendencies within the data. These, in turn, enhance the robustness of ML algorithms and the accuracy of the derived SST_{skin}.

Water vapor has a profound impact on the atmospheric absorption and emission of infrared radiation. Consequently, fluctuations in specific humidity directly alter the atmospheric transmittance of radiation originating from the sea surface [75], leading to potential inaccuracies in the satellite-derived SST_{skin} [27,28,76,77]. As depicted in Figure 4F, a distinct trend of SST biases relative to specific humidity emerges for the NLSST; as specific humidity increases, the bias becomes increasingly negative. Regarding the ML models, all exhibit negative biases when the specific humidity exceeds 20 g/kg, but this is found in a small proportion of the conditions represented in the MUDBs. To address this issue, a more-extensive atmospheric dataset to capture a wider range of environmental conditions is needed.

The patterns of the SST_{skin} biases presented in Figure 3 (right) do not reveal temporal changes. Figure 5 presents Hovmöller diagrams that outline both the temporal and spatial evolution of SST_{skin} biases using the NLSST and the four ML algorithms. Such visualization aids in understanding spatio-temporal variations and identifying patterns. There is a significant data gap between July 2019 and January 2020 resulting from a lack of available in situ SST measurements from iQuam. Prominent negative biases in the NLSST retrievals in the study area from July to October are apparent.

From the Hovmöller diagrams, substantial seasonal fluctuations, especially in July each year, and notable latitudinal variations are found in the SST_{skin} biases for all the ML models used. Despite the presence of large and negative biases at the lower latitudes, the ML models demonstrated a superior performance compared to that of the SST_{skin} derived from the standard NLSST algorithm. Specifically, the XGBoost and RF models exhibited enhanced robustness to noise, a characteristic of their ensemble nature, which aggregates predictions from multiple trees, thereby reducing the risk of overfitting and noise sensitivity.

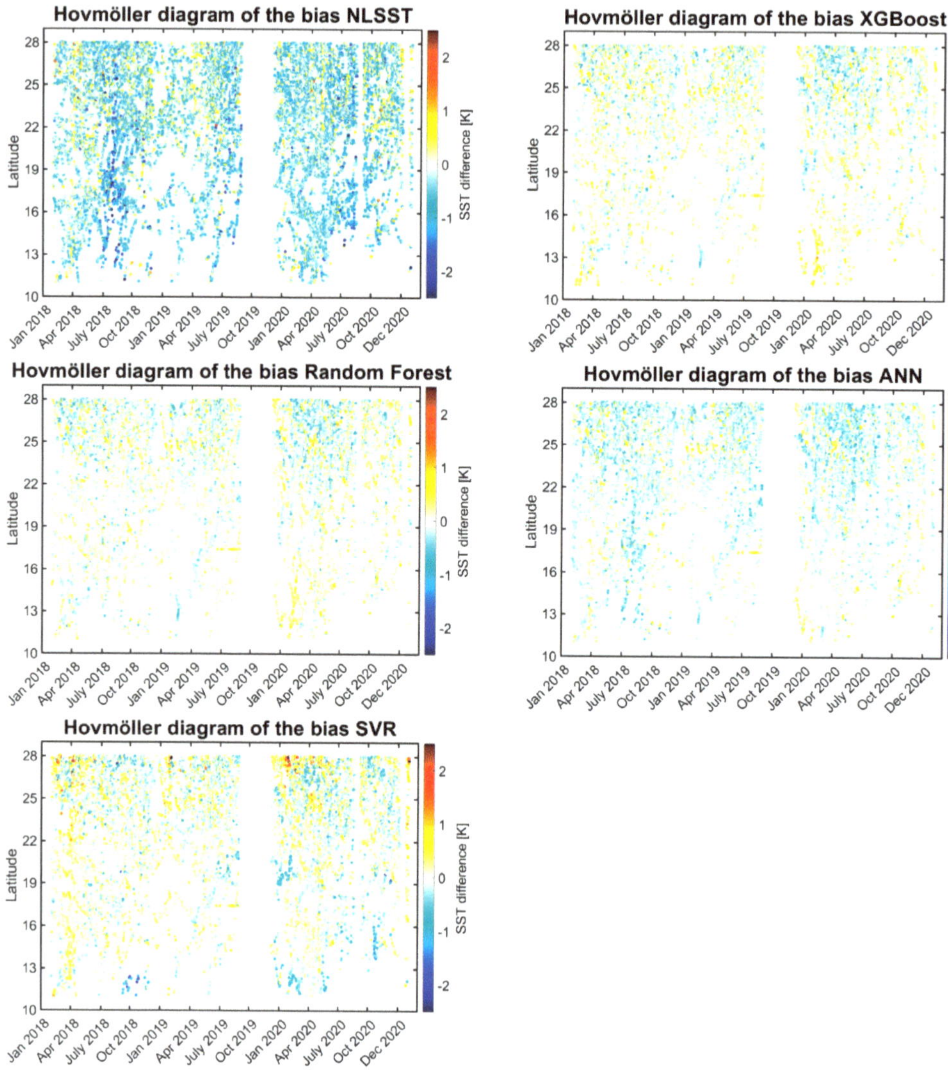

Figure 5. Hovmöller diagrams show the time and latitude evolutions of the biases from NLSST retrievals and four ML predictions compared with iQuam SSTs for 2018 to 2020. Since the in-situ data from July 2019 to Jan 2020 were not available, a gap exists between them. The color indicates SST difference in K, according to the scale on the right.

4.2. Comparisons with M-AERI SST_{skin}

To further assess each ML method's ability to predict accurate SST_{skin}, the SST_{skin} data derived from the M-AERIs mounted on ships were utilized as the reference. This approach is well established, with several published studies using the M-AERI SST_{skin} to assess the accuracy of retrievals from MODIS [20], GOES-ABI [47], and Sentinel-3 SLSTR [48], and also from the MERRA-2 and ERA-5 fields [54,78]. The ML-derived SST_{skin} was also compared with the current NLSST retrieved SST_{skin} [79] using the MUDBs to quantify accuracy improvements and to understand under which conditions improvements are minimal or absent, and thus direct further algorithm development.

Figure 6 includes scatter plots, histograms, and maps of differences of the NLSST retrievals and the four ML predictions against M-AERI SST_{skin} in the Caribbean region. Figure 6 has the same structure as Figure 3, but using SSTskin derived from M-AERIs mounted on ships to assess the performance of the NLSST SSTskin and ML algorithms. Such ship-based, self-calibrating radiometers provide data for a like-with-like SST_{skin} comparison, thereby avoiding the need for corrections of the diurnal heating and thermal skin layer effects.

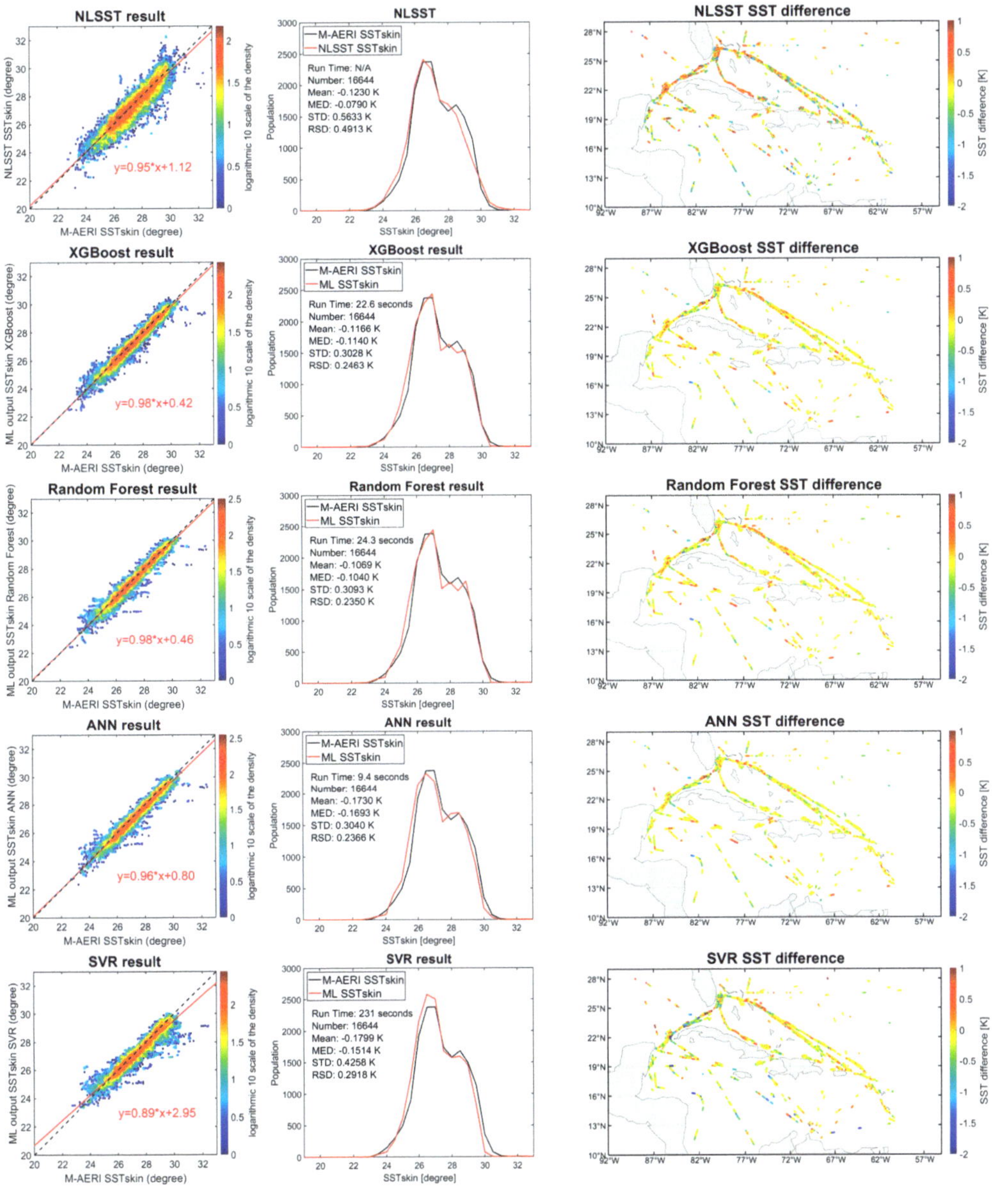

Figure 6. Like Figure 3, but utilizes the SST_{skin} values derived from the M-AERIs mounted on ships as comparisons for the assessment of the NLSST retrievals and ML prediction errors.

The left panel of Figure 6 presents scatter plots showing the correlation between NLSST retrieval and the ML predictions and the SST_{skin} derived by using the ship-mounted M-AERIs. The dashed line indicates an ideal one-to-one correspondence, the central panel histograms show the distribution of both the satellite- and M-AERI-derived SST_{skin} data. The NLSST algorithm exhibits an average discrepancy of −0.123 K, with an STD of 0.563 K. The ML models present varying performance metrics. Specifically, the XGBoost and RF models produce mean biases of −0.117 K and −0.107 K, with STDs of 0.303 K and 0.309 K, respectively. However, larger mean discrepancies of −0.173 K and −0.180 K are found in the ANN and SVR results. Notably, the SVR model demonstrates the highest STD of the ML results at 0.426 K. The right panel provides a visualization of the spatial distribution of temperature differences between the satellite-derived SST_{skin} and the M-AERI data; the NLSST and SVR models SST_{skin} values show pronounced discrepancies from the M-AERI data in specific regions. Such significant biases contribute to larger STDs, emphasizing the need for the refinement in those methods. Overall, validation with the M-AERI data demonstrates the potential of using ML approaches to produce accurate atmospheric corrections for SST_{skin} retrieval and gain new insight into the source of inaccuracies in the MODIS-derived SST_{skin} fields under different conditions.

Figure 7 displays the discrepancies between NLSST retrievals and the ML predictions and the M-AERI-determined SST_{skin} under various environmental factors, similar to Figure 4. The satellite-derived SST_{skin} data align favorably at U10 spanning from 3 to 8 m/s or at surface air temperature between 24 and 29 °C, maintaining a small bias below 0.2 K. In extreme conditions, such as strong wind speeds and high air temperatures, there are pronounced inaccuracies in both the SVR and NLSST outputs. Ideally, all the SST_{skin} values should exhibit negligible wind speed and air temperature dependences; however, the performance of SVR is highly impacted at air temperatures poorly represented in the data. As shown in Figure 7C, the influence of latitude on the NLSST algorithm and the ML models is evident; the SVR and the NLSST display a discernible negative bias of up to −0.75 K south of 15°N, likely due to the sparsity of M-AERI data, as illustrated in Figure 1. The lack of comprehensive data lowers the ML models' training efficacy. Note that the behavior at latitudes south of 15°N is more extreme than when validating that of the NLSST and ML outputs with the iQuam in situ SST. The deviations in all the ML models increasing with the rise in SST_{skin}, as shown in Figure 7D, is primarily attributed to the lack of sufficient training data at a warmer SST_{skin}. As the ML models heavily rely on the patterns present in the training data, the dataset we used for model training has more SST_{skin} values in the range of 24 °C to 30 °C compared to those at a warmer SST_{skin}, which limits the ability of the ML models to accurately learn and predict patterns when SST_{skin} > 31 °C. The monthly variations shown in Figure 7E indicate a larger training dataset is needed, especially for the SVR algorithm for summertime conditions. The ML models display significant variance at low humidity and high temperatures, underscoring potential data inadequacies for those specific conditions.

Considering that the M-AERIs provide SST_{skin} data, there is a need to improve the subsurface-temperature-to-SST_{skin} conversion when using the iQuam SST from drifting buoy measurements taken at depth to derive the coefficients for the NLSST algorithm. There is also a need for better optimization for regional applications.

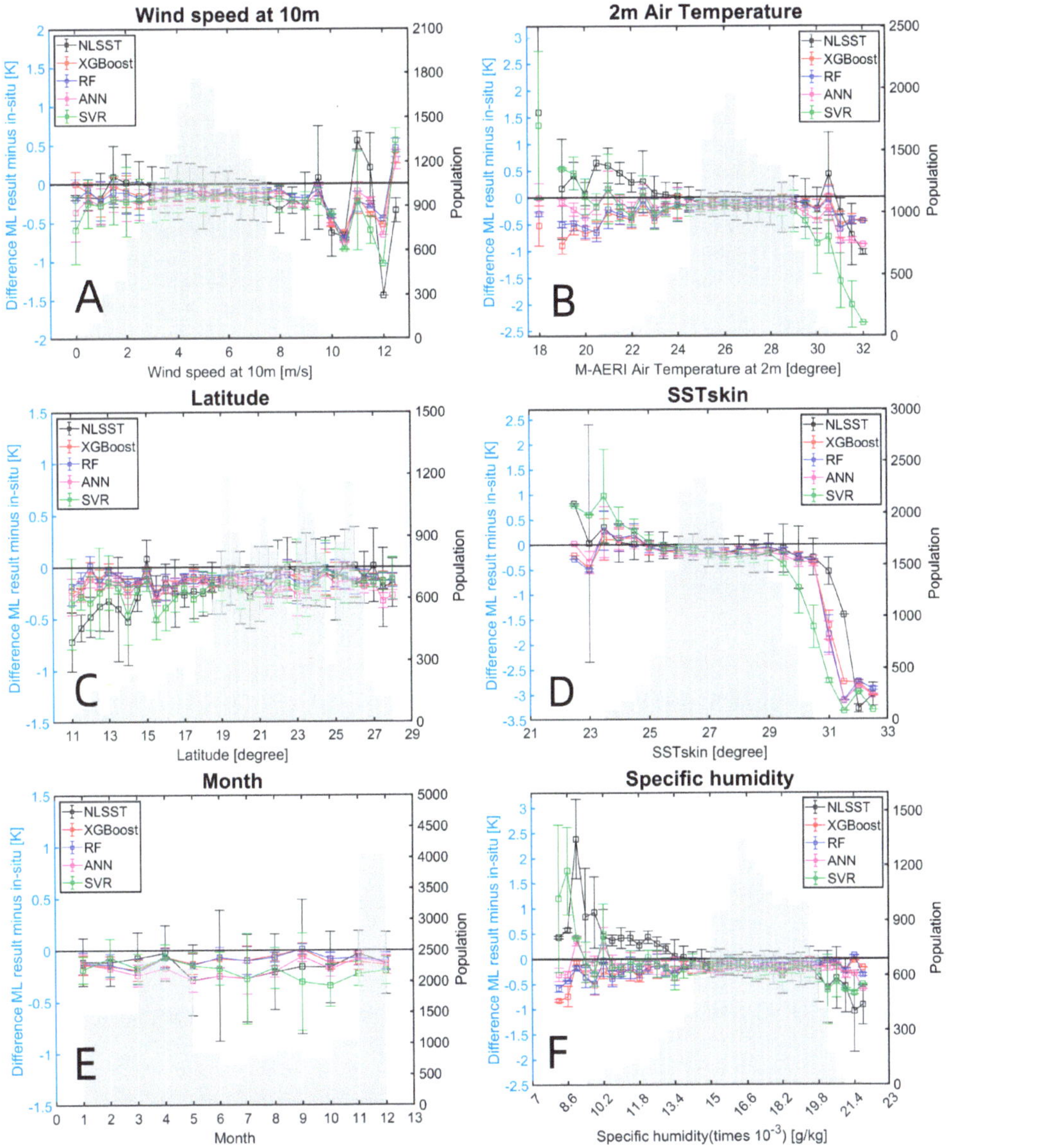

Figure 7. As Figure 4, but for NLSST and ML SST_{skin} values compared with M-AERI SST_{skin} data.

5. Summary

This paper reports the application of four ML models, XGBoost, RF, the ANN, and SVR, in enhancing the accuracy of SST_{skin} retrieval from the MODIS onboard NASA Aqua satellite. Research demonstrates the ML methods compared well with the conventional NLSST algorithm in retrieving SST_{skin} in the Caribbean region. The ML models were trained using an extensive MUDB containing both satellite and in situ measurements and related meteorological variables. The NLSST algorithm coefficients were derived from very large datasets in latitude bands that span the entire range of longitudes. The NLSST SST_{skin} values were taken from NASA data servers. The primary sources of validation data for the NLSST- and ML-derived SSTs were iQuam and M-AERI datasets. When the outputs from the various models were assessed against M-AERI SST_{skin}, XGBoost and RF demonstrated a superior performance, while SVR fell behind due to issues, including overfitting and higher computational requirements. Some factors, such as wind speed, surface air temperature, latitude, and specific humidity, were found to influence the accuracy of the ML models' SST_{skin} predictions. There were larger discrepancies in the ML models' retrievals under conditions, which were often associated with sparse training data.

The ML models, especially XGBoost, provides a promising avenue for improving the accuracy and robustness of SST_{skin} retrieval, with mean differences of -0.001 K compared with iQuam SST and -0.117 K compared with M-AERI SST_{skin}, suggesting potential benefits over the current NLSST algorithm. Overall, this research advocates for the further investigation of ML models to refine the atmospheric correction processes for SST_{skin}; this research also emphasizes the need for more comprehensive datasets by sampling the global range of marine conditions and meticulous validation to ensure the optimal performance of the NLSST algorithm and ML models and the confident assessment of the accuracies of the derived SST_{skin} values on a global basis.

Future work should focus on reducing the discrepancies observed, particularly in regions or temperature ranges that challenge the current atmospheric correction algorithms, including high latitudes [76,77], equatorial and tropical conditions with a high water vapor content, aerosol-burdened atmospheres, and coastal regions influence by terrestrial atmospheres. The focus of the research reported here was primarily in the Caribbean region; this was a choice driven by the abundant in situ data available in this area. As we continue to gather more data for training and validating these algorithms, we plan to broaden the scope of our study to encompass larger geographical regions. Moreover, applying the ML techniques to data from newer satellite sensors, such as the VIIRS on the Suomi-NPP, and subsequent NOAA satellites [80], and the SLSTR onboard the Copernicus Sentinel-3 satellites [48], should be explored.

Author Contributions: Conceptualization, B.L.; data curation, P.J.M.; formal analysis, B.L.; supervision, P.J.M.; writing—original draft preparation, B.L. and C.J.; writing—review and editing, B.L., C.J. and P.J.M.; visualization, B.L.; supervision, P.J.M.; project administration, P.J.M.; funding acquisition, B.L. and P.J.M. All authors have read and agreed to the published version of the manuscript.

Funding: This work was supported by the Future Investigators in NASA Earth and Space Science and Technology (FINESST) award (80NSSC19K1326) and benefited from research supported by NASA grants over several years, most recently 80NSSC21K1514 and 80NSSC24K1825. Support from the Royal Caribbean Group is gratefully acknowledged for allowing us to install our equipment on three of their ships.

Data Availability Statement: The MUDB databases used here are accessible through the SeaBASS validation system at the NASA Ocean Biology Distributed Active Archive Center (OB.DAAC). https://seabass.gsfc.nasa.gov/archive/SSTVAL (accessed on 21 June 2021).

Conflicts of Interest: The authors declare no conflicts of interest.

References

1. Minnett, P.J.; Alvera-Azcárate, A.; Chin, T.M.; Corlett, G.K.; Gentemann, C.L.; Karagali, I.; Li, X.; Marsouin, A.; Marullo, S.; Maturi, E.; et al. Half a century of satellite remote sensing of sea-surface temperature. *Remote Sens. Environ.* **2019**, *233*, 111366. [CrossRef]
2. Guenther, B.; Xiong, X.; Salomonson, V.V.; Barnes, W.L.; Young, J. On-orbit performance of the Earth Observing System Moderate Resolution Imaging Spectroradiometer; first year of data. *Remote Sens. Environ.* **2002**, *83*, 16–30. [CrossRef]
3. Esaias, W.E.; Abbott, M.R.; Barton, I.; Brown, O.B.; Campbell, J.W.; Carder, K.L.; Clark, D.K.; Evans, R.H.; Hoge, F.E.; Gordon, H.R.; et al. An Overview of MODIS Capabilities for Ocean Science Observations. *IEEE Trans. Geosci. Remote Sens.* **1998**, *36*, 1250–1265. [CrossRef]
4. Donlon, C.J.; Robinson, I.; Casey, K.S.; Vazquez-Cuervo, J.; Armstrong, E.; Arino, O.; Gentemann, C.; May, D.; LeBorgne, P.; Piollé, J.; et al. The Global Ocean Data Assimilation Experiment High-resolution Sea Surface Temperature Pilot Project. *Bull. Am. Meteorol. Soc.* **2007**, *88*, 1197–1213. [CrossRef]
5. Kumar, C.; Podestá, G.; Kilpatrick, K.; Minnett, P. A machine learning approach to estimating the error in satellite sea surface temperature retrievals. *Remote Sens. Environ.* **2021**, *255*, 112227. [CrossRef]

6. O'Carroll, A.G.; Armstrong, E.M.; Beggs, H.M.; Bouali, M.; Casey, K.S.; Corlett, G.K.; Dash, P.; Donlon, C.J.; Gentemann, C.L.; Høyer, J.L.; et al. Observational Needs of Sea Surface Temperature. *Front. Mar. Sci.* **2019**, *6*, 1–27. [CrossRef]
7. Le Traon, P.Y.; Antoine, D.; Bentamy, A.; Bonekamp, H.; Breivik, L.A.; Chapron, B.; Corlett, G.; Dibarboure, G.; DiGiacomo, P.; Donlon, C.; et al. Use of satellite observations for operational oceanography: Recent achievements and future prospects. *J. Oper. Oceanogr.* **2015**, *8*, s12–s27. [CrossRef]
8. Domingues, R.; Kuwano-Yoshida, A.; Chardon-Maldonado, P.; Todd, R.E.; Halliwell, G.; Kim, H.-S.; Lin, I.-I.; Sato, K.; Narazaki, T.; Shay, L.K.; et al. Ocean Observations in Support of Studies and Forecasts of Tropical and Extratropical Cyclones. *Front. Mar. Sci.* **2019**, *6*, 1–23. [CrossRef]
9. Kolstad, E.W.; Bracegirdle, T.J. Sensitivity of an apparently hurricane-like polar low to sea-surface temperature. *Q. J. R. Meteorol. Soc.* **2017**, *143*, 966–973. [CrossRef]
10. Hallam, S.; Marsh, R.; Josey, S.A.; Hyder, P.; Moat, B.; Hirschi, J.J.-M. Ocean precursors to the extreme Atlantic 2017 hurricane season. *Nat. Commun.* **2019**, *10*, 896. [CrossRef]
11. Hu, L.; Ritchie, E.A.; Tyo, J.S. Short-term tropical cyclone intensity forecasting from satellite imagery based on the deviation angle variance technique. *Weather Forecast.* **2020**, *35*, 285–298. [CrossRef]
12. Malkus, J.S.; Riehl, H. On the dynamics and energy transformations in steady-state hurricanes. *Tellus* **1960**, *12*, 1–20. [CrossRef]
13. Shay, L.K.; Goni, G.J.; Black, P.G. Effects of a Warm Oceanic Feature on Hurricane Opal. *Mon. Weather Rev.* **2000**, *128*, 1366–1383. [CrossRef]
14. Frame, D.J.; Wehner, M.F.; Noy, I.; Rosier, S.M. The economic costs of Hurricane Harvey attributable to climate change. *Clim. Chang.* **2020**, *160*, 271–281. [CrossRef]
15. Klotzbach, P.J.; Bowen, S.G.; Pielke, R.; Bell, M. Continental U.S. hurricane landfall frequency and associated damage: Observations and future risks. *Bull. Am. Meteorol. Soc.* **2018**, *99*, 1359–1376. [CrossRef]
16. Dinan, T. Projected Increases in Hurricane Damage in the United States: The Role of Climate Change and Coastal Development. *Ecol. Econ.* **2017**, *138*, 186–198. [CrossRef]
17. Garner, A.J. Observed increases in North Atlantic tropical cyclone peak intensification rates. *Sci. Rep.* **2023**, *13*, 16299. [CrossRef]
18. Manikanta, N.D.; Joseph, S.; Naidu, C.V. Recent global increase in multiple rapid intensification of tropical cyclones. *Sci. Rep.* **2023**, *13*, 15949. [CrossRef]
19. Walton, C.C.; Pichel, W.G.; Sapper, J.F.; May, D.A. The development and operational application of nonlinear algorithms for the measurement of sea surface temperatures with the NOAA polar-orbiting environmental satellites. *JGR Oceans.* **1998**, *103*, 27999–28012. [CrossRef]
20. Kilpatrick, K.A.; Podestá, G.; Walsh, S.; Williams, E.; Halliwell, V.; Szczodrak, M.; Brown, O.B.; Minnett, P.J.; Evans, R. A decade of sea surface temperature from MODIS. *Remote Sens. Environ.* **2015**, *165*, 27–41. [CrossRef]
21. GSFC. Sea Surface Temperature (SST) R2019 Algorithm Theoretical Basis Documents. 2020. Available online: https://oceancolor.gsfc.nasa.gov/resources/atbd/sst/ (accessed on 21 June 2021).
22. Brasnett, B. The impact of satellite retrievals in a global sea-surface-temperature analysis. *Q. J. R. Meteorol. Soc.* **2008**, *134*, 1745–1760. [CrossRef]
23. Luo, B.; Minnett, P.J.; Gentemann, C.; Szczodrak, G. Improving satellite retrieved night-time infrared sea surface temperatures in aerosol contaminated regions. *Remote Sens. Environ.* **2019**, *223*, 8–20. [CrossRef]
24. Dunion, J.P.; Velden, C.S. The impact of the Saharan air layer on Atlantic tropical cyclone activity. *Bull. Am. Meteorol. Soc.* **2004**, *85*, 353–366. [CrossRef]
25. Jia, C.; Minnett, P.J. Effects of the Hunga Tonga-Hunga Ha'apai Eruption on MODIS-Retrieved Sea Surface Temperatures. *Geophys. Res. Lett.* **2023**, *50*, e2023GL104297. [CrossRef]
26. Luo, B.; Minnett, P.J.; Nalli, N.R. Infrared satellite-derived sea surface skin temperature sensitivity to aerosol vertical distribution–Field data analysis and model simulations. *Remote Sens. Environ.* **2021**, *252*, 112151. [CrossRef]
27. Merchant, C.J.; Filipiak, M.J.; Le Borgne, P.; Roquet, H.; Autret, E.; Piollé, J.F.; Lavender, S. Diurnal warm-layer events in the western Mediterranean and European shelf seas. *Geophys. Res. Lett.* **2008**, *35*, L04601. [CrossRef]
28. Szczodrak, M.; Minnett, P.J.; Evans, R.H. The effects of anomalous atmospheres on the accuracy of infrared sea-surface temperature retrievals: Dry air layer intrusions over the tropical ocean. *Remote Sens. Environ.* **2014**, *140*, 450–465. [CrossRef]
29. Luo, B.; Minnett, P.J.; Szczodrak, M.; Akella, S. Regional and Seasonal Variability of the Oceanic Thermal Skin Effect. *J. Geophys. Res. Ocean.* **2022**, *127*, e2022JC018465. [CrossRef]
30. Olsen, A.; Triñanes, J.; Wanninkhof, R. Sea-air flux of CO_2 in the Caribbean Sea estimated using in situ and remote sensing data. *Remote Sens. Environ.* **2004**, *89*, 309–325. [CrossRef]
31. Williams, E.; Prager, E.; Wilson, D. Research Combines with Public Outreach on a Cruise Ship. *EOS Trans. AGU* **2002**, *83*, 590–596. [CrossRef]
32. Wickramaratna, K.; Kubat, M.; Minnett, P. Discovering numeric laws, a case study: CO_2 fugacity in the ocean. *Intell. Data Anal.* **2008**, *12*, 379–391. [CrossRef]
33. Dendamrongvit, S.; Kubat, M.; Minnett, P. Regression Trees in Sea-Surface Temperature Measurements. In Proceedings of the Ninth International Conference on Soft Computing Applied in Computer and Economic Environments, Hodonín, Czech Republic, 21 January 2011; pp. 49–53.

34. Kilpatrick, K.A.; Podestá, G.; Williams, E.; Walsh, S.; Minnett, P.J. Alternating decision trees for cloud masking in MODIS and VIIRS NASA sea surface temperature products. *J. Atmos. Ocean. Technol.* **2019**, *36*, 387–407. [CrossRef]
35. Chen, T.; Guestrin, C. XGBoost: A Scalable Tree Boosting System. In Proceedings of the 22nd ACM SIGKDD International Conference on Knowledge Discovery and Data Mining, San Francisco, CA, USA, 13–17 August 2016; pp. 785–794.
36. Cao, Z.; Ma, R.; Duan, H.; Pahlevan, N.; Melack, J.; Shen, M.; Xue, K. A machine learning approach to estimate chlorophyll-a from Landsat-8 measurements in inland lakes. *Remote Sens. Environ.* **2020**, *248*, 111974. [CrossRef]
37. Chen, S.; Hu, C.; Barnes, B.B.; Xie, Y.; Lin, G.; Qiu, Z. Improving ocean color data coverage through machine learning. *Remote Sens. Environ.* **2019**, *222*, 286–302. [CrossRef]
38. Aradpour, S.; Deng, Z. Remote sensing algorithm for retrieving global-scale sea surface solar irradiance. *Environ. Monit. Assess.* **2023**, *195*, 1355. [CrossRef] [PubMed]
39. Hrisko, J.; Ramamurthy, P.; Gonzalez, J.E. Estimating heat storage in urban areas using multispectral satellite data and machine learning. *Remote Sens. Environ.* **2021**, *252*, 112125. [CrossRef]
40. Frouin, R.J.; Franz, B.A.; Ibrahim, A.; Knobelspiesse, K.; Ahmad, Z.; Cairns, B.; Chowdhary, J.; Dierssen, H.M.; Tan, J.; Dubovik, O.; et al. Atmospheric Correction of Satellite Ocean-Color Imagery During the PACE Era. *Front. Earth Sci.* **2019**, *7*, 1–43. [CrossRef]
41. Balasubramanian, S.V.; Pahlevan, N.; Smith, B.; Binding, C.; Schalles, J.; Loisel, H.; Gurlin, D.; Greb, S.; Alikas, K.; Randla, M.; et al. Robust algorithm for estimating total suspended solids (TSS) in inland and nearshore coastal waters. *Remote Sens. Environ.* **2020**, *246*, 111768. [CrossRef]
42. Liu, H.; Li, Q.; Bai, Y.; Yang, C.; Wang, J.; Zhou, Q.; Hu, S.; Shi, T.; Liao, X.; Wu, G. Improving satellite retrieval of oceanic particulate organic carbon concentrations using machine learning methods. *Remote Sens. Environ.* **2021**, *256*, 112316. [CrossRef]
43. Kilpatrick, K.A.; Minnett, P.; Luo, B. Validation of NASA MODIS R2019.0 reprocessed SST Products. In Proceedings of the 20th GHRSST International Science Team Meeting (GHRSST XX), Frascati, Italy, 3–7 June 2019.
44. Jia, C.; Minnett, P.J. High latitude sea surface temperatures derived from MODIS infrared measurements. *Remote Sens. Environ.* **2020**, *251*, 112094. [CrossRef]
45. Minnett, P.J.; Knuteson, R.O.; Best, F.A.; Osborne, B.J.; Hanafin, J.A.; Brown, O.B. The Marine-Atmospheric Emitted Radiance Interferometer (M-AERI), a high-accuracy, sea-going infrared spectroradiometer. *J. Atmos. Ocean. Technol.* **2001**, *18*, 994–1013. [CrossRef]
46. Minnett, P.J.; Maillet, K.A.; Hanafin, J.A.; Osborne, B.J. Infrared interferometric measurements of the near-surface air temperature over the oceans. *J. Atmos. Ocean. Technol.* **2005**, *22*, 1019–1032. [CrossRef]
47. Luo, B.; Minnett, P. Skin Sea Surface Temperatures from the GOES-16 ABI Validated with Those of the Shipborne M-AERI. *IEEE Trans. Geosci. Remote Sens.* **2021**, *59*, 9902–9913. [CrossRef]
48. Luo, B.; Minnett, P.J.; Szczodrak, M.; Kilpatrick, K.; Izaguirre, M. Validation of Sentinel-3A SLSTR derived Sea-Surface Skin Temperatures with those of the shipborne M-AERI. *Remote Sens. Environ.* **2020**, *244*, 111826. [CrossRef]
49. Xu, F.; Ignatov, A. In situ SST Quality Monitor (iQuam). *J. Atmos. Ocean. Technol.* **2014**, *31*, 164–180. [CrossRef]
50. Xu, F.; Ignatov, A. Evaluation of in situ sea surface temperatures for use in the calibration and validation of satellite retrievals. *J. Geophys. Res. Ocean.* **2010**, *115*, C09022. [CrossRef]
51. Centurioni, L.R.; Turton, J.; Lumpkin, R.; Braasch, L.; Brassington, G.; Chao, Y.; Charpentier, E.; Chen, Z.; Corlett, G.; Dohan, K.; et al. Global in situ Observations of Essential Climate and Ocean Variables at the Air–Sea Interface. *Front. Mar. Sci.* **2019**, *6*, 1–23. [CrossRef]
52. Elipot, S.; Sykulski, A.; Lumpkin, R.; Centurioni, L.; Pazos, M. A dataset of hourly sea surface temperature from drifting buoys. *Sci. Data* **2022**, *9*, 567. [CrossRef]
53. Gelaro, R.; McCarty, W.; Suárez, M.J.; Todling, R.; Molod, A.; Takacs, L.; Randles, C.A.; Darmenov, A.; Bosilovich, M.G.; Reichle, R.; et al. The Modern-Era Retrospective Analysis for Research and Applications, Version 2 (MERRA-2). *J. Clim.* **2017**, *30*, 5419–5454. [CrossRef]
54. Luo, B.; Minnett, P.J.; Szczodrak, M.; Nalli, N.R.; Morris, V.R. Accuracy assessment of MERRA-2 and ERA-Interim sea surface temperature, air temperature, and humidity profiles over the Atlantic Ocean using AEROSE measurements. *J. Clim.* **2020**, *33*, 6889–6909. [CrossRef]
55. Breiman, L. Random forests. *Mach. Learn.* **2001**, *45*, 5–32. [CrossRef]
56. Roy, M.-H.; Larocque, D. Robustness of random forests for regression. *J. Nonparametr. Stat.* **2012**, *24*, 993–1006. [CrossRef]
57. Kianian, B.; Liu, Y.; Chang, H.H. Imputing Satellite-Derived Aerosol Optical Depth Using a Multi-Resolution Spatial Model and Random Forest for PM 2.5 Prediction. *Remote Sens.* **2021**, *13*, 126. [CrossRef]
58. Pelletier, C.; Valero, S.; Inglada, J.; Champion, N.; Dedieu, G. Assessing the robustness of Random Forests to map land cover with high resolution satellite image time series over large areas. *Remote Sens. Environ.* **2016**, *187*, 156–168. [CrossRef]
59. Hassoun, M.H. *Fundamentals of Artificial Neural Networks*; MIT Press: Cambridge, MA, USA, 1995.
60. Gross, L.; Thiria, S.; Frouin, R. Applying artificial neural network methodology to ocean color remote sensing. *Ecol. Model.* **1999**, *120*, 237–246. [CrossRef]
61. Drucker, H.; Burges, C.J.; Kaufman, L.; Smola, A.; Vapnik, V. Support vector regression machines. *Adv. Neural Inf. Process. Syst.* **1997**, *9*, 155–161.
62. Mountrakis, G.; Im, J.; Ogole, C. Support vector machines in remote sensing: A review. *ISPRS J. Photogramm. Remote Sens.* **2011**, *66*, 247–259. [CrossRef]

63. Su, H.; Wu, X.; Yan, X.-H.; Kidwell, A. Estimation of subsurface temperature anomaly in the Indian Ocean during recent global surface warming hiatus from satellite measurements: A support vector machine approach. *Remote Sens. Environ.* **2015**, *160*, 63–71. [CrossRef]
64. Corlett, G.K.; Merchant, C.J.; Minnett, P.J.; Donlon, C.J. Assessment of Long-Term Satellite Derived Sea Surface Temperature Records. In *Experimental Methods in the Physical Sciences, Vol 47, Optical Radiometry for Ocean Climate Measurements*; Zibordi, G., Donlon, C.J., Parr, A.C., Eds.; Academic Press: Cambridge, MA, USA, 2014; Volume 47, pp. 639–677.
65. Donlon, C.J.; Minnett, P.J.; Gentemann, C.; Nightingale, T.J.; Barton, I.J.; Ward, B.; Murray, J. Toward improved validation of satellite sea surface skin temperature measurements for climate research. *J. Clim.* **2002**, *15*, 353–369. [CrossRef]
66. Minnett, P.J.; Smith, M.; Ward, B. Measurements of the oceanic thermal skin effect. *Deep Sea Res. Part II Top. Stud. Oceanogr.* **2011**, *58*, 861–868. [CrossRef]
67. Akella, S.; Todling, R.; Suarez, M. Assimilation for skin SST in the NASA GEOS atmospheric data assimilation system. *Q. J. R. Meteorol. Soc.* **2017**, *143*, 1032–1046. [CrossRef] [PubMed]
68. Gentemann, C.L.; Minnett, P.J.; Ward, B. Profiles of Ocean Surface Heating (POSH): A new model of upper ocean diurnal warming. *JGR Oceans.* **2009**, *114*, C07017. [CrossRef]
69. Jia, C.; Minnett, P.J.; Luo, B. Significant Diurnal Warming Events Observed by Saildrone at High Latitudes. *J. Geophys. Res. Ocean.* **2023**, *128*, e2022JC019368. [CrossRef]
70. Fairall, C.W.; Bradley, E.F.; Hare, J.E.; Grachev, A.A.; Edson, J.B. Bulk parameterization of air-sea fluxes: Updates and verification for the COARE algorithm. *J. Clim.* **2003**, *16*, 571–591. [CrossRef]
71. Gentemann, C.L.; Akella, S. Evaluation of NASA GEOS-ADAS Modeled Diurnal Warming Through Comparisons to SEVIRI and AMSR2 SST Observations. *J. Geophys. Res. Ocean.* **2018**, *123*, 1364–1375. [CrossRef]
72. Dash, P.; Ignatov, A.; Martin, M.; Donlon, C.; Brasnett, B.; Reynolds, R.W.; Banzon, V.; Beggs, H.; Cayula, J.-F.; Chao, Y.; et al. Group for High Resolution Sea Surface Temperature (GHRSST) analysis fields inter-comparisons—Part 2: Near real time web-based level 4 SST Quality Monitor (L4-SQUAM). *Deep Sea Res. Part II Top. Stud. Oceanogr.* **2012**, *77–80*, 31–43. [CrossRef]
73. Minnett, P.J. Sea surface temperature measurements from the MODerate-resolution Imaging Spectroradiometer (MODIS) on AQUA. In Proceedings of the AGU Fall Meeting, San Francisco, CA, USA, 8–12 December 2003.
74. May, D.A.; Holyer, R.J. Sensitivity of satellite multichannel sea surface temperature retrievals to the air-sea temperature difference. *J. Geophys. Res. Ocean.* **1993**, *98*, 12567–12577. [CrossRef]
75. Saunders, P.M. The temperature at the ocean-air interface. *J. Atmos. Sci.* **1967**, *24*, 269–273. [CrossRef]
76. Jia, C.; Minnett, P.J.; Szczodrak, M. Assessment of Accuracy of Moderate-Resolution Imaging Spectroradiometer Sea Surface Temperature at High Latitudes Using Saildrone Data. *Remote Sens.* **2024**, *16*, 2008. [CrossRef]
77. Jia, C.; Minnett, P.J.; Szczodrak, M. Characteristics of R2019 Processing of MODIS Sea Surface Temperature at High Latitudes. *Remote Sens.* **2024**, *16*, 4102. [CrossRef]
78. Luo, B.; Minnett, P. Evaluation of the ERA5 Sea Surface Skin Temperature with Remotely-Sensed Shipborne Marine-Atmospheric Emitted Radiance Interferometer Data. *Remote Sens.* **2020**, *12*, 1873. [CrossRef]
79. Minnett, P.J.; Kilpatrick, K.; Szczodrak, G.; Izaguirre, M.; Luo, B.; Jia, C.; Proctor, C.; Bailey, S.W.; Armstrong, E.; Vazquez-Cuervo, J.; et al. MODIS Sea-Surface Temperatures: Characteristics of the R2019.0 Reprocessing of the Terra And Aqua Missions. In Proceedings of the XXI Science Team Meeting of the Group for High-Resolution Sea-Surface Temperature, Online, 1–4 June 2020; Available online: www.ghrsst.org (accessed on 21 June 2021).
80. Minnett, P.J.; Kilpatrick, K.A.; Podestá, G.P.; Evans, R.H.; Szczodrak, M.D.; Izaguirre, M.A.; Williams, E.J.; Walsh, S.; Reynolds, R.M.; Bailey, S.W.; et al. Skin Sea-Surface Temperature from VIIRS on Suomi-NPP—NASA Continuity Retrievals. *Remote Sens.* **2020**, *12*, 3369. [CrossRef]

remote sensing

MDPI

Article

Shallow Water Bathymetry Inversion Based on Machine Learning Using ICESat-2 and Sentinel-2 Data

Mengying Ye [1], Changbao Yang [1], Xuqing Zhang [1,*], Sixu Li [2], Xiaoran Peng [1], Yuyang Li [1] and Tianyi Chen [1]

[1] College of Geo-Exploration Science and Technology, Jilin University, Changchun 130026, China; yemy22@mails.jlu.edu.cn (M.Y.); yangcb@jlu.edu.cn (C.Y.); pengxr22@mails.jlu.edu.cn (X.P.); yuyangl22@mails.jlu.edu.cn (Y.L.); chenty22@mails.jlu.edu.cn (T.C.)
[2] College of Instrumentation and Electrical Engineering, Jilin University, Changchun 130021, China; lsx22@mails.jlu.edu.cn
* Correspondence: zxq@jlu.edu.cn

Abstract: Shallow water bathymetry is essential for maritime navigation, environmental monitoring, and coastal management. While traditional methods such as sonar and airborne LiDAR provide high accuracy, their high cost and time-consuming nature limit their application in remote and sensitive areas. Satellite remote sensing offers a cost-effective and rapid alternative for large-scale bathymetric inversion, but it still relies on significant in situ data to establish a mapping relationship between spectral data and water depth. The ICESat-2 satellite, with its photon-counting LiDAR, presents a promising solution for acquiring bathymetric data in shallow coastal regions. This study proposes a rapid bathymetric inversion method based on ICESat-2 and Sentinel-2 data, integrating spectral information, the Forel-Ule Index (FUI) for water color, and spatial location data (normalized X and Y coordinates and polar coordinates). An automated script for extracting bathymetric photons in shallow water regions is provided, aiming to facilitate the use of ICESat-2 data by researchers. Multiple machine learning models were applied to invert bathymetry in the Dongsha Islands, and their performance was compared. The results show that the XG-CID and RF-CID models achieved the highest inversion accuracies, 93% and 94%, respectively, with the XG-CID model performing best in the range from −10 m to 0 m and the RF-CID model excelling in the range from −15 m to −10 m.

Keywords: ICESat-2; Sentinel-2; satellite-derived bathymetry; shallow water

Citation: Ye, M.; Yang, C.; Zhang, X.; Li, S.; Peng, X.; Li, Y.; Chen, T. Shallow Water Bathymetry Inversion Based on Machine Learning Using ICESat-2 and Sentinel-2 Data. *Remote Sens.* **2024**, *16*, 4603. https://doi.org/10.3390/rs16234603

Academic Editors: Xiao-Hai Yan, Hua Su and Wenfang Lu

Received: 12 October 2024
Revised: 30 November 2024
Accepted: 5 December 2024
Published: 7 December 2024

1. Introduction

Shallow bays and areas around islands and reefs are hotspots for human marine activities, and information on bathymetry is crucial for the study of these shallow seas. With the growth of the ocean economy and the increasing demand for the exploitation of resources such as fisheries, oil and gas, and marine tourism, knowledge of bathymetry is essential for safe navigation, harbor planning, and fishery resource assessments. These areas are also often important components of ecosystems, such as coral reefs and seagrass beds, and accurate bathymetric data can help to study their distribution and growth [1,2]. In addition, changes in bathymetry are closely linked to climate change, with rising sea levels due to global warming raising concerns about coastline retreat. Monitoring changes in bathymetry can also help to predict the extent of natural disasters such as tsunamis and hurricanes, supporting the mitigation of potential damage. The in-depth study of bathymetric information in shallow waters has significant scientific, economic, and social value [3–5].

Traditional shallow water bathymetric methods are divided into five main categories. (1) Sonar echo sounding techniques based on shipboard systems: These include single-beam echo sounding (SBES [6]) and multibeam echo sounding (MBES [7,8]). The former has a small coverage and low spatial resolution, while the latter provides detailed underwater topography through complete insonification of the area. (2) Bathymetry using airborne

LiDAR data from non-imaging active remote sensing and satellite radar altimetry (e.g., SEASAT altimetry), with the former obtaining accurate measured depths in near-shore waters and the latter being suitable for only coarse, large-scale monitoring of seafloor topography changes. (3) Synthetic Aperture Radar (SAR) based on imaging active remote sensing for bathymetric inversion: When extracting seafloor features, this technique only identifies features with wavelengths similar to those of the local swell. If there is a large gap between the scale of the seabed topographic features and the wavelength of the waves, the effectiveness of the SAR technique will be limited [3]. (4) Bathymetric techniques based on imaging passive remote sensing: The use of satellite remote sensing images to establish bathymetric inversion models is divided into statistical and physics-based methods. Physics-based methods usually have higher accuracy, but need to consider complex optical properties; statistics-based methods analyze the relationship between spectral properties and depth through regression [9–16]. (5) Bathymetric inversion based on photogrammetry: This technique uses high-resolution satellite or airborne imagery to extract submerged features through advanced photogrammetric techniques. By applying stereo-matching and disparity estimation, this method reconstructs three-dimensional underwater topography. However, accurate refraction correction is essential for precise bathymetric data, and the technique is primarily applicable to shallow and ultra-shallow waters with depths up to 10 m [17–21]. Sonar and airborne LiDAR bathymetry are expensive and have limitations in remote and sensitive areas that are difficult to reach by ships and drones [22]. Photogrammetry can achieve high-resolution shallow water topography, but its depth capability is limited (typically up to 10 m). Satellite bathymetry allows for fast and cost-effective large-scale bathymetric inversion.

The research of satellite-derived empirical-based bathymetry (SDB) can be traced back to the 1970s and 1980s, when Polcyn et al. [23–25] proposed the SDB algorithm based on the band ratio, which gradually enabled the estimation of shallow water depths up to 5 m. Lyzenga et al. [26–28] simplified the classical radiative transfer equation to establish a quantitative relationship between the surface radiant energy and the water depth, thus simplifying the multispectral bathymetric inversion model and successfully estimated the water depth up to 15 m. Subsequently, Lyzenga et al. [29] proposed a multi-band linear model to correct optical attenuation and bottom reflection changes by log-transforming the blue and green band radiance combinations to improve the bathymetry accuracy, and Stumpf et al. [30] proposed an empirical ratio formula with only two unknown parameters, improving upon previous bathymetry inversion models. Experimental results demonstrated that the dual-band ratio model not only requires fewer parameters, but also performs well for low bottom depths and low reflectivity conditions. This model has become one of the classical approaches and forms the basis for many current studies. In recent years, scholars have made significant progress on the basis of these classical models. Pacheco et al. [31] improved the linear transformation algorithm of Lyzenga and inverted the nearshore SDB maps from Landsat 8 imagery. Hedley et al. [32] compared the ability of Sentinel-2 and Landsat 8 imagery in shallow water bathymetry and seabed mapping. With the development of machine learning technology, many researchers applied it to bathymetric inversion. Sandidge et al. [33] proposed a BP neural network for bathymetric inversion for the first time, and the effect exceeded traditional linear regression. Manessa et al. [34] used the random forest algorithm to carry out the bathymetric inversion of shallow coral reefs based on WorldView-2 imagery. Wang L et al. [35] used IKONOS-2 imagery and airborne LiDAR samples to implement bathymetric inversion by support vector machine model, while Wang Y et al. [5] improved the inversion accuracy by integrating spectral and spatial features through multilayer perceptron. Leng Z et al. [36] used the GRU deep learning model to carry out segmented bathymetric inversion of turbidity in Liaodong Bay. Ji X et al. [9] proposed an adaptive empirical method for different substrate types based on WorldView-2 imagery and multibeam echo sounding, airborne laser bathymetry (ALB) system. Knudby A et al. [12] compared five SDB models and discussed the importance of local neighborhood information for optimizing the effectiveness of bathymetric inversion.

These studies have promoted the continuous development and application of SDB field, but satellite bathymetry still needs a large amount of in situ measured data to construct the mapping relationship between spectra and depth.

ICESat-2 (Ice, Cloud, and land Elevation Satellite 2) [22,37–41] was launched in September 2018 with the first on-board photon-counting lidar system, known as ATLAS (Advanced Topographic Laser Altimeter System). As a novel source of a priori bathymetry data, it makes up for the shortcomings of traditional satellite bathymetry that requires a large amount of measured data, and has been widely used in the field of Satellite-Derived Bathymetry (SDB) in recent years. Parrish et al. [42] successfully achieved 40 m bathymetry in clear waters using ICESat-2 data. Hsu H J et al. [43] combined ICESat-2 and Sentinel-2 data to achieve shallow water bathymetry of six islands in the South China Sea based on a semi-empirical model [30]. Chen Y et al. [44] proposed a photon-counting LIDAR bathymetry method based on adaptive variable ellipsoid filtering (AVEBM) and verified the accuracy in Yongle Atoll and Chilianyu Archipelago. Xie C et al. [45] applied the density clustering algorithm (DBSCAN) to remove noise from ICESat-2 raw photons and combined them with Sentinel-2 data to perform bathymetric inversion, demonstrating the potential of combining data from multiple sources. Peng K et al. [46] proposed a physically assisted convolutional neural network (PACNN) model based on convolutional neural networks (CNNs) by linking Sentinel-2 and ICESat-2 data for shallow water bathymetry. Guo X et al. [47] performed bathymetric inversion by integrating ICESat-2 and Sentinel-2 data using a BP neural network model, which effectively enhanced the bathymetric inversion results. Xie C et al. [11] fused ICESat-2 and Sentinel-2 data and incorporated a radiative transfer-based model into a convolutional neural network (CNN) for bathymetric inversion, significantly improving inversion accuracy and further validating the effectiveness of multi-source data fusion.

This study aims to propose a simple and convenient method for shallow water depth inversion based on satellite datasets, enhancing the performance of the water depth inversion model through the integration of various types of information. First, the feasibility of applying ICESat-2 data in shallow water bathymetry is explored and improved. To this end, we developed a fully automated script capable of extracting water depth photons, where users only need to define the study area and select high-quality ICESat-2 data tracks and dates. Second, this study uses the information from the red, green, and blue bands of Sentinel-2 data as spectral feature information, the Forel-Ule index (FUI) [48–50] as water color information, and normalized latitude and longitude coordinates, along with polar coordinates, as spatial information. These are combined with the extracted ICESat-2 water depth point data to train the traditional Stumpf model, Polynomial Regression model, Random Forest model, Gradient Boosting model, and XGBoost model for water depth inversion. Through a comprehensive analysis of the accuracy and applicability of each model in shallow water bathymetry, this study provides new perspectives and methodologies for the effective application of ICESat-2 and Sentinel-2 data in shallow water depth inversion.

2. Materials and Methods

2.1. Study Area and Data

2.1.1. Study Area and In Situ Bathymetric Data

The first study area is located in the shallow coastal regions of Clearwater Bay, Haitang Bay, and Yalong Bay (Lingshui-Sanya Bay) in Hainan Province, China, situated in the southeastern part of the province within a low-latitude coastal zone, as shown in Figure 1a. The in situ data consist of 24 bathymetric points collected in 2020, which are used to evaluate the bathymetric capability of ICESat-2 data. The locations of these measurement points are indicated by red dots in Figure 1c.

The second study area is located in the Dongsha Islands (Figure 1b), a group of islands and reefs in the northern part of the South China Sea. It consists of 11 coral reefs and 35 islands with a total area of about 0.57 km^2. The Dongsha Islands are the furthest group of islands in the South China Sea from Hainan, about 350 km from Hainan Island and about 460 km from the Leizhou Peninsula in Guangdong Province. These islands consist mainly of coral reefs and sandy islands with low reliefs and small islands. The natural environment of the Dongsha Islands remains relatively pristine, with a well-preserved ecosystem. Covering a total sea area of approximately 5000 km^2, the Dongsha Islands feature a comprehensive topography that includes unique natural formations such as reef flats, lagoons, sandbars, shoals, channels, and islands, making it a quintessential example of an atoll landform.

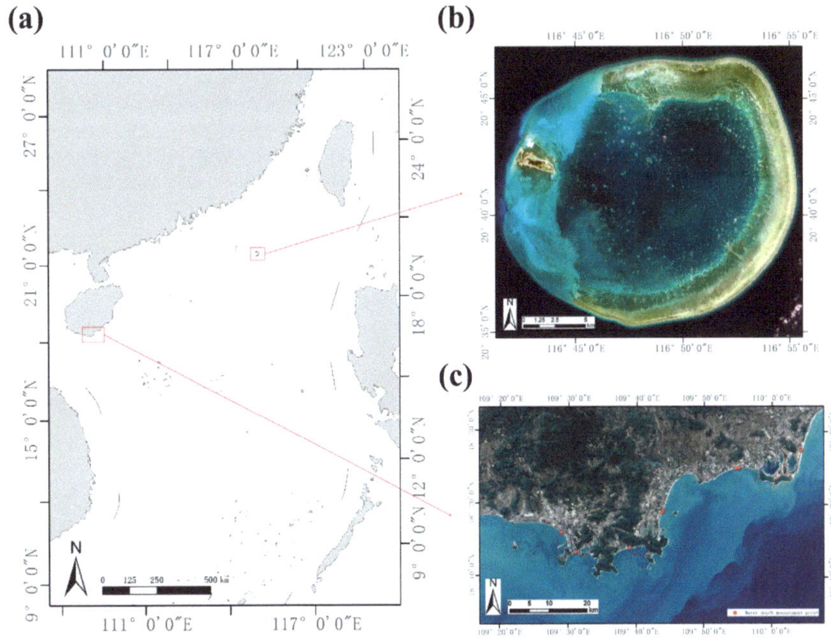

Figure 1. Map of the study area for this study. (**a**) Location of the study area for this study. (**b**) Sentinel-2 image map of Dongsha Islands. (**c**) Sentinel-2 image map of Lingshui-Sanya Bay; the red dots are the actual measurement points of water depth.

2.1.2. ICESat-2 Data

ICESat-2 (Ice, Cloud, and Land Elevation Satellite-2) is an Earth observation satellite launched by NASA in September 2018, designed to accurately measure changes in surface elevation through laser altimetry to support global environmental monitoring and climate change research. ICESat-2 carries the Advanced Topographic Laser Altimeter System (ATLAS), one of the most advanced laser altimeters in Earth's orbit to date. ICESat-2's primary mission includes assessing volumetric changes in the polar ice caps in order to establish an active monitoring system related to sea level change and ocean circulation impacts. In addition, ICESat-2 is used to measure global vegetation characteristics, land topography, and the backscattering properties of molecules, clouds, and aerosols in the atmosphere. These data are critical to understanding global change and supporting environmental protection [51–58]. The ATLAS uses six laser beams divided into three pairs, each consisting of a strong beam and a weak beam. The strong beam has four times the energy of the weak beam, a design that helps to obtain stable data under varying albedo conditions. The distance between each pair of laser pulses is 90 m, while the distance between each pair and the next is 3.3 km. This spatial configuration strikes a balance between high-resolution sampling and wide area coverage, enabling ICESat-2 to capture detailed altimetry data across diverse global surfaces with improved accuracy.

ICESat-2 provides a variety of data products, and the Level 2 data of ICESat-2, ATL03 (Global Geopositioning Photonics Data), was used in this study (as shown in Table 1). The

ATL03 dataset consists of all the raw photon data recorded in six different trajectories (three strong beams and three weak beams), each with unique latitude, longitude, and elevation angles based on the WGS84 ellipsoidal datum with unique latitude, longitude, and elevation angles. The dataset is corrected for atmospheric delays, solid tides, and systematic pointing biases, but does not correct for bathymetric errors such as water surface fluctuations, tilted surfaces, and water column effects. Although the ATL03 dataset provides detailed photon data, due to the high sensitivity of the detector, the data contain a large number of noise photons, especially in the daytime solar background. In order to distinguish between signal and noise photons, the ATL03 dataset introduces a 'confidence' parameter ranging from 0 to 4, where the higher the confidence, the more likely the photon is a signal. However, due to attenuation and scattering effects in the water column, the distributions of signal and noise photons are different from those in the atmosphere, resulting in poor performance of the confidence parameter in undersea signal photon detection [42]. Therefore, this study proposes a density-based signal detection algorithm to filter the photons and identify the water depth signal photons.

Table 1. Data table of ICESat-2 and Sentinel-2 in the study area.

Site	Lingshui-Sanya Bay	Dongsha Islands
Latitude **Longitude**	18°3.84′N–18°33.3′N 109°17.1′E–110°7.8′E	20°34.75′N–20°47.20′N 116°41.34′E–116°55.61′E
ICESat-2 Data	ATL03_20200130213819_05370607_006_01 ATL03_20200427052044_04840701_006_02 ATL03_20200430171804_05370707_006_02 ATL03_20200530034826_09870701_006_01 ATL03_20200727010031_04840801_006_01 ATL03_20200828232813_09870801_006_01 ATL03_20200901112535_10400807_006_02 ATL03_20200930100134_00950907_006_02 ATL03_20210502234906_05981107_006_01 ATL03_20210524103607_09261101_006_01 ATL03_20210524103607_09261101_006_01 ATL03_20220624154329_00421601_006_01 ATL03_20220628034053_00951607_006_01 ATL03_20220829004446_10401607_006_01	ATL03_20190129144159_04910207_006_02 ATL03_20190730060118_04910407_006_02 ATL03_20191021135212_03770501_006_02 ATL03_20191029014115_04910507_006_01 ATL03_20200420051144_03770701_006_02 ATL03_20200427170047_04910707_006_02
Sentinel-2 Data	S2A_MSIL2A_20201203T031109_N0500_R075_ T49QCA_20230303T030821	S2A_MSIL2A_20240222T023721_N0510_R089_ T50QMH_20240222T061746

2.1.3. Sentinel-2 Data

Sentinel-2 [59] is a key satellite in the European Space Agency's (ESA) Copernicus program, designed to monitor the Earth's surface through high-resolution optical imaging. The Sentinel-2 satellites, comprising Sentinel-2A and Sentinel-2B, were launched in June 2015 and March 2017, respectively. Sentinel-2's L2A-level data are radiometrically calibrated and atmospherically corrected surface reflectance images specifically designed for detailed surface analyses. The L2A class data are processed from the original Level-1C data (geometrically corrected orthophotos). It is characterized by high spatial resolution and multi-spectral coverage. Sentinel-2 L2A level data provides resolutions of 10, 20, and 60 m. The resolution varies by band, with the 10-meter resolution band being suitable for detailed surface analysis. Coverage of 13 spectral bands, ranging from visible to near-infrared (VNIR) and short-wave infrared (SWIR), provides rich spectral information to support a wide range of applications.

The Sentinel-2 L2A level data for this study was obtained from the European Space Agency's (ESA) Copernicus Open Access Hub. The data are projected using the UTM/WGS84 (Universal Transverse Mercator/World Geodetic System 84) projection, which facilitates its use in conjunction with other Geographic Information System (GIS) data (data table shown in Table 1).

2.2. Methodology

The main work of this study involves the following aspects: First, bathymetric measured data from Lingshui-Sanya Bay, as well as ICESat-2 data and Sentinel-2 images from both Lingshui-Sanya Bay and Dongsha Islands, were obtained. Second, Sentinel-2 images of the two study areas were preprocessed, and ICESat-2 bathymetric photon signals were extracted using a fully automated script. In the Lingshui-Sanya Bay area, ICESat-2 bathymetric photon data were matched with measured bathymetric data in terms of coordinates to evaluate the feasibility of ICESat-2 data for bathymetric applications. Subsequently, ICESat-2 bathymetric photon data from Dongsha Islands were resampled to a 10 m resolution and matched with Sentinel-2 images to obtain the red, green, and blue band reflectance values of the ICESat-2 bathymetric points. Additionally, the dataset was augmented with the FUI to represent water color information and spatial information, including normalized latitude and longitude coordinates as well as polar coordinates (radius and angle). Using this comprehensive dataset, the Stumpf model, Polynomial Regression model, Random Forest model, Gradient Boosting model, and XGBoost model were trained to invert the bathymetry of the Dongsha Islands. Finally, the accuracy and applicability of each model were comparatively evaluated. Figure 2 illustrates the technical workflow of this study.

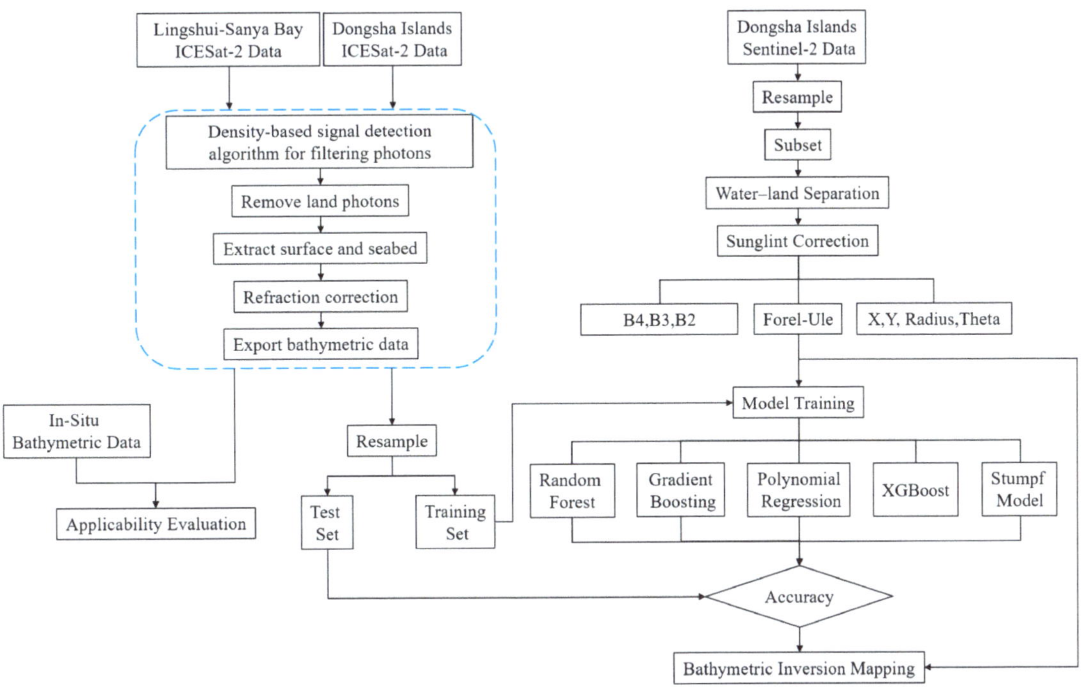

Figure 2. The technical flowchart of this study. The blue dashed box illustrates the key steps in the ICESat-2 bathymetric photon extraction process.

2.2.1. Lingshui-Sanya Bay Measured Data Acquisition

On 15–16 August 2020, our team carried out a field collection of seawater depths in Lingshui-Sanya Bay. This study used the bathymetric rod method, which measures the distance from the seafloor to the water surface by inserting a bathymetric rod vertically into the seawater and measuring the distance from the seafloor to the water surface through the scale marked on the rod. The bathymetric rod detection method is widely used in marine scientific research because of its simplicity and practicality. We used this method to detect a total of 24 discrete points of seawater bathymetry data in Lingshui-Sanya Bay, and tidally corrected the measured data by checking the tide tables of the harbors near the measured points, as shown in Table 2.

Table 2. In situ bathymetry data sheet.

Longitude	Latitude	Distance from Shore (m)	Measured Water Depth (m)	Tide-Corrected Water Depth (m)	Time
110.07672	18.45667	23.7	0.1	0.5	11:49
110.07676	18.45665	27.4	0.6	1	11:49
110.07686	18.45660	39.1	1.1	1.5	11:52
109.91875	18.41580	57.8	0.1	1.02	16:50
109.91877	18.41569	70.5	0.46	1.38	16:50
109.91878	18.41562	78.1	1.1	2.02	16:53
109.91078	18.41475	85.8	0.1	1.02	17:12
109.91079	18.41464	97.5	0.4	1.32	17:13
109.91082	18.41454	109.4	1.1	2.02	17:15
109.73072	18.31802	11.8	0.1	0.78	18:24
109.73077	18.31799	18.9	0.4	1.08	18:25
109.73118	18.31777	67.4	1.1	1.78	18:30
109.72836	18.31361	32.1	0.1	0.78	18:41
109.72848	18.31354	46.1	0.4	1.08	18:42
109.72884	18.31329	93.2	1.1	1.78	18:47
109.65143	18.23303	15.2	0.1	0	10:17
109.65143	18.23300	18.5	0.5	0.4	10:18
109.65144	18.23284	36.4	1	0.9	10:22
109.51883	18.22201	15.8	0.1	0.97	14:24
109.51884	18.22172	48.2	0.3	1.17	14:25
109.51886	18.22095	133.9	0.9	1.77	14:32
109.48227	18.26732	40.2	0.1	1.01	15:07
109.48217	18.26719	57.2	0.4	1.31	15:08
109.48197	18.26691	95.5	1	1.91	15:13

2.2.2. ICESat-2 Data Preprocessing

Our main objective was to extract the bathymetric photon signals from ICESat-2 satellites and compare them with the measured data of Sanya Bay, and then to evaluate the accuracy and reliability of the extraction method of ICESat-2 data bathymetric photon signals used in this study, taking into account the hydrographic characteristics of Sanya Bay. The specific data processing steps include data acquisition, signal filtering, land photon removal, water surface and seafloor extraction, refraction correction, and the exportation of bathymetric data, as shown in the blue box in Figure 2.

In this study, we referred to the methodology provided by the 2023 ICESat-2 Hackweek (https://icesat-2-2023.hackweek.io/tutorials/bathymetry/bathymetry_tutorial.html (accessed on 24 February 2024)) [60] to write a script that automatically extracts water depths in batch based on the date and orbit number of ICESat-2 data within the study area. Before running the script, we just needed to determine the study area extent and filter out the orbits and dates with good data quality. Using OpenAltimetry ICESat-2 Webpage (https://openaltimetry.earthdatacloud.nasa.gov/data/icesat2/ (accessed on 24 February 2024)), we could select the region of interest. By modifying the date and orbit number, we could select photon data orbits with good quality, density, and regular point clouds. By inputting the selected orbit, date, and the latitude and longitude of the study area into the script and running it, the water depth signal photons for the study area were automatically filtered. The following describes the main workflow and theoretical methods of the script.

Our study utilized the Python library "Sliderule" and an EarthData account to download the ATL03 data corresponding to specific latitude, longitude, date, and orbit numbers of the study area. ICESat-2, equipped with a laser altimeter system, conducts high-precision measurements of the Earth's surface, generating point cloud data that include ground and water surface elevations. To ensure that the ICESat-2 orbital data acquired covers the target area's laser detection information, we employed the distribution preview feature of the ATL08 dataset to identify orbits that potentially contain high-quality signals. After the data was downloaded, we proceeded with the filtering of photon signals. The ATL03 product provided by ICESat-2, along with the "Sliderule" tool, encompasses a variety of photon signal measurement and processing techniques. In this study, we utilized the YAPC (Yet Another Photon Classifier) algorithm, which was developed by NASA researchers [61]. The YAPC algorithm is a density-based signal detection method that identifies valid signals by analyzing the spatial distribution of photon signals. Compared to traditional photon classification approaches, YAPC exhibits heightened sensitivity to environmental variations, enabling it to adapt more accurately to the characteristics of diverse water bodies. Utilizing the YAPC algorithm, we filtered and identified effective photon signals. Taking the processing of ATL03_20190129144159_04910207_006_02 data as an example, Figure 3a presents a photon signal density confidence map based on the YAPC algorithm, illustrating the spatial distribution of photon signals along the track. Different colors in the figure represent photon signals of varying density levels, with signals of higher confidence indicated by more prominent colors. To determine the minimum threshold for valid signals, we employed the Otsu [62] thresholding method for automatic acquisition (Figure 4a). Photons with YAPC signal scores above this threshold are considered valid signals. Subsequently, we excluded land photons from the valid signals. To achieve this, we constructed a histogram to tally the frequency of photon occurrences across various height intervals, ranging from −50 m to 50 m with a step size of 0.1 m. By identifying the height value with the highest frequency in the histogram (i.e., the most common photon height), we estimated the water surface height. To account for the effects of waves or surface undulations, we added a 1 m buffer to this height. The vertical black line in Figure 4b represents the estimated water surface height; photons above this height were considered land photons and were removed, while valid photons below this height were classified as water area photons (Figure 3b). Building on this foundation, we extracted the water surface and seafloor from the remaining water area photons. We performed binning on the spatial distribution of photon signals, setting the resolution along the track to 20 m. Based on the distribution of photon density, we adaptively adjusted the height resolution and generated a two-dimensional histogram. The binning operation aimed to divide the photon data into multiple intervals along both the height and track dimensions, facilitating the analysis of photon height distribution characteristics. To enhance the detection of signal peaks, we applied adaptive filtering to the generated two-dimensional histogram. The filtering strength was dynamically adjusted based on local variance to smooth the signal and reduce the impact of noise, thereby highlighting the main peaks of the signal. Subsequently, based on the peak values of each waveform, we assumed that the topmost return signal represents the water surface. After removing the water surface peak, we selected the prominent peak as an indicator of seafloor depth, extracting water surface and seafloor return information from the photon height histogram. After acquiring the water surface and seabed depth, we proceeded with refraction correction based on the research outcomes of Parrish et al. [42] in 2019, which effectively enhances the precision of bathymetric measurements. Subsequently, we iteratively traversed each waveform, extracted the water surface and seabed information, applied refraction correction (Figure 5), calculated the water depth, and compiled the ICESat-2 bathymetric signal extraction results for the area.

Suitability assessment of processed ICESat-2 bathymetric data. The bathymetric performance of the ICESat-2 satellite data were evaluated by matching and comparing with the existing Lingshui-Sanya Bay bathymetry data. Firstly, the bathymetric real measurements were coordinate-matched with the extracted Lingshui-Sanya Bay ICESat-2 bathymetry data. For each measured point, the K Nearest Neighbours algorithm (KNN) was used to find the five nearest ICESat-2 data points, and the difference in bathymetry between these points and the measured points was computed, using the Root Mean Squared Error (RMSE), MeanDepthDiff, Variance of Depth Difference (VarDepthDiff), and Mean Squared Error (MSE) to quantify the differences to assess the consistency and error of the data.

Additionally, the average distance to the five nearest ICESat-2 points was calculated to further assess the spatial distribution and proximity of the ICESat-2 points relative to the measured locations.

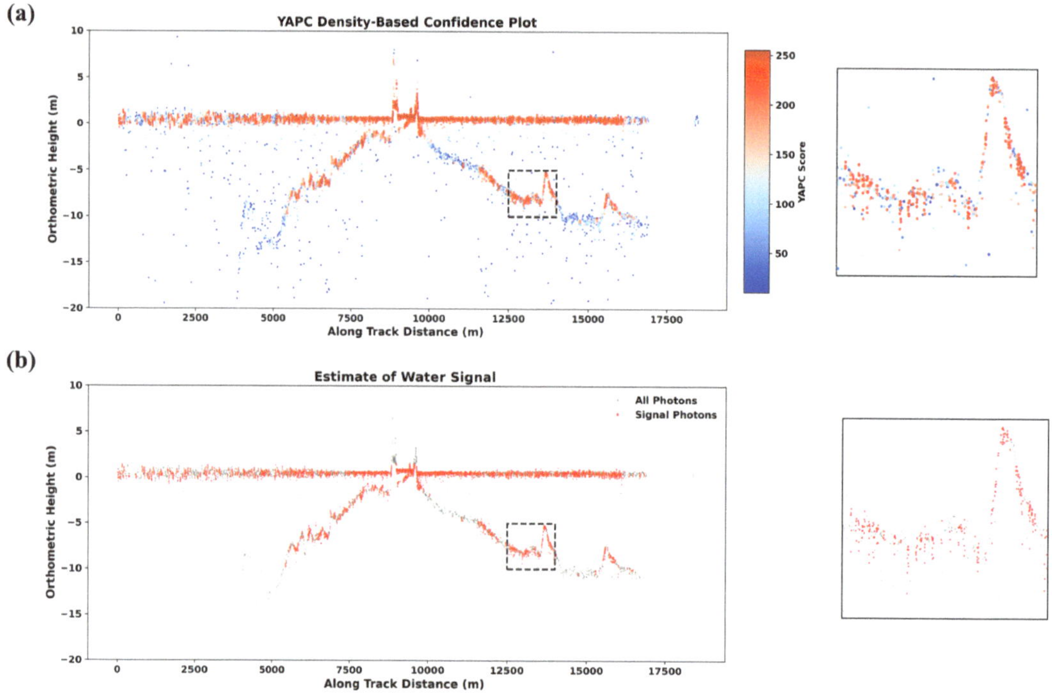

Figure 3. Noise and land photons were filtered using the YAPC algorithm. The right panel shows a zoomed-in view of the rectangular area in the image. (**a**) Photon signal density confidence map based on the YAPC algorithm, with red areas indicating high-confidence regions. (**b**) Water signal estimation map, where the red dots represent valid photon signals from the water surface and below.

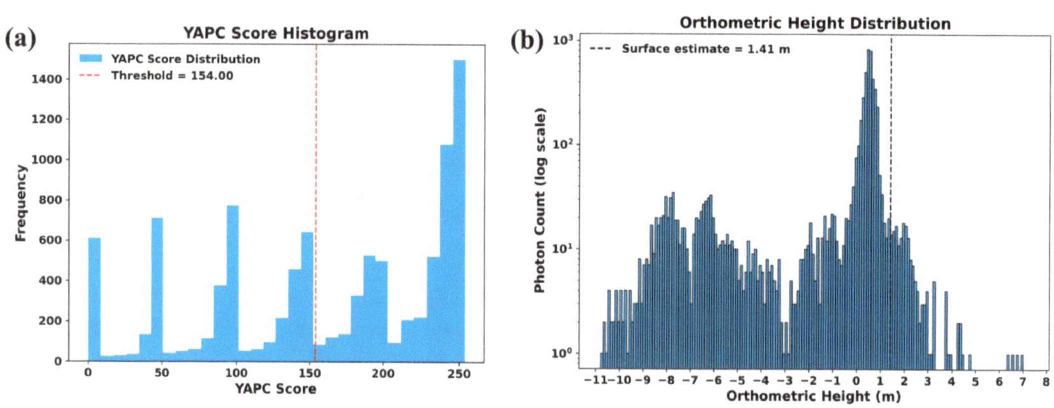

Figure 4. Reference lines for filtering noise and land photons. (**a**) The Otsu threshold method is used to automatically determine the minimum threshold for valid photon signals, with the red vertical line representing the threshold line for valid photon signals. (**b**) The estimated water surface height is obtained based on the histogram statistics of valid photon signals along the track, with the vertical black line indicating the estimated water surface height.

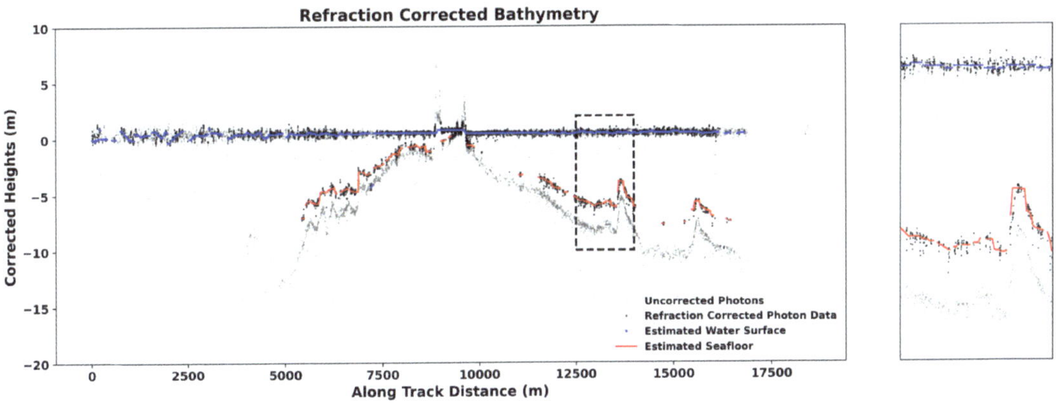

Figure 5. Water depth map after refraction correction. The gray points represent uncorrected photon data, while the black points indicate refraction-corrected photon data. The blue points denote estimated water surface photons, and the red line represents the estimated seafloor.

2.2.3. Sentinel-2 Image Preprocessing

Pre-processing the Sentinel-2 L2A data was an important step in constructing models for remote sensing analyses. Although the L2A data have been atmospherically corrected to generate surface reflectance data, there are still some necessary preprocessing steps before specific analyses can be performed.

Sentinel-2 L2A data were acquired from the Copernicus Open Access Hub (https://dataspace.copernicus.eu/ (accessed on 24 February 2024)), selecting high-quality imagery with minimal cloud cover. After resampling to 10 m resolution using SNAP, the images were cropped to the study area. Water bodies were extracted using a mask based on the near-infrared band (B8) with the formula (If B8 > 0.05, then NaN, else 1) [63]. Sunglint correction [64] was applied using the Deglint processor in the Sen2Cor plugin to reduce surface reflections, enhancing water body analysis accuracy.

Finally, consistency between ICESat-2 bathymetric data and Sentinel-2 imagery was ensured. To achieve spatial consistency, the ICESat-2 bathymetric data points were first mapped to the nearest grid cell in a 10 m resolution raster coordinate system. Data cleaning was then performed to ensure consistency: each RGB combination was checked against its corresponding depth value to ensure a unique depth value for each RGB combination. Additionally, it was verified that each depth value corresponded to a unique RGB combination, preventing the association of a single depth value with multiple RGB combinations. This process ensured that each raster cell was associated with only one depth value, providing a consistent data foundation for subsequent analysis.

2.2.4. Bathymetric Inversion Model for the Dongsha Islands

With the development of satellite remote sensing technology, bathymetric inversion using satellite images has become a fast and economical alternative to traditional bathymetry. Machine learning algorithms are good at learning complex nonlinear relationships from large amounts of data, which can significantly improve the accuracy of bathymetric inversion. This section describes the use of four machine learning methods, Random Forest, Gradient Boosting, XGBoost, and Polynomial regression, to train bathymetric inversion models based on spectral feature information, water body color information, and spatial location data. The performance of these models is compared with the improved logarithmic band ratio algorithm proposed by Stumpf et al. [30] in 2003 to explore the application of machine learning in satellite-derived bathymetry.

Creation of a Comprehensive Information Dataset

In the previous section, we obtained the ICESat-2 bathymetric dataset after data consistency processing. The ICESat-2 bathymetric points were matched with Sentinel-2 imagery, and the reflectance values of the red, green, and blue bands corresponding to each bathymetric point were extracted. These reflectance values were used as spectral feature information.

The Forel-Ule Index (FUI) is a classic index used to characterize the color of water bodies, primarily assessing the optical properties and water quality. In this study, we adopted the FUI algorithm developed by Van der Woerd H. J. and Wernand M. R. in 2018 for Sentinel-2 imagery [48], using the obtained FUI values as water color information.

Previous studies [5,65] have shown that incorporating spatial location information can improve the accuracy of bathymetric inversion using machine learning. However, these studies typically considered only the X and Y coordinates of the pixels, without accounting for polar coordinates. Polar coordinates provide additional spatial features, such as distance and angle, which are more sensitive to areas with non-uniform data distribution. In this study, spatial location information was enhanced by introducing polar coordinates alongside the traditional normalized pixel coordinates (X, Y). Specifically, for each pixel's normalized coordinates, the distance (R) and angle (θ) from the bottom left corner of the image were calculated.

The integrated dataset includes the three aforementioned components of feature information. Before model training, data standardization [Equation (1)] was applied to address potential issues arising from discrepancies in the scale and range of different features. Without standardization, features with larger numerical ranges could dominate the training process, overshadowing other important variables. Additionally, significant differences in feature scales could impede the convergence of gradient-based optimization algorithms, ultimately reducing training efficiency. This standardization ensured consistent value ranges across the different features, thereby improving the training performance and predictive accuracy of the model.

$$X_{\text{standardized}} = \frac{X - \mu}{\sigma} \quad (1)$$

where X is the original data, μ is the mean of the feature, σ is the standard deviation of the feature, and $X_{\text{standardized}}$ is the standardized data. Through this standardization process, the data distribution is adjusted to a normal distribution with a mean of 0 and a standard deviation of 1.

Model Training

In this study, we use the integrated information as features and the ICESat-2 depth values as labels for model training. The dataset is divided into 80% for training and 20% for testing. During the hyperparameter optimization process, we employ the FLAML framework for automated tuning. In this process, FLAML defines a hyperparameter space for each model and utilizes a Bayesian optimization algorithm to search for the optimal combination of hyperparameters. At each step of Bayesian optimization, FLAML evaluates the performance of each hyperparameter combination using ten-fold cross-validation. Specifically, we partition the 80% training data into 10 subsets, using 9 subsets for training and 1 subset for validation, repeating this process 10 times to comprehensively assess the model's performance. This method allows us to calculate performance metrics for each hyperparameter combination, including Root Mean Square Error (RMSE), Mean Absolute Error (MAE), R^2 score, and Explained Variance Score. The mean and standard deviation of these metrics demonstrate the stability of the model across different folds and help us assess its fitting ability and predictive performance. After hyperparameter optimization, FLAML returns the best hyperparameter configuration, and, based on this configuration, the final model is trained on the entire training set to maximize performance. The five models used in this study are described in detail below:

(1) Random Forest algorithm: Random Forest [66] is an integrated learning algorithm that performs classification and regression tasks by constructing multiple decision trees and combining their predictions. Its core idea is to use diversity by randomly sampling data with replacement to obtain a training subset and randomly selecting some of the features when training each decision tree, so as to introduce diversity and reduce the risk of overfitting. In the regression task, it gives the final prediction by taking the average value [Equation (2)]. Random Forest has the advantages of high prediction accuracy, resistance to overfitting, handling high-dimensional data, and strong robustness to noise and outliers.

$$\hat{y} = \frac{1}{B}\sum_{b=1}^{B} h_b(x) \tag{2}$$

where $\hat{y}$ is the predicted value obtained by averaging the predictions from B decision trees, $h_b(x)$ is the prediction function of the b-th tree for the input data, and B is the total number of decision trees in the random forest model. The summation $\sum_{b=1}^{B}$ indicates the accumulation of the prediction results from all B trees, and dividing by B gives the average prediction, which is the final predicted value $\hat{y}$.

(2) Gradient Boosting algorithm: Gradient Boosting [67] is a commonly used machine learning method for classification and regression tasks. It constructs a high-performance predictive model by iteratively combining multiple weak learners, typically decision trees. The algorithm starts with an initial model to predict the target variable [Equation (3)]. Then, new weak learners are trained to fit the residuals of the current model [Equations (4) and (5)], progressively optimizing the model's performance. The predictions of the new learner are weighted and added to the current model to form the updated model [Equation (6)]. This process aims to minimize the loss function, using gradient descent to guide each optimization step. The final model is the weighted sum of multiple weak learners.

$$F_0(x) = \underset{\gamma}{\operatorname{argmin}} \sum_{i=1}^{N} L(y_i, \gamma) \tag{3}$$

where $F_0(x)$ is the initial model obtained by minimizing the loss function L over all samples, $L(y_i, \gamma)$ is the loss function that measures the difference between the predicted value γ and the actual value y_i, and γ is the parameter of the initial model that we aim to optimize. The summation $\sum_{i=1}^{N}$ indicates the accumulation of the loss over all N samples, and the argument of the minimum $\underset{\gamma}{\operatorname{argmin}}$ tells us the value of γ that minimizes this total loss.

$$r_{im} = -\left[\frac{\partial L(y_i, F(x_i))}{\partial F(x_i)}\right]_{F(x)=F_{m-1}(x)} \tag{4}$$

where r_{im} is the residual for the i-th observation at the m-th iteration of the Gradient Boosting algorithm. The true value for the i-th observation is denoted by y_i. The predicted value generated by the model for the i-th observation is represented by $F(x_i)$. The model's predictions at the end of the $(m-1)$-th iteration for all observations are given by $F_{m-1}(x_i)$. The partial derivative $\left[\frac{\partial L(y_i, F(x_i))}{\partial F(x_i)}\right]$ is the rate at which the loss function L changes with respect to the predicted value $F(x_i)$, evaluated at the current model $F_{m-1}(x_i)$. The residual r_{im} is calculated as the negative of this partial derivative and is used to guide the training of the next weak learner in the Gradient Boosting process.

$$h_m(x) = \underset{h}{\operatorname{argmin}} \sum_{i=1}^{N} (r_{im} - h(x_i))^2 \tag{5}$$

where $h_m(x)$ is the weak learner function that is being optimized during the m-th iteration of the Gradient Boosting algorithm. The goal is to find a function h that minimizes the sum of the squared differences between the residuals r_{im} and the predictions of the weak learner $h(x_i)$ across all N training samples. The residual r_{im} represents the difference between the true value y_i and the prediction of the model from the previous iteration $F_{m-1}(x_i)$. The notation $\underset{h}{\operatorname{argmin}}$ indicates that we are looking for the function h (a weak learner, typically a decision tree) that results in the smallest possible sum of squared errors.

$$F_m(x) = F_{m-1}(x) + v \cdot h_m(x) \tag{6}$$

where $F_m(x)$ represents the predictive model function after the m-th iteration of the Gradient Boosting algorithm. This updated model is obtained by adding the contribution of the newly trained weak learner $h_m(x)$, scaled by the learning rate, to the predictive function from the previous iteration $F_{m-1}(x_i)$. The weak learner $h_m(x)$ is typically a decision tree that has been optimized to predict the residuals from the previous iteration. The learning rate v is a hyperparameter that controls the impact of each weak learner on the final prediction.

(3) Polynomial Regression algorithm: Polynomial Regression is an extended regression analysis method for modeling nonlinear relationships between dependent variables and multiple independent variables. Unlike multivariate linear regression, Polynomial Regression captures complex patterns in the data by introducing higher-order and interaction terms for the independent variables. The core idea is to use polynomial functions to describe the relationship between dependent and independent variables [Equation (7)]. To balance the model's expressive power and generalization, we select a second-order polynomial as the model form. This choice effectively captures nonlinear relationships while reducing model complexity to mitigate the risk of overfitting.

$$y = \beta_0 + \sum_{i=1}^{n} \beta_i x_i + \sum_{i=1}^{n} \sum_{j=1}^{n} \beta_{ij} x_i x_j + \varepsilon \tag{7}$$

where y is the dependent variable, β_0 is the intercept, β_i is the coefficient for the independent variable x_i, β_{ij} is the coefficient for the interaction term between the independent variables x_i and x_j, and ε is the error term.

(4) XGBoost algorithm: XGBoost [68] (Extreme Gradient Boosting) is an efficient machine learning algorithm widely used for classification and regression tasks. It enhances traditional gradient boosting methods through several key optimizations aimed at improving model performance and computational efficiency. XGBoost introduces L1 and L2 regularization to control model complexity and reduce overfitting, and utilizes second-order gradient information (Hessian matrix) to accelerate convergence and improve precision. The algorithm supports column sampling and optimized tree splitting, which randomly selects feature subsets to train decision trees, thereby increasing computational efficiency and mitigating overfitting. Parallelization is also employed to speed up the training process, making XGBoost particularly effective for large datasets. These optimizations lead to significant improvements in both model performance and training speed compared to traditional Gradient Boosting algorithms.

(5) Stumpf logarithmic band ratio algorithm: The improved logarithmic band ratio algorithm proposed by Stumpf et al. [30] is widely used in satellite bathymetric inversion (SDB). The method is based on the differential absorption and scattering properties of various wavelengths of light in the water column. In general, short-wavelength blue light penetrates deeper, while long-wavelength green light penetrates shallower. However, this relationship can vary depending on water quality, such as turbidity and other factors. In clear shallow water areas, the reflectance ratio between blue and green wavelengths changes with increasing depth. This nonlinear relationship is linearized by applying a logarithmic transformation to the reflectance of each band, enabling the development of

a mathematical model to describe water depth. The model's constant parameters can be determined through regression analysis of known depth data.

$$Z = m_1 \frac{\ln(nR_{rs}(\lambda_i))}{\ln(nR_{rs}(\lambda_j))} - m_0 \tag{8}$$

where Z represents the water depth, $R_{rs}(\lambda_i)$ and $R_{rs}(\lambda_j)$ are the reflectances of the respective bands i and j, and m_1 and m_0 are the constants derived from the regression analysis of calibration data. The constant m_1 is used to scale the ratio of the band reflectances to the water depth, while m_0 represents the offset at a depth of 0 m ($Z = 0$). The variable n is a predetermined fixed value that ranges between 500 and 1500, ensuring that the logarithmic ratio is always positive and varies linearly with depth.

3. Results

3.1. ICESat-2 Bathymetric Photon Extraction Results and Bathymetric Performance Evaluation

The ICESat-2 bathymetric photon data of Lingshui-Sanya Bay and Dongsha Islands were extracted using the fully automated bathymetric photon extraction algorithm constructed in this study (shown in Figure 6). Among them, 11,144 ICESat-2 bathymetry points were extracted from Lingshui-Sanya Bay and 10,581 bathymetry points were extracted from Dongsha Islands. In order to evaluate the accuracy of the ICESat-2 bathymetry data, we coordinate-matched the measured bathymetry data of Lingshui-Sanya Bay with the ICESat-2 bathymetry data. During the matching process, the K Nearest Neighbour (KNN) algorithm was used to search for the five nearest ICESat-2 bathymetry points around each measured point. The average distance to the five nearest ICESat-2 points from each measured point was calculated to be 38.88 m. Using this method, we obtained a total of 120 data pairs and calculated the depth difference of each pair. In order to show the data differences more intuitively, different colors were used to classify the data points according to the absolute value of the depth differences, and a difference distribution map (shown in Figure 7) was drawn to visualize the error distribution between the measured bathymetry and the ICESat-2 bathymetry. The statistical analysis results are shown in Table 3, where MeanDepthDiff represents the mean depth difference and VarDepthDiff represents the variance of depth differences. Overall, the ICESat-2 bathymetry data are slightly higher compared to the measured data, and the fluctuation of the bathymetry difference is small, the error is more stable, and all of them are within the acceptable range. This indicates that ICESat-2 data can obtain high-precision shallow water bathymetry data, which has good potential for bathymetric applications.

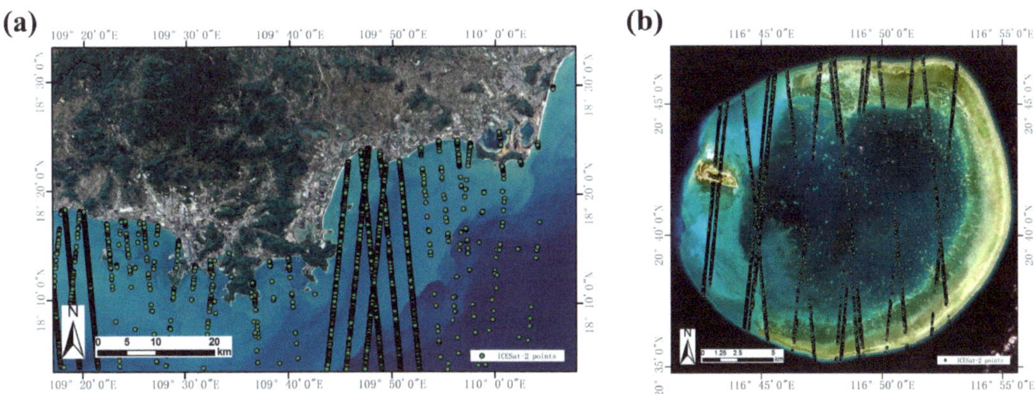

Figure 6. Plot of ICESat-2 bathymetric photon extraction data results. (**a**) Lingshui-Sanya Bay. (**b**) Dongsha Islands.

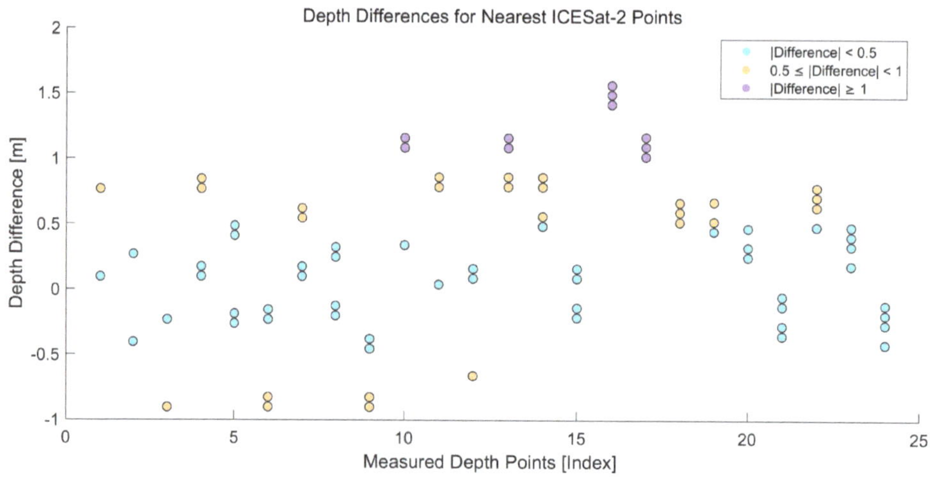

Figure 7. Difference distribution map showing the distribution of depth differences between the measured points and the nearest ICESat-2 bathymetry points. The *X*-axis represents the sequence number of the measured points.

Table 3. Comparison of ICESat-2 bathymetry data with in situ measurements.

MeasuredPointIndex	MeanDepthDiff (m)	VarDepthDiff (m)	MSE (m)	RMSE (m)
1	0.50	0.14	0.36	0.60
2	0.00	0.14	0.11	0.33
3	−0.50	0.14	0.36	0.60
4	0.55	0.14	0.41	0.64
5	0.19	0.14	0.15	0.39
6	−0.45	0.14	0.32	0.57
7	0.40	0.06	0.20	0.45
8	0.10	0.06	0.06	0.24
9	−0.60	0.06	0.41	0.64
10	0.98	0.13	1.06	1.03
11	0.68	0.13	0.57	0.75
12	−0.02	0.13	0.10	0.32
13	0.94	0.03	0.90	0.95
14	0.64	0.03	0.43	0.65
15	−0.06	0.03	0.03	0.17
16	1.49	0.00	2.23	1.49
17	1.09	0.00	1.19	1.09
18	0.59	0.00	0.35	0.59
19	0.52	0.01	0.28	0.53
20	0.32	0.01	0.11	0.33
21	−0.22	0.02	0.06	0.24
22	0.62	0.02	0.39	0.63
23	0.32	0.02	0.11	0.34
24	−0.28	0.02	0.09	0.31
ALL	0.32	0.33	0.43	0.65

3.2. Bathymetric Inversion Based on Sentinel-2 Data

Through the preprocessing of Sentinel-2 imagery over the Dongsha Islands (Figure 1), we extracted the spectral characteristics of the region and performed data consistency processing, ultimately obtaining 9562 ICESat-2 bathymetry points. Using the computed FUI (shown in Figure 8k), we extracted the corresponding FUI values for the ICESat-2 bathymetry points and calculated the spatial information for each bathymetry point.

Based on these data, we constructed a comprehensive information dataset and used it for model training. The trained model was then applied to perform bathymetry inversion for the Dongsha Islands (Figure 8). The inversion results from all machine learning models exhibited similar overall trends and were consistent with previous bathymetry inversion results for this region [43,69]. Therefore, our automated script for extracting ICESat-2 bathymetry points demonstrates good feasibility in shallow, clear-water areas, providing effective support for rapid bathymetric inversion.

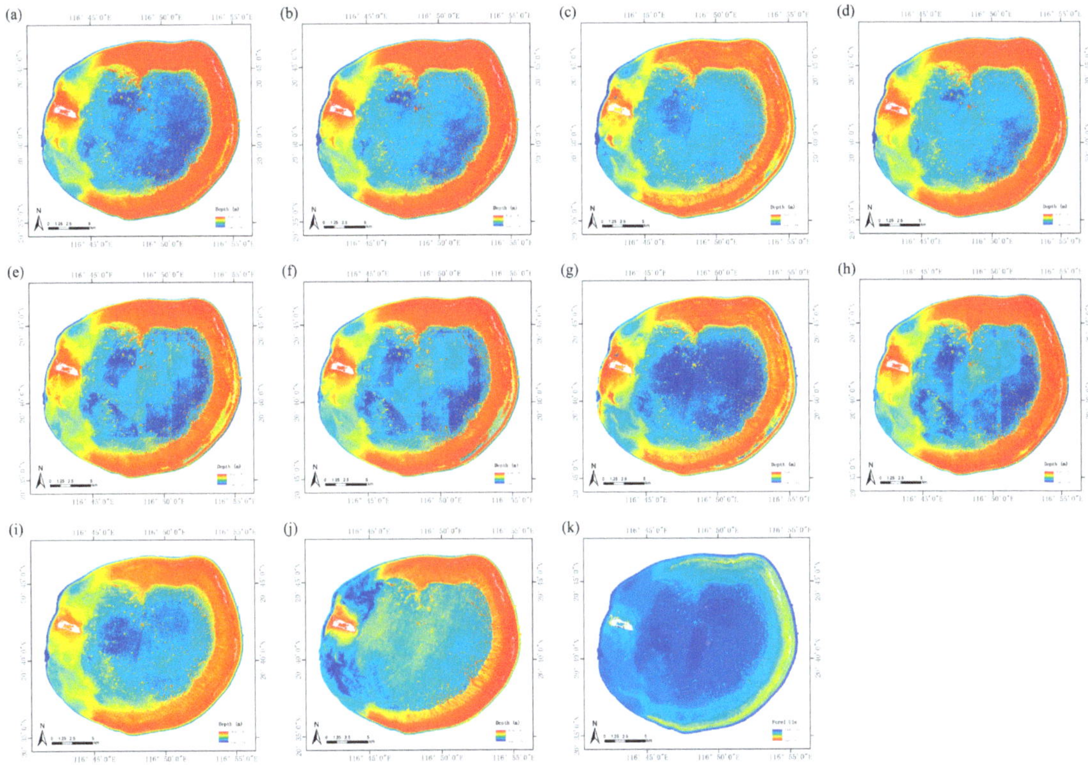

Figure 8. Plot of inversion results based on four models for Dongsha Islands, where '-Bands' represents bathymetric images inverted using spectral characteristic information, and '-CID' represents bathymetric images inverted using comprehensive information. (**a**) Random Forest-Bands. (**b**) Gradient Boosting-Bands. (**c**) Polynomial Regression-Bands. (**d**) XGBoost-Bands. (**e**) Random Forest-CID. (**f**) Gradient Boosting-CID. (**g**) Polynomial Regression-CID. (**h**) XGBoost-CID. (**i**) Stumpf-BG. (**j**) Stumpf-BR. (**k**) Forel-Ule Index.

3.3. Evaluation of Model Accuracy

The trained models were evaluated on a 20% test set to assess their generalization. Scatter plots comparing predicted water depth with ICESat-2 depth values were generated to visually demonstrate the model's prediction performance (as shown in Figure 9). Additionally, the R^2 and RMSE for each model were calculated on the test set to quantify the correlation between the predicted and true values, providing further validation of the model's performance. The results indicate that the bathymetric inversion models using integrated features significantly improved R^2 and reduced RMSE compared to models with single features. When only spectral information was used as the input feature, the prediction performance of Random Forest, Gradient Boosting, and XGBoost models was similar. However, after incorporating water color information and spatial data, the Random Forest model performed the best, achieving an R^2 of 0.94 and an RMSE of 0.84 m.

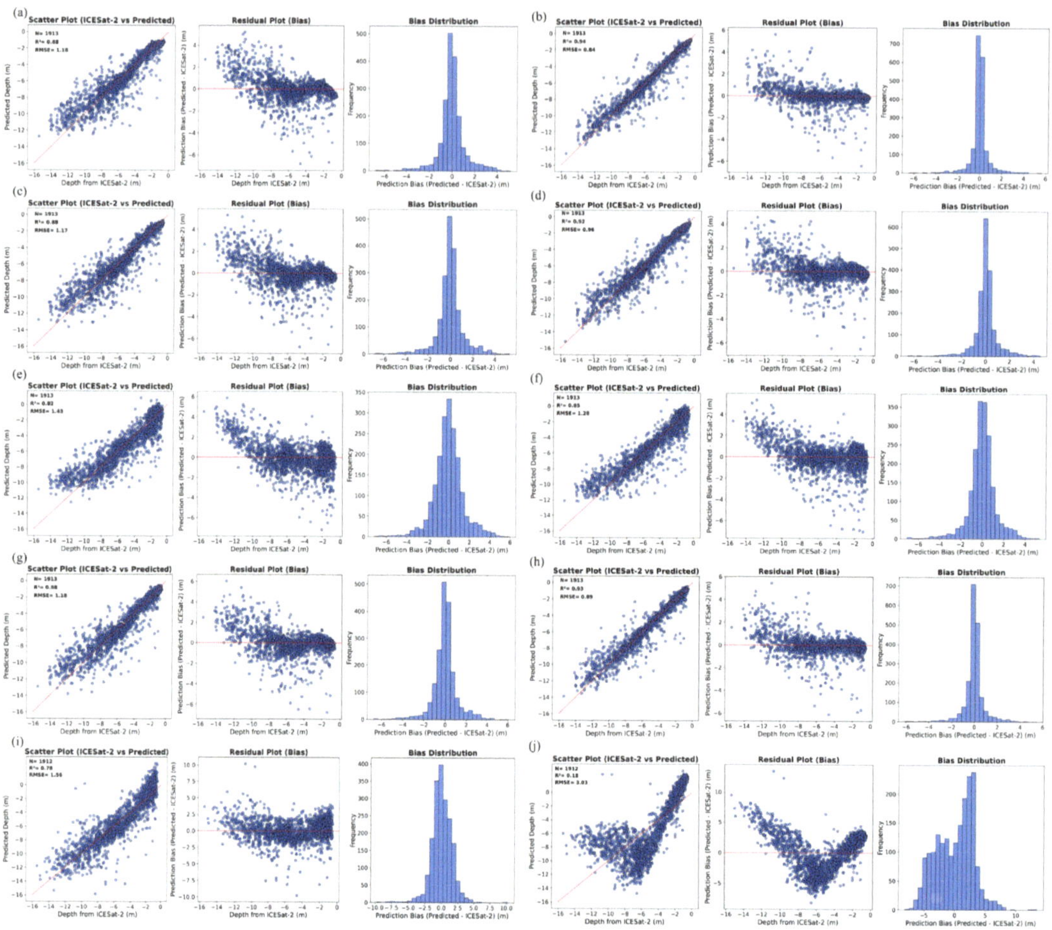

Figure 9. Scatter plots, residual plots, and deviation distributions of predicted bathymetry versus ICESat-2 bathymetry values. (**a**) Random Forest-Bands. (**b**) Gradient Boosting-Bands. (**c**) Polynomial Regression-Bands. (**d**) XGBoost-Bands. (**e**) Random Forest-CID. (**f**) Gradient Boosting-CID. (**g**) Polynomial Regression-CID. (**h**) XGBoost-CID. (**i**) Stumpf-BG. (**j**) Stumpf-BR.

4. Discussion

4.1. The Rationality of Feature Selection

In this study, we selected spectral information, spatial data (including normalized X and Y coordinates, as well as polar coordinates), and water color information (Forel-Ule Index, FUI) as model features. Spectral information is the core variable for bathymetric inversion, as water depth directly influences the absorption and scattering properties of light within the water column. Despite minimal changes in water quality within the study area, FUI, as a proxy for water color, provides complementary optical features. In clear waters, FUI assists in capturing subtle optical characteristics of the water, thereby enhancing the model's ability to detect depth-related variations. The inclusion of spatial information, particularly normalized X and Y coordinates and polar coordinates, effectively captures spatial patterns in bathymetric distribution. The use of polar coordinates simplifies spatial calculations and improves the model's ability to learn depth variations. By integrating these features, the model comprehensively accounts for optical, spatial, and water-related characteristics, ultimately improving the accuracy of bathymetric inversion and enhancing model performance in clear water environments.

4.2. Model Evaluation

To further evaluate the predictive performance of the models, residual and bias distribution plots were generated (as shown in Figure 9). Statistical analysis was performed for three depth ranges: from −5 m to 0 m, from −10 m to −5 m, and from −15 m to −10 m. The root mean square error (RMSE), mean absolute error (MAE), bias average (BIAS_AVG), and bias standard deviation (BIAS_STD) for each depth range were calculated (as shown in Table 4). Additionally, bar charts of the performance evaluation metrics for each model across different depth ranges were plotted (as shown in Figure 10) to provide a comprehensive comparison of model performance at various depth intervals. RMSE and MAE reflect the predictive accuracy of the models, with lower values indicating smaller prediction errors within the depth ranges. BIAS_AVG and BIAS_STD reveal the bias in model predictions, with a lower BIAS_AVG indicating predictions closer to the true water depths and a smaller BIAS_STD suggesting higher stability in the model's performance across different depth ranges.

Table 4. Statistics for different depth bands for different models.

Model	Segment	N	RMSE	MAE	BIAS_AVG	BIAS_STD
	−5~0 m	1006	0.82	0.46	−0.22	0.79
RF-Bands	−10~−5 m	727	1.23	0.86	−0.08	1.23
	−15~−10 m	179	2.20	1.88	1.75	1.34
	−5~0 m	1006	0.82	0.46	−0.18	0.79
GB-Bands	−10~−5 m	727	1.26	0.89	−0.12	1.26
	−15~−10 m	179	2.07	1.77	1.58	1.33
	−5~0 m	1006	1.30	0.91	−0.41	1.23
PR-Bands	−10~−5 m	727	1.12	0.83	−0.01	1.12
	−15~−10 m	179	2.65	2.37	2.32	1.28
	−5~0 m	1006	0.84	0.48	−0.18	0.82
XG-Bands	−10~−5 m	727	1.23	0.87	−0.11	1.23
	−15~−10 m	179	2.15	1.81	1.60	1.43
	−5~0 m	1006	0.68	0.35	−0.13	0.66
RF-CID	−10~−5 m	727	0.89	0.55	−0.05	0.89
	−15~−10 m	179	1.51	1.15	0.94	1.18
	−5~0 m	1006	0.70	0.36	−0.13	0.69
GB-CID	−10~−5 m	727	1.00	0.67	−0.05	1.00
	−15~−10 m	179	1.74	1.37	1.15	1.31
	−5~0 m	1006	1.19	0.78	−0.34	1.15
PR-CID	−10~−5 m	727	1.07	0.77	−0.01	1.07
	−15~−10 m	179	2.36	1.91	1.81	1.52
	−5~0 m	1006	0.66	0.31	−0.12	0.65
XG-CID	−10~−5 m	727	0.88	0.53	−0.04	0.88
	−15~−10 m	179	1.54	1.12	0.79	1.32
	−5~0 m	998	1.45	1.10	0.21	1.44
Stumpf-BG	−10~−5 m	737	1.56	1.10	−0.35	1.51
	−15~−10 m	175	2.06	1.54	0.99	1.81
	−5~0 m	998	2.32	2.05	0.87	2.16
Stumpf-BR	−10~−5 m	737	3.34	2.89	−1.92	2.73
	−15~−10 m	175	4.65	4.18	4.17	2.05

The analysis indicates that incorporating comprehensive information as input features improves model accuracy across all depth intervals. In the range from −15 m to −10 m, prediction errors were significantly reduced, suggesting that the inclusion of comprehensive information enhances the accuracy of predictions for the deeper segments of the shallow water zone (from −15 m to −10 m). The XGBoost model with Comprehensive Information (XG-CID) as input features performed best across all depth intervals, especially in the range from −10 m to 0 m, where both RMSE and MAE remained low. The Random Forest model with Comprehensive Information (RF-CID) inputs followed closely, demonstrating stable performance across all depth intervals, particularly maintaining strong predictive capability in the range from −15 m to −10 m.

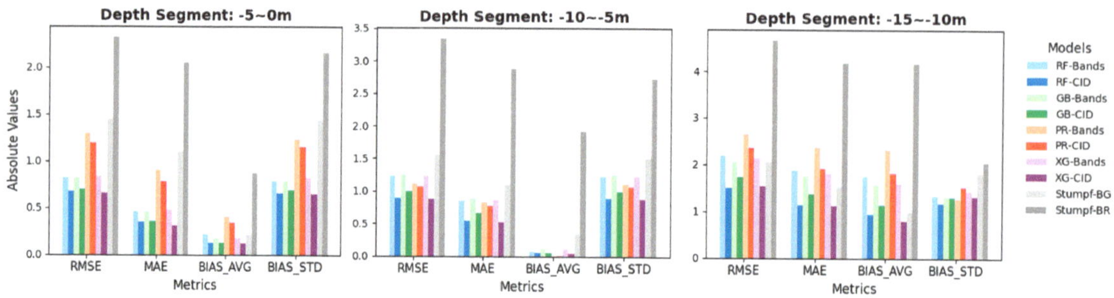

Figure 10. The bar charts of performance evaluation metrics for each model across different depth ranges.

However, in the range from −15 m to −10 m, all models exhibited systematic positive bias, with prediction errors generally exceeding 1 m. This bias is likely related to the sparse data in this depth range. The ICESat-2 data in this region is relatively sparse, which limits the model's ability to accurately capture the complex variation in water depth, leading to an overestimation of the water depth and resulting in positive bias.

Furthermore, to gain deeper insights into the contribution of each feature to the model's predictions, we employed SHAP (Shapley Additive Explanations) plots to analyze the impact of each feature on the model's outputs across different depth intervals. SHAP values were calculated for each feature, highlighting their importance in the model's predictions for various depth ranges (Figure 11). The results indicate that spectral information contributed the most to the model's depth predictions. Despite the relatively minor changes in water quality, the Forel-Ule Index (FUI), which represents water color, played a significant role in capturing the optical properties of the water. Additionally, spatial information also contributed to the model's predictions. Through these analyses, we gained a clearer understanding of the model's behavior, which provides a basis for further optimization.

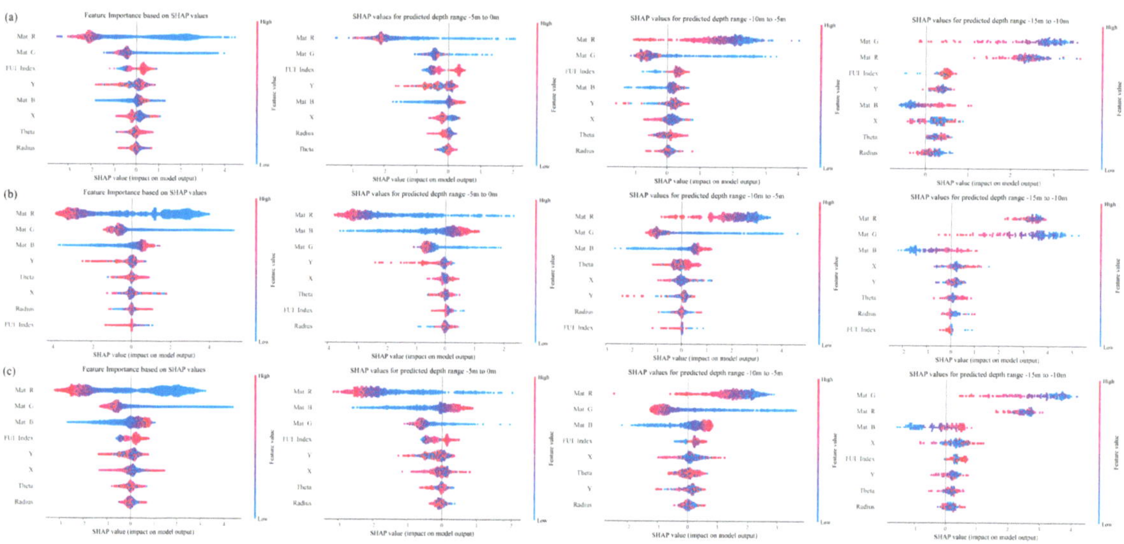

Figure 11. SHAP analysis of feature contributions across depth intervals. The leftmost plot in each group represents the overall analysis, covering feature contribution analysis across all depth intervals, while the remaining plots correspond to different depth intervals. (**a**) Random Forest-CID. (**b**) Gradient Boosting-CID. (**c**) XGBoost-CID.

4.3. Limitations and Directions for Improvement

Due to practical limitations, the in situ water depth data in Lingshui-Sanya Bay in this study reached a maximum depth of only 2 m. This depth restriction hindered a more in-depth analysis of the script's ability to extract water depth photons in deeper waters, and as a result, a comprehensive validation of the script's performance in deeper regions was not possible. However, the water depth inversion results obtained in this study align with previous experimental findings, indicating that the water depth photon extraction script can still effectively provide water depth information in shallow areas, thus offering a convenient tool for water depth inversion in shallow waters. It is worth noting that the portability of this script requires further investigation, particularly in regions with complex water characteristics, which will be a key focus for future research improvements. Further validation of ICESat-2's performance in deeper waters will require more extensive and deeper in situ data.

Additionally, ICESat-2 faces certain technical limitations in ultra-shallow water areas, particularly in regions where the water depth is less than 2 m. Due to the similarity between the water surface and seabed echo signals, the lidar system struggles to effectively distinguish between the reflections from the water surface and the seabed, thereby affecting depth measurement accuracy. Consequently, the depth accuracy of ICESat-2 in this depth range is relatively low, limiting its application potential in ultra-shallow water zones.

Furthermore, the water depth photon data provided by ICESat-2 has relatively low spatial resolution, leading to data sparsity and uneven distribution in certain areas. The discontinuity of the data may impact the accuracy of water depth inversion, especially in regions where water body characteristics are complex or data are sparse. Future research could optimize the inversion process by improving data fusion methods, integrating additional high-resolution remote sensing data, and considering factors such as water depth spatial distribution and water body environments. This could enhance the accuracy and applicability of the model.

5. Conclusions

This study proposes a rapid bathymetric inversion method based on ICESat-2 and Sentinel-2 data, integrating spectral information, the Forel-Ule Index (FUI) as water color information, and spatial location data (normalized X and Y coordinates and polar coordinates). Building upon previous work, an automated script for extracting bathymetric photon data was developed, enabling users to easily obtain the required photon data by simply inputting the study area, photon orbit number, and date. This aims to facilitate the use of ICESat-2 data for a wider range of researchers.

Although the in situ water depth data in Sanya Bay only reached 2 m, the bathymetric inversion results for the Dongsha Islands in this study are consistent with previous research, validating the script's effectiveness in shallow water regions. The performance evaluation of several machine learning models showed that the XGBoost model with comprehensive input features (XG-CID) performed best across all depth intervals, particularly in the range from -10 m to 0 m, where its prediction accuracy was especially notable. The Random Forest model with comprehensive input features (RF-CID) also demonstrated strong predictive capability in the range from -15 m to -10 m.

Through SHAP analysis, this study enhanced the model's interpretability, visually illustrating the influence of each feature on the predictions across different depth intervals. Spectral information contributed the most to the depth predictions, while FUI and spatial data also played a significant role in improving prediction accuracy.

Future research will focus on improving the extraction of bathymetric photons, incorporating higher-resolution remote sensing data, and considering additional factors such as spatial distribution of water depths and water body environment to further enhance the accuracy and applicability of the model.

Author Contributions: Conceptualization, M.Y. and C.Y.; methodology, M.Y., C.Y., X.Z., and S.L.; software, M.Y.; validation, M.Y.; formal analysis, M.Y.; investigation, M.Y.; resources, C.Y. and X.Z.; data curation, M.Y. and C.Y.; writing—original draft preparation, M.Y.; writing—review and editing, M.Y., S.L., C.Y., X.Z., X.P., Y.L., and T.C.; visualization, M.Y.; supervision, C.Y. and X.Z. All authors have read and agreed to the published version of the manuscript.

Funding: This research was funded by the Natural Science Foundation of China, grant numbers 42130805 and 42074154.

Data Availability Statement: The ICESat-2 ATL03 data are available at https://nsidc.org/data/atl03/versions/6 (accessed on 22 February 2024). The Sentinel-2 Level-2A (L2A) imagery products are available at https://dataspace.copernicus.eu/ (accessed on 22 February 2024). The script code for extracting ICESat-2 bathymetric signals is available in the following repository on GitHub: https://github.com/luzhu-star/ICESat_2 (accessed on 22 February 2024).

Acknowledgments: The authors gratefully acknowledge the NASA National Snow and Ice Data Center (NSIDC) for providing ICESat-2 data and the European Space Agency (ESA) for Sentinel-2 imagery. We also thank the ICESat-2 Hackweek for inspiring our approach.

Conflicts of Interest: The authors declare no conflicts of interest.

References

1. Wang, Y.; He, X.; Bai, Y.; Wang, D.; Zhu, Q.; Gong, F.; Yang, D.; Li, T. Satellite retrieval of benthic reflectance by combining lidar and passive high-resolution imagery: Case-I water. *Remote Sens. Environ.* **2022**, *272*, 112955. [CrossRef]
2. da Silveira, C.B.L.; Strenzel, G.M.R.; Maida, M.; Araujo, T.C.M.; Ferreira, B.P. Multiresolution Satellite-Derived Bathymetry in Shallow Coral Reefs: Improving Linear Algorithms with Geographical Analysis. *J. Coast. Res.* **2020**, *36*, 1247–1265. [CrossRef]
3. Kutser, T.; Hedley, J.; Giardino, C.; Roelfsema, C.; Brando, V.E. Remote sensing of shallow waters—A 50 year retrospective and future directions. *Remote Sens. Environ.* **2020**, *240*, 111619. [CrossRef]
4. Ma, S.; Tao, Z.; Yang, X.; Yu, Y.; Zhou, X.; Li, Z. Bathymetry Retrieval from Hyperspectral Remote Sensing Data in Optical-Shallow Water. *Ieee Trans. Geosci. Remote Sens.* **2014**, *52*, 1205–1212. [CrossRef]
5. Wang, Y.; Zhou, X.; Li, C.; Chen, Y.; Yang, L. Bathymetry Model Based on Spectral and Spatial Multifeatures of Remote Sensing Image. *Ieee Geosci. Remote Sens. Lett.* **2020**, *17*, 37–41. [CrossRef]
6. Kulbacki, A.; Lubczonek, J.; Zaniewicz, G. Acquisition of Bathymetry for Inland Shallow and Ultra-Shallow Water Bodies Using PlanetScope Satellite Imagery. *Remote Sens.* **2024**, *16*, 3165. [CrossRef]
7. Bannari, A.; Kadhem, G. MBES-CARIS Data Validation for Bathymetric Mapping of ShallowWater in the Kingdom of Bahrain on the Arabian Gulf. *Remote Sens.* **2017**, *9*, 385. [CrossRef]
8. Costa, B.M.; Battista, T.A.; Pittman, S.J. Comparative evaluation of airborne LiDAR and ship-based multibeam SoNAR bathymetry and intensity for mapping coral reef ecosystems. *Remote Sens. Environ.* **2009**, *113*, 1082–1100. [CrossRef]
9. Ji, X.; Ma, Y.; Zhang, J.; Xu, W.; Wang, Y. A Sub-Bottom Type Adaption-Based Empirical Approach for Coastal Bathymetry Mapping Using Multispectral Satellite Imagery. *Remote Sens.* **2023**, *15*, 3570. [CrossRef]
10. Ashphaq, M.; Srivastava, P.K.; Mitra, D. Review of near-shore satellite derived bathymetry: Classification and account of five decades of coastal bathymetry research. *J. Ocean Eng. Sci.* **2021**, *6*, 340–359. [CrossRef]
11. Xie, C.; Chen, P.; Zhang, S.; Huang, H. Nearshore Bathymetry from ICESat-2 LiDAR and Sentinel-2 Imagery Datasets Using Physics-Informed CNN. *Remote Sens.* **2024**, *16*, 511. [CrossRef]
12. Knudby, A.; Richardson, G. Incorporation of neighborhood information improves performance of SDB models. *Remote Sens. Appl. -Soc. Environ.* **2023**, *32*, 101033. [CrossRef]
13. He, C.L.; Jiang, Q.G.; Wang, P. An Improved Physics-Based Dual-Band Model for Satellite-Derived Bathymetry Using SuperDove Imagery. *Remote Sens.* **2024**, *16*, 3801. [CrossRef]
14. Klotz, A.N.; Almar, R.; Quenet, Y.; Bergsma, E.W.J.; Youssefi, D.; Artigues, S.; Rascle, N.; Sy, B.A.; Ndour, A. Nearshore satellite-derived bathymetry from a single-pass satellite video: Improvements from adaptive correlation window size and modulation transfer function. *Remote Sens. Environ.* **2024**, *315*, 114411. [CrossRef]
15. Richardson, G.; Foreman, N.; Knudby, A.; Wu, Y.L.; Lin, Y.W. Global deep learning model for delineation of optically shallow and optically deep water in Sentinel-2 imagery. *Remote Sens. Environ.* **2024**, *311*, 114302. [CrossRef]
16. Wu, Z.Q.; Zhao, Y.C.; Wu, S.L.; Chen, H.D.; Song, C.H.; Mao, Z.H.; Shen, W. Satellite-Derived Bathymetry Using a Fast Feature Cascade Learning Model in Turbid Coastal Waters. *J. Remote Sens.* **2024**, *4*. [CrossRef]
17. Dietrich, J.T. Bathymetric Structure-from-Motion: Extracting shallow stream bathymetry from multi-view stereo photogrammetry. *Earth Surf. Process. Landf.* **2017**, *42*, 355–364. [CrossRef]
18. Lubczonek, J.; Kazimierski, W.; Zaniewicz, G.; Lacka, M. Methodology for Combining Data Acquired by Unmanned Surface and Aerial Vehicles to Create Digital Bathymetric Models in Shallow and Ultra-Shallow Waters. *Remote Sens.* **2022**, *14*, 105. [CrossRef]
19. Hodúl, M.; Chénier, R.; Faucher, M.A.; Ahola, R.; Knudby, A.; Bird, S. Photogrammetric Bathymetry for the Canadian Arctic. *Mar. Geod.* **2020**, *43*, 23–43. [CrossRef]

20. Del Savio, A.A.; Torres, A.L.; Olivera, M.A.V.; Rojas, S.R.L.; Ibarra, G.T.U.; Neckel, A. Using UAVs and Photogrammetry in Bathymetric Surveys in Shallow Waters. *Appl. Sci.* **2023**, *13*, 3420. [CrossRef]
21. Bandini, F.; Sunding, T.P.; Linde, J.; Smith, O.; Jensen, I.K.; Köppl, C.J.; Butts, M.; Bauer-Gottwein, P. Unmanned Aerial System (UAS) observations of water surface elevation in a small stream: Comparison of radar altimetry, LIDAR and photogrammetry techniques. *Remote Sens. Environ.* **2020**, *237*, 111487. [CrossRef]
22. Ma, Y.; Xu, N.; Liu, Z.; Yang, B.; Yang, F.; Wang, X.H.; Li, S. Satellite-derived bathymetry using the ICESat-2 lidar and Sentinel-2 imagery datasets. *Remote Sens. Environ.* **2020**, *250*, 112047. [CrossRef]
23. Polcyn, F.C.; Rollin, R.A. Remote sensing techniques for the location and measurement of shallow-water features. Available online: https://deepblue.lib.umich.edu/handle/2027.42/7114 (accessed on 6 June 2024).
24. Polcyn, F.C.; Brown, W.L.; Sattinger, I.J. The Measurement of Water Depth by Remote Sensing Techniques. Available online: https://agris.fao.org/search/en/providers/122415/records/647368ca53aa8c89630d65ca (accessed on 6 June 2024).
25. Polcyn, F.C. Calculations of water depth from ERTS-MSS data. Available online: https://ntrs.nasa.gov/citations/19730019626 (accessed on 6 June 2024).
26. Lyzenga, D.R. Passive remote sensing techniques for mapping water depth and bottom features. *Appl. Opt.* **1978**, *17*, 379–383. [CrossRef]
27. Lyzenga, D.R. Remote sensing of bottom reflectance and water attenuation parameters in shallow water using aircraft and Landsat data. *Int. J. Remote Sens.* **1981**, *2*, 71–82. [CrossRef]
28. Lyzenga, D.R. Shallow-water bathymetry using combined lidar and passive multispectral scanner data. *Int. J. Remote Sens.* **1985**, *6*, 115–125. [CrossRef]
29. Lyzenga, D.R.; Malinas, N.P.; Tanis, F.J. Multispectral bathymetry using a simple physically based algorithm. *IEEE Trans. Geosci. Remote Sens.* **2006**, *44*, 2251–2259. [CrossRef]
30. Stumpf, R.P.; Holderied, K.; Sinclair, M. Determination of water depth with high-resolution satellite imagery over variable bottom types. *Limnol. Oceanogr.* **2003**, *48*. [CrossRef]
31. Pacheco, A.; Horta, J.; Loureiro, C.; Ferreira, Ó. Retrieval of nearshore bathymetry from Landsat 8 images: A tool for coastal monitoring in shallow waters. *Remote Sens. Environ.* **2015**, *159*, 102–116. [CrossRef]
32. Hedley, J.D.; Roelfsema, C.; Brando, V.; Giardino, C.; Kutser, T.; Phinn, S.; Mumby, P.J.; Barrilero, O.; Laporte, J.; Koetz, B. Coral reef applications of Sentinel-2: Coverage, characteristics, bathymetry and benthic mapping with comparison to Landsat 8. *Remote Sens. Environ.* **2018**, *216*, 598–614. [CrossRef]
33. Sandidge, J.; Holyer, R.J. Coastal bathymetry from hyperspectral observations of water radiance. *Remote Sens. Environ.* **1998**, *65*, 341–352. [CrossRef]
34. Manessa, M.D.M.; Kanno, A.; Sekine, M.; Haidar, M.; Yamamoto, K.; Imai, T.; Higuchi, T. Satellite-Derived Bathymetry Using Random Forest Algorithm and Worldview-2 Imagery. *Geoplanning J. Geomat. Plan.* **2016**, *3*, 117–126. [CrossRef]
35. Wang, L.; Liu, H.; Su, H.; Wang, J. Bathymetry retrieval from optical images with spatially distributed support vector machines. *Giscience Remote Sens.* **2019**, *56*, 323–337. [CrossRef]
36. Leng, Z.; Zhang, J.; Ma, Y.; Zhang, J. Underwater Topography Inversion in Liaodong Shoal Based on GRU Deep Learning Model. *Remote. Sens.* **2020**, *12*, 4068. [CrossRef]
37. Neumann, T.A.; Martino, A.J.; Markus, T.; Bae, S.; Bock, M.R.; Brenner, A.C.; Brunt, K.M.; Cavanaugh, J.; Fernandes, S.T.; Hancock, D.W.; et al. The Ice, Cloud, and Land Elevation Satellite-2 mission: A global geolocated photon product derived from the Advanced Topographic Laser Altimeter System. *Remote Sens. Environ.* **2019**, *233*, 111325. [CrossRef]
38. Markus, T.; Neumann, T.; Martino, A.; Abdalati, W.; Brunt, K.; Csatho, B.; Farrell, S.; Fricker, H.; Gardner, A.; Harding, D.; et al. The Ice, Cloud, and land Elevation Satellite-2 (ICESat-2): Science requirements, concept, and implementation. *Remote Sens. Environ.* **2017**, *190*, 260–273. [CrossRef]
39. Abdalati, W.; Zwally, H.J.; Bindschadler, R.; Csatho, B.; Farrell, S.L.; Fricker, H.A.; Harding, D.; Kwok, R.; Lefsky, M.; Markus, T.; et al. The ICESat-2 Laser Altimetry Mission. *Proc. IEEE* **2010**, *98*, 735–751. [CrossRef]
40. Smith, B.; Fricker, H.A.; Holschuh, N.; Gardner, A.S.; Adusumilli, S.; Brunt, K.M.; Csatho, B.; Harbeck, K.; Huth, A.; Neumann, T.; et al. Land ice height-retrieval algorithm for NASA's ICESat-2 photon-counting laser altimeter. *Remote Sens. Environ.* **2019**, *233*, 111352. [CrossRef]
41. Wang, C.; Zhu, X.; Nie, S.; Xi, X.; Li, D.; Zheng, W.; Chen, S. Ground elevation accuracy verification of ICESat-2 data: A case study in Alaska, USA. *Opt. Express* **2019**, *27*, 38168–38179. [CrossRef]
42. Parrish, C.E.; Magruder, L.A.; Neuenschwander, A.L.; Forfinski-Sarkozi, N.; Alonzo, M.; Jasinski, M. Validation of ICESat-2 ATLAS Bathymetry and Analysis of ATLAS's Bathymetric Mapping Performance. *Remote Sens.* **2019**, *11*, 1634. [CrossRef]
43. Hsu, H.-J.; Huang, C.-Y.; Jasinski, M.; Li, Y.; Gao, H.; Yamanokuchi, T.; Wang, C.-G.; Chang, T.-M.; Ren, H.; Kuo, C.-Y.; et al. A semi-empirical scheme for bathymetric mapping in shallow water by ICESat-2 and Sentinel-2: A case study in the South China Sea. *Isprs J. Photogramm. Remote Sens.* **2021**, *178*, 1–19. [CrossRef]
44. Chen, Y.; Le, Y.; Zhang, D.; Wang, Y.; Qiu, Z.; Wang, L. A photon-counting LiDAR bathymetric method based on adaptive variable ellipse filtering. *Remote Sens. Environ.* **2021**, *256*, 112326. [CrossRef]
45. Xie, C.; Chen, P.; Pan, D.; Zhong, C.; Zhang, Z. Improved Filtering of ICESat-2 Lidar Data for Nearshore Bathymetry Estimation Using Sentinel-2 Imagery. *Remote Sens.* **2021**, *13*, 4303. [CrossRef]

46. Peng, K.; Xie, H.; Xu, Q.; Huang, P.; Liu, Z. A Physics-Assisted Convolutional Neural Network for Bathymetric Mapping Using ICESat-2 and Sentinel-2 Data. *IEEE Trans. Geosci. Remote Sens.* **2022**, *60*, 3213248. [CrossRef]
47. Guo, X.; Jin, X.; Jin, S. Shallow Water Bathymetry Mapping from ICESat-2 and Sentinel-2 Based on BP Neural Network Model. *Water* **2022**, *14*, 3862. [CrossRef]
48. van der Woerd, H.J.; Wernand, M.R. Hue-Angle Product for Low to Medium Spatial Resolution Optical Satellite Sensors. *Remote Sens.* **2018**, *10*, 180. [CrossRef]
49. Fronkova, L.; Greenwood, N.; Martinez, R.; Graham, J.A.; Harrod, R.; Graves, C.A.; Devlin, M.J.; Petus, C. Can Forel-Ule Index Act as a Proxy of Water Quality in Temperate Waters? Application of Plume Mapping in Liverpool Bay, UK. *Remote Sens.* **2022**, *14*, 2375. [CrossRef]
50. Nie, Y.F.; Guo, J.T.; Sun, B.N.; Lv, X.Q. An evaluation of apparent color of seawater based on the in-situ and satellite-derived Forel-Ule color scale. *Estuar. Coast. Shelf Sci.* **2020**, *246*, 107032. [CrossRef]
51. Zhang, J.; Tian, J.; Li, X.; Wang, L.; Chen, B.; Gong, H.; Ni, R.; Zhou, B.; Yang, C. Leaf area index retrieval with ICESat-2 photon counting LiDAR. *Int. J. Appl. Earth Obs. Geoinf.* **2021**, *103*, 102488. [CrossRef]
52. Xing, Y.; Huang, J.; Gruen, A.; Qin, L. Assessing the Performance of ICESat-2/ATLAS Multi-Channel Photon Data for Estimating Ground Topography in Forested Terrain. *Remote Sens.* **2020**, *12*, 2084. [CrossRef]
53. Xiang, J.; Li, H.; Zhao, J.; Cai, X.; Li, P. Inland water level measurement from spaceborne laser altimetry: Validation and comparison of three missions over the Great Lakes and lower Mississippi River. *J. Hydrol.* **2021**, *597*, 126312. [CrossRef]
54. Magruder, L.; Neumann, T.; Kurtz, N. ICESat-2 Early Mission Synopsis and Observatory Performance. *Earth Space Sci.* **2021**, *8*, e2020EA001555. [CrossRef] [PubMed]
55. Liu, X.; Su, Y.; Hu, T.; Yang, Q.; Liu, B.; Deng, Y.; Tang, H.; Tang, Z.; Fang, J.; Guo, Q. Neural network guided interpolation for mapping canopy height of China's forests by integrating GEDI and ICESat-2 data. *Remote Sens. Environ.* **2022**, *269*, 112844. [CrossRef]
56. Li, Y.; Gao, H.; Zhao, G.; Tseng, K.-H. A high-resolution bathymetry dataset for global reservoirs using multi-source satellite imagery and altimetry. *Remote Sens. Environ.* **2020**, *244*, 111831. [CrossRef]
57. Li, Y.; Gao, H.; Jasinski, M.F.; Zhang, S.; Stoll, J.D. Deriving High-Resolution Reservoir Bathymetry From ICESat-2 Prototype Photon-Counting Lidar and Landsat Imagery. *Ieee Trans. Geosci. Remote Sens.* **2019**, *57*, 7883–7893. [CrossRef]
58. Gwenzi, D.; Lefsky, M.A.; Suchdeo, V.P.; Harding, D.J. Prospects of the ICESat-2 laser altimetry mission for savanna ecosystem structural studies based on airborne simulation data. *Isprs J. Photogramm. Remote Sens.* **2016**, *118*, 68–82. [CrossRef]
59. Drusch, M.; Del Bello, U.; Carlier, S.; Colin, O.; Fernandez, V.; Gascon, F.; Hoersch, B.; Isola, C.; Laberinti, P.; Martimort, P.; et al. Sentinel-2: ESA's Optical High-Resolution Mission for GMES Operational Services. *Remote Sens. Environ.* **2012**, *120*, 25–36. [CrossRef]
60. Markel, J. Shallow Water Bathymetry with ICESat-2 (Tutorial Led by Jonathan Markel at the 2023 ICESat-2 Hackweek). Available online: https://icesat-2-2023.hackweek.io/tutorials/bathymetry/bathymetry_tutorial.html (accessed on 24 February 2024).
61. Sutterley, T. Python Interpretation of the NASA Goddard Space Flight Center YAPC ("Yet Another Photon Classifier") Algorithm. Available online: https://yapc.readthedocs.io/en/latest/ (accessed on 24 February 2024).
62. Xu, X.; Xu, S.; Jin, L.; Song, E. Characteristic analysis of Otsu threshold and its applications. *Pattern Recognit. Lett.* **2011**, *32*, 956–961. [CrossRef]
63. Bernardis, M.; Nardini, R.; Apicella, L.; Demarte, M.; Guideri, M.; Federici, B.; Quarati, A.; De Martino, M. Use of ICEsat-2 and Sentinel-2 Open Data for the Derivation of Bathymetry in Shallow Waters: Case Studies in Sardinia and in the Venice Lagoon. *Remote Sens.* **2023**, *15*, 2944. [CrossRef]
64. Harmel, T.; Chami, M.; Tormos, T.; Reynaud, N.; Danis, P.-A. Sunglint correction of the Multi-Spectral Instrument (MSI)-SENTINEL-2 imagery over inland and sea waters from SWIR bands. *Remote Sens. Environ.* **2018**, *204*, 308–321. [CrossRef]
65. He, C.L.; Jiang, Q.G.; Tao, G.F.; Zhang, Z.C. A Convolutional Neural Network with Spatial Location Integration for Nearshore Water Depth Inversion. *Sensors* **2023**, *23*, 8493. [CrossRef]
66. Breiman, L. Random forests. *Mach. Learn.* **2001**, *45*, 5–32. [CrossRef]
67. Friedman, J.H. Greedy function approximation: A gradient boosting machine. *Ann. Stat.* **2001**, *29*, 1189–1232. [CrossRef]
68. Chen, T.Q.; Guestrin, C.; Assoc Comp, M. XGBoost: A Scalable Tree Boosting System. In Proceedings of the 22nd ACM SIGKDD International Conference on Knowledge Discovery and Data Mining (KDD), San Francisco, CA, USA, 13–17 August 2016; pp. 785–794.
69. Xu, N.; Wang, L.; Zhang, H.-S.; Tang, S.; Mo, F.; Ma, X. Machine Learning Based Estimation of Coastal Bathymetry From ICESat-2 and Sentinel-2 Data. *Ieee J. Sel. Top. Appl. Earth Obs. Remote Sens.* **2024**, *17*, 1748–1755. [CrossRef]

remote sensing

Article

Detecting Ocean Eddies with a Lightweight and Efficient Convolutional Network

Haochen Sun, Hongping Li *, Ming Xu, Tianyu Xia and Hao Yu

College of Marine Technology, Faculty of Information Science and Engineering, Ocean University of China, Qingdao 266100, China; sunhaochen@stu.ouc.edu.cn (H.S.); xuming@stu.ouc.edu.cn (M.X.); xiatianyu@stu.ouc.edu.cn (T.X.); yuhao1004@stu.ouc.edu.cn (H.Y.)
* Correspondence: lhp@ouc.edu.cn

Abstract: As a ubiquitous mesoscale phenomenon, ocean eddies significantly impact ocean energy and mass exchange. Detecting these eddies accurately and efficiently has become a research focus in ocean remote sensing. Many traditional detection methods, rooted in physical principles, often encounter challenges in practical applications due to their complex parameter settings, while effective, deep learning models can be limited by the high computational demands of their extensive parameters. Therefore, this paper proposes a new approach to eddy detection based on the altimeter data, the Ghost Attention Deeplab Network (GAD-Net), which is a lightweight and efficient semantic segmentation model designed to address these issues. The encoder of GAD-Net consists of a lightweight ECA+GhostNet and an Atrous Spatial Pyramid Pooling (ASPP) module. And the decoder integrates an Efficient Attention Network (EAN) module and an Efficient Ghost Feature Integration (EGFI) module. Experimental results show that GAD-Net outperforms other models in evaluation indices, with a lighter model size and lower computational complexity. It also outperforms other segmentation models in actual detection results in different sea areas. Furthermore, GAD-Net achieves detection results comparable to the Py-Eddy-Tracker (PET) method with a smaller eddy radius and a faster detection speed. The model and the constructed eddy dataset are publicly available.

Keywords: eddy detection; deep learning; semantic segmentation; lightweight; ghost attention deeplab network (GAD-Net)

Citation: Sun, H.; Li, H.; Xu, M.; Xia, T.; Yu, H. Detecting Ocean Eddies with a Lightweight and Efficient Convolutional Network. *Remote Sens.* **2024**, *16*, 4808. https://doi.org/10.3390/rs16244808

Academic Editors: Xiao-Hai Yan, Hua Su and Wenfang Lu

Received: 5 November 2024
Revised: 20 December 2024
Accepted: 21 December 2024
Published: 23 December 2024

1. Introduction

Ocean eddies, a significant component of the oceanic mesoscale phenomenon, are characterized by irregular egg-shaped contours in the ocean [1]. The eddies have a large spatial extent, ranging from tens of kilometers to hundreds of kilometers, and a long lifetime, ranging from a few days to a few years [2]. Based on the rotational direction, ocean eddies can be categorized as cyclonic eddies and anticyclonic eddies [3,4]. Cyclonic eddies are distinguished by their outward water displacement from the central eddy, which promotes the ascent of colder, deeper waters to the surface. This phenomenon results in lower temperatures in the eddy's center than in the surrounding waters. In contrast, the anticyclonic eddies flow inward, drawing warmer surface water to the depths, thus keeping the temperature at the eddy's center higher than the surrounding waters. These vertical mixing and diffusion movements from the surface to the depths influence energy exchange, mass transfer, and climate change [5–9]. Consequently, how to detect ocean eddies accurately and efficiently has become a hot topic in ocean remote sensing. Satellite remote sensing is characterized by all-weather monitoring, large-area coverage, and time-continuous acquisition, so this technique has become one of the preferred approaches to ocean eddy detection [10,11].

Over the past decades, ocean eddy detection has mainly relied on traditional physical principles. These traditional methods are the mainstay of ocean eddy detection and consist of four main principles: temperature anomaly detection, material motion tracking, rotating flow field analysis, and closed topology checking [12]. Temperature anomaly detection leverages sea surface temperature data as a foundation for detection [13]. It operates by calculating the gradient of satellite temperature data or sea surface temperature imagery, thus accurately recognizing the classification, size, and intensity of eddies [14,15]. Material tracking is a technique for eddy detection that utilizes the rotation and migration patterns of matter such as chlorophyll and plankton [16]. This method utilizes the natural movement of ocean constituents to track and detect oceanic eddies, which provides great help in marine ecological studies. The basic principle of the rotating flow fields method is to distinguish the eddy center and boundary based on the geometric characteristics of seawater flow velocity [17,18]. The closed topography method determines the boundaries of ocean eddies by analyzing closed profiles in altimeter data [19]. This technique is highly regarded for its accuracy and efficiency in eddy detection and has been widely used [20,21].

With the emergence of artificial intelligence, this technology has gradually entered into ocean remote sensing [22], and scholars have begun to use deep learning models in detecting ocean eddies. Lguensat et al. [23] proposed EddyNet by adopting the model structure of UNet [24]. This innovative model allows for pixel-level eddy species classification based on sea surface height (SSH) data. Similarly, Sun et al. [25] introduced a novel convolutional neural network using sea surface height (SSH) data. The model structure is very similar to Deeplabv3+ [26] and demonstrates strong detection capabilities in practical applications. After that, Xu et al. [27] used PSPNet [28], Deeplabv3+, and BiSeNet [29] for ocean eddy detection, respectively, and compared the number, size, and lifetime of these eddies. From the results, PSPNet detected more ocean eddies and BiSeNet detected eddies with larger sizes. Santana et al. [30] upgraded their REDN1 model [31] to its successor model, REDN2, both of which are based on the UNet architecture. Compared to the RDEN1 model, authors optimized the model's structure and reconstructed the residual module. These improvements give the REDN2 stronger detection capabilities and better results in different areas. The following year, Saida et al. [32] proposed a new model based on eddy features. In this model, the authors proposed an attention mechanism module, a feature enhancement module, and a serial Atrous Spatial Pyramid module. These modules greatly enhance the eddy feature extraction. In the same year, Hammoud et al. [33] used ResNet50-FCN [34] and UNet-PPM [24,28] for eddy detection in the Arabian Sea and investigated the various factors affecting the model training results.

Currently, deep learning models designed for eddy detection often have excessive parameters and complex computational processes. For example, the deep framework model [25] has a high parameter size of 157.36 MB and its Giga Floating-point Operations Per Second (GFLOPS) is 87.93 G, while the automatic eddy detection model [35] has a parameter size of 90.14 MB, corresponding to the GFLOPS of 61.86 G. Due to the large number of parameters, these models consume a large amount of computational resources and require more advanced computer hardware support. Consequently, this situation, to a certain extent, impacts the practical application for everyday users. Moreover, the more complex the computational process, the longer the detection time required, which also reduces the detection efficiency of the model. In these regards, we propose a lightweight deep learning model, the Ghost Attention Deeplab Network (GAD-Net), for eddy detection based on sea level anomaly (SLA) data. This model provides a more concise and efficient way for eddy detection, greatly reducing the actual cost for users. GAD-Net uses the encoder–decoder framework, which can efficiently extract eddy features by concatenating low-level and high-level features. The encoder consists of an ECA+GhostNet network and an Atrous Spatial Pyramid Pooling (ASPP) module. Among them, ECA+GhostNet is a lightweight backbone network mainly used for eddy feature extraction, while the ASPP module is used for eddy feature enhancement. Considering the inadequacy of lightweight backbone networks for feature extraction, GAD-Net proposes two new modules in the decoder: Efficient Attention Network (EAN) module and Efficient Ghost Feature Integration (EGFI) module. These designed modules increase the depth of the network, which, in turn, enhances the eddy feature extraction and integration. Moreover, to make the training process more balanced, the model uses the Dice loss [36] and the Focal loss [37] as the loss functions. The main work of our study can be summarized as follows:

- We built an ocean eddy dataset, which was constructed from manually labeled SLA data. This dataset covers the ocean area from 10° to 30°N latitude and 120° to 150°E longitude, providing an eddy data spanning the years 2017 to 2020.
- We proposed a lightweight eddy detection model, GAD-Net, which uses an encoder–decoder model architecture. The encoder consists of the ECA+GhostNet network and ASPP module. The decoder mainly comprises the EAN, EGFI, and upsampling modules.
- We compared the detection results of GAD-Net with other deep learning models. The results show that GAD-Net exhibits better advantages in model lightness, accuracy, and efficiency. And we compared the detection results of GAD-Net with the Py-Eddy-Tracker (PET) [21] method. The results show that both methods have a similar eddy detection performance, but GAD-Net detects smaller eddy radius and detects faster.

2. Methodology

2.1. Overall Model Structure

Following the encoder–decoder model framework, this paper proposes GAD-Net to implement accurate eddy detection for SLA data. The encoder is mainly responsible for downsampling and eddy feature extraction and is mainly composed of an ECA+GhostNet network and an ASPP. The decoder focuses on upsampling and integrating eddy features, which is primarily achieved by the EAN module, EGFI module, and upsampling module. The complete model structure of GAD-Net is shown in Figure 1. This model can detect ocean eddies in L4-level SLA data from altimetry satellites.

In the encoder phase, GAD-Net first receives dataset images, which are input into the ECA+GhostNet for eddy feature extraction. Next, ECA+GhostNet outputs two layers with different eddy feature layers: shapes of (40, 40, 112) and (20, 20, 160). In this case, the feature layer of (40, 40, 112) represents the low-level eddy feature information and is transmitted to the decoder. The feature layer of (20, 20, 160) is the high-level eddy feature layer and is transmitted to the ASPP module. And to enhance the characterization of the small-size eddies, we set the ASPP to small dilation rates of 2, 3, and 4. After ASPP feature enhancement, the encoder performs feature integration and channel tuning through a 1×1 convolution and passes the feature layer to the decoder. At this point, the encoder's work is complete.

The decoder receives the high-level eddy feature layer from the encoder and uses an upsampling module to scale up the feature size for concatenation with the low-level feature layer. When the low-level feature layer is input to the decoder, the model uses a 1×1 convolution for channel tuning to ensure that it can be fused with the high-level feature layer. Then, these two eddy feature layers are input into the EAN module to calculate their importance weights. After the calculation, the decoder concatenates different feature layers according to these weights. And these concatenated features are input into the EGFI module for further enhancement. Finally, the decoder performs the final channel tuning using a 1×1 convolution and upsampling module.

In addition, to make the training process more stable and balanced, GAD-Net uses Dice loss combined with Focal loss as the loss function.

Overall, the model takes SLA data as input and employs a convolutional neural network (CNN) to extract multi-level and multi-scale data features. Subsequently, the model learns from the annotated label information based on these data features, including the labeled positions, shapes, and directions of eddies. Finally, through continuous learning and correction via the loss function, the model achieves ocean eddy detection in SLA data.

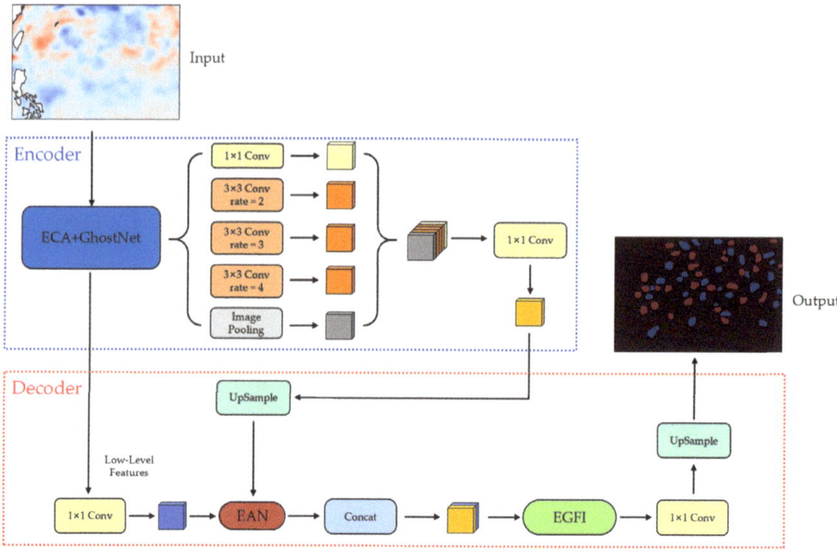

Figure 1. GAD-Net's overall structure.

2.2. ECA+GhostNet

When the model extracts image feature extraction, the traditional convolution produces much redundant feature information, which greatly reduces the efficiency of the model operation. GhostNet [38] provides an efficient and lightweight solution that effectively alleviates this problem by using the Ghost module. Specifically, the Ghost module first builds a base feature layer using a 1×1 convolution. Next, it transforms the base feature layer into a linear feature layer by linear transformation. Then, it generates a convolutional feature layer using multiple depth-wise convolutions for the base feature layer. Finally, the two generated feature layers are concatenated and output. Its structure is shown in Figure 2.

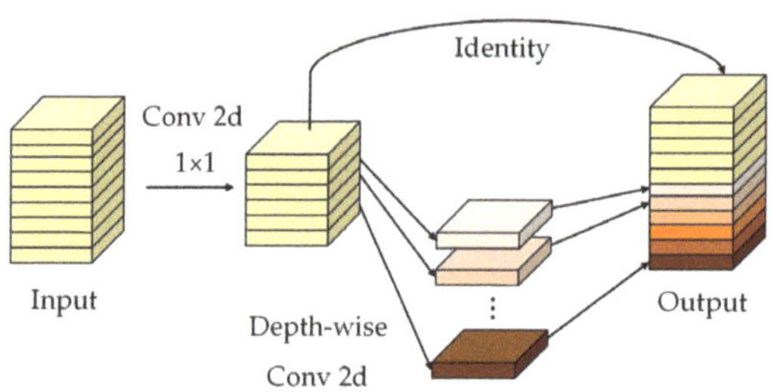

Figure 2. Ghost module's structure.

The Ghost bottleneck, as the basic module for feature extraction, is formed by stacking two Ghost modules. And, for different downsampling steps, the Ghost bottleneck is realized by adding a depth-wise convolution, as shown in Figure 3. Moreover, to improve the computational efficiency, the Ghost bottleneck add the residual block [39] are used.

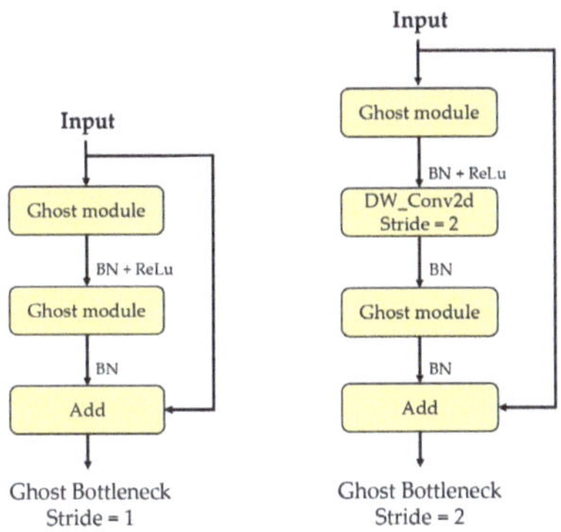

Figure 3. Ghost bottleneck with different step sizes.

GhostNet utilizes the attention mechanism to enhance its feature extraction. The initial Squeeze-Excitation Network (SENet) [40] had more parameters that reduce the efficiency. For this reason, we choose the Efficient Channel Attention Plus network (ECA+Net) [41] as an alternative, which is an upgraded version of the Efficient Channel Attention network (ECA-Net) [42] and provides a more comprehensive channel attention. The structure of ECA+Net is shown in Figure 4.

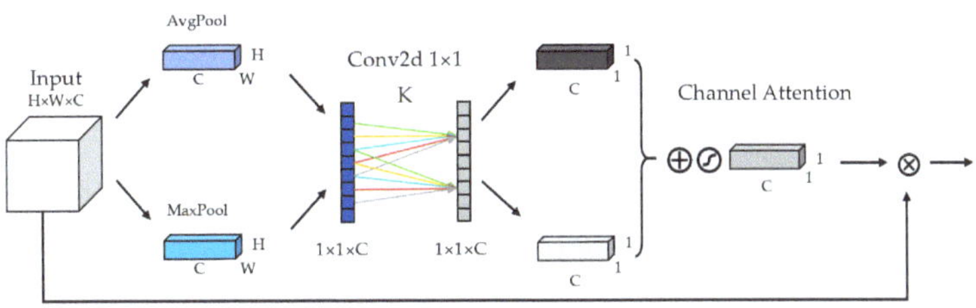

Figure 4. ECA+Net's structure.

As shown above, ECA+Net first performs global average pooling and global maximum pooling respectively. Second, it uses a one-dimensional convolution with a convolution kernel of K to learn the channel information. This design greatly improves the processing efficiency and running speed. And the convolution kernel K is the same as ECA-Net with an adaptive size. After that, ECA+Net sums up the learning results and uses a sigmoid function for importance weight calculation. Finally, these importance weights are multiplied with the input feature layer. The adaptive size convolutional kernel K and sigmoid function are calculated as follows:

$$K = \left| \frac{\log_2(C)}{\gamma} + \frac{b}{\gamma} \right| odd \tag{1}$$

$$Sigmoid(x) = \frac{1}{1 + e^{-x}} \tag{2}$$

where C is the input channel number, $|x|odd$ is the odd number closest to x, and b and γ are the relationship coefficients. Referring to the ECA-Net structure, we set b to 1 and γ to 2.

The composition of ECA+GhostNet is shown in Table 1.

Table 1. Composition of ECA+GhostNet.

	#exp [1]	#out [2]	ECA+Net	Stride
Conv 2d 3 × 3	-	16	-	2
G-bneck [3]	16	16	-	1
G-bneck	48	24	-	2
G-bneck	72	24	-	1
G-bneck	72	40	1	2
G-bneck	120	40	1	1
G-bneck	240	80	-	2
G-bneck	200	80	-	1
G-bneck	200	80	-	1
G-bneck	184	80	-	1
G-bneck	184	80	-	1
G-bneck	480	112	1	1
G-bneck	672	112	1	1
G-bneck	672	160	1	2
G-bneck	960	160	-	1
G-bneck	960	160	1	1
G-bneck	960	160	-	1
G-bneck	960	160	1	1

[1] "#exp" is the expansion channel size. [2] "#out" is the output channel size. [3] "G-bneck" is the Ghost bottleneck.

2.3. EAN Module

For more effective eddy feature fusion of different levels, this paper proposes a lightweight and efficient attention mechanism, the EAN module, as shown in Figure 5. This attention mechanism can calculate attention in channel and spatial dimensions, realizing more comprehensive eddy feature fusion. Therefore, the EAN module is mainly composed of two parts: the channel feature extraction part and the spatial feature extraction part.

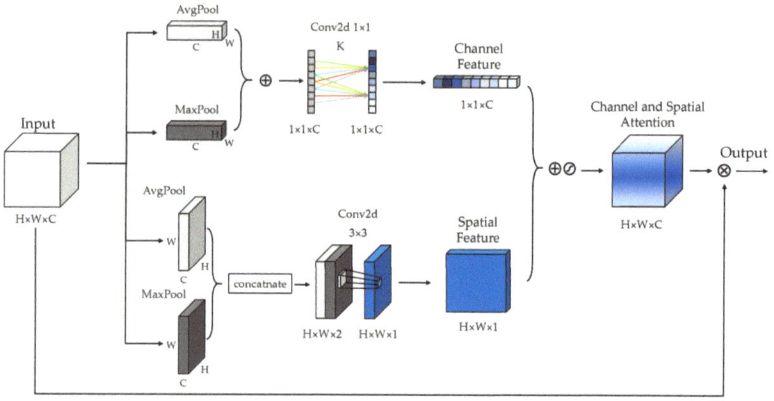

Figure 5. EAN module's structure.

In the channel feature extraction part, the EAN module uses one-dimensional convolution to learn channel features. Specifically, the EAN module first performs average channel pooling and maximum channel pooling on the input feature layer and sums the pooling results. Then, the EAN module uses one-dimensional convolution to learn the cross-channel feature information. The convolution kernel of this one-dimensional convolution is the same as that of ECA-Net, which is also an adaptive-size. After that, the EAN module outputs the learned channel features and fuses them with the spatial features.

In the spatial feature extraction part, the EAN module references the Convolutional Block Attention Module (CBAM) [43], which also uses convolution to extract spatial features. The EAN module's spatial feature extraction is performed as follows: first, perform spatial average pooling and spatial maximum pooling on the input feature layer, and concatenate these pooled results. Then, extract the spatial features through a regular 3 × 3 convolution. Finally, fuse the spatial feature layer obtained by convolution with the channel features.

The EAN module employs the feature fusion approach similar to the Bottleneck Attention Module (BAM) [44]. It sums the channel features and spatial features in three dimensions to effectively fuse the different kinds of feature information. This computational process not only improves the extraction of channel features, but also improves the understanding of spatial information.

2.4. EGFI Module

The original model's decoder design was relatively simple compared to the encoder, leading to suboptimal performance in fusing eddy features. Consequently, this study proposes a lightweight feature integration module, the EGFI module, as shown in Figure 6.

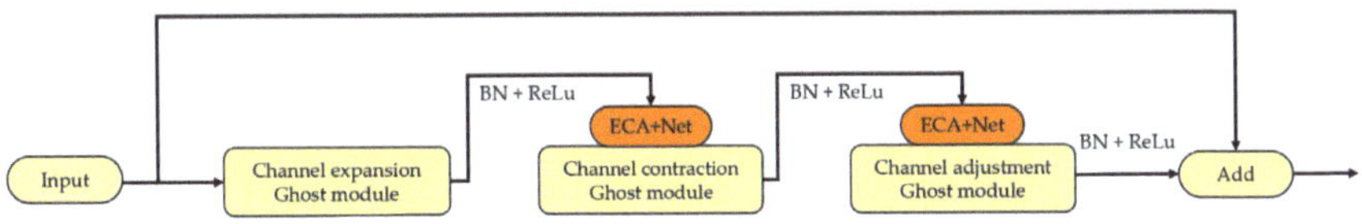

Figure 6. EGFI module's structure.

The EGFI module is mainly composed of three Ghost modules and two ECA+Nets. In this configuration, the Ghost modules are dedicated to eddy feature integration, while the ECA+Nets provide supplementary support. To enhance the fusion of eddy features, the EGFI module uses three Ghost modules with different functions. The first is the channel expansion Ghost module, which aims to increase the feature channel number to capture richer eddy information. The second is the channel contraction Ghost module, which realizes a more compact and diverse eddy feature extraction by reducing the feature channel numbers. The third is the channel adjustment Ghost module, which adjusts the feature channel numbers to the output channel numbers, ensuring that the model can be concatenated to the shortcut.

Specifically, the EGFI module first inputs the eddy feature layer into the channel expansion Ghost module for eddy feature integration. At this point, the feature channels are twice as many as the number of output channels. Then, the EGFI module continues the eddy feature integration using an ECA+Net and a channel contraction Ghost module. At this time, the number of feature channels is half the number of output channels. Finally, the EGFI module uses a channel adjustment Ghost module to match the feature channels to the output channels and concatenates this feature layer to the shortcut. By stretching and shrinking the feature channel number to integrate the eddy features, more comprehensive and diverse eddy feature information can be obtained, which, in turn, improves the model feature fusion performance.

2.5. Loss Function

For a more stable and balanced training process, GAD-Net uses Dice loss and Focal loss as the loss function. By using Focal loss, the training imbalance caused by the different numbers of different eddy annotations can be effectively mitigated. Using Dice loss can better compensate for the large difference in the number of pixel points between the eddy and background, making the training process more balanced. The loss function of GAD-Net is calculated as follows:

$$Loss = L_{Dice} + L_{Focal} \tag{3}$$

where L_{Dice} denotes the Dice loss and L_{Focal} denotes the Focal loss. The Dice loss solves the problem of an unbalanced sample distribution using the Dice coefficient, a function that measures the similarity of samples and is used to assess the degree of overlap between two samples. Its value is between 0 and 1 and is calculated as follows:

$$\text{Dice coefficient} = \frac{2|X \cap Y|}{|X| + |Y|} \tag{4}$$

where X is the ground truth, and Y is the predicted pixel. Therefore, the formula for Dice loss is as follows:

$$\text{Dice loss} = 1 - \frac{2|X \cap Y|}{|X| + |Y|} \tag{5}$$

The Focal loss balances the training process by increasing the weight of difficult-to-classify samples and reducing the weight of easy-to-classify samples. It reduces the sample imbalance of different eddies, and the calculation formula is as follows:

$$\text{Focal loss} = -\alpha_t (1 - p_t)^{\gamma} \log(p_t) \tag{6}$$

where α_t is the weight balancing factor that can control the contribution of positive and negative samples to the loss. And the $(1 - p_t)^{\gamma}$ is the modulating factor that can be used to control the difficulty of sample classification. When $\gamma = 0$, the Focal loss is the cross-entropy function, and when $\gamma > 0$, the loss function reduces the calculation of the easy-to-classify sample and increases the calculation of the hard-to-classify sample, which, in turn, makes the training process more balanced.

3. Datasets and Evaluation Index

3.1. Eddy Dataset

In this study, the 2017–2020 L4-level SLA data in the 10°–30°N and 120°–150°E ocean areas were selected as the dataset data, as shown in the black box in Figure 7. The data are published by the Copernicus Marine Environment Monitoring Service (CMEMS) (https://marine.copernicus.eu/, accessed on 5 November 2024) and encompass a variety of SLA information across different years and seasons. These data were obtained through optimal interpolation, which amalgamates L3-level along-track measurements from diverse altimeter missions. Additionally, a segment of this processing is specifically calibrated for the Global Ocean.

To highlight the eddy morphology features, we plotted the raw gridded data as color contour images. We used the Labelme software to accurately and manually label these 1461 SLA images, with cyclonic eddy labeled as "CE" and anticyclonic eddy labeled as "AE". Following this, experts reviewed and validated the annotated eddies in conjunction with the velocity data within the annotated areas. Any incorrectly annotated eddies were manually removed. The labeled labels contained information such as the position, shape, and direction of eddies, which can provide data support for deep learning models. After expert validation, the dataset was divided in the ratio of 8:1:1, resulting in 1169 training images, 146 validation images, and 146 testing images. The training images were used for model training. The validation images were used for model hyperparameter tuning. The testing images were used to judge the model's generalizability.

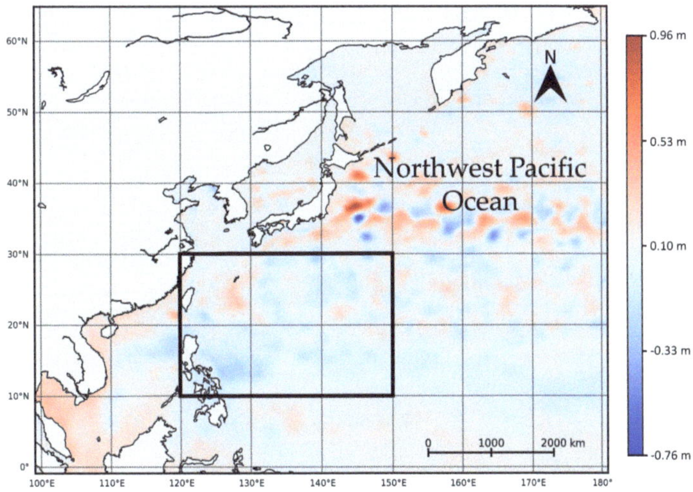

Figure 7. The dataset area is shown in a black box (10°N−30°N, 120°E−150°E).

3.2. Model Evaluation Index

In this study, we used Recall, Precision and Mean Intersection over Union (MIoU) as model evaluation indices. Recall is the ratio of correct pixel points detected by the model to all correct pixel points. Precision is the ratio of correct pixel points detected by the model to all pixel points detected by the model. And MIoU is the average of the intersection and concatenation ratios between the true and predicted values for each category, which is a key index for measuring the generalizability of the segmentation model. The specific calculation process for each index is as follows:

$$\text{Recall} = \frac{TP}{TP + FN} \tag{7}$$

$$\text{Precision} = \frac{TP}{TP + FP} \tag{8}$$

$$\text{MIoU} = \frac{1}{k+1} \sum_{i=0}^{k} \frac{TP}{FN + FP + TP} \tag{9}$$

where TP is the number of correct pixels detected as correct, FN is the number of incorrect pixels detected as incorrect, and FP is the number of incorrect pixels detected as correct.

In addition, this study evaluated the efficiency of model eddy detection by calculating the GFLOPS, which is an important measure of the model's computational complexity. In general, the larger the GFLOPS value, the higher the model complexity, and the lower the detection efficiency.

4. Experiment

4.1. Ablation Experiment

To assess the influence of each enhancement on model performance, we utilized Deeplabv3+ with GhostNet as our base model and performed a sequential ablation study. The model was initialized with a learning rate of 1×10^{-3} and a minimum learning rate of 1×10^{-5}. We used the Adamw [45] optimizer, which is widely recognized for its effectiveness in training deep neural networks. The momentum parameter was configured at 0.9 to enhance the stability of the optimization process, while the weight decay was set to 1×10^{-2} to prevent overfitting. To further refine the learning dynamics, a cosine annealing learning rate [46] was implemented. The training was conducted for 100 epochs to ensure thorough model convergence.

Given the relatively small size of the eddy in the image, we chose the larger input size of 640 × 640 pixels to capture more detailed features. Additionally, considering the important role of batch size in model training [47], we set the batch size to 8 to balance computational efficiency and model performance. Table 2 presents the ablation experiment results.

Table 2. Ablation experiment results.

Model	ECA+GhostNet	EAN	EGFI	MIoU (%)	Params Size (MB)	GFLOPS (G)
DeepLabv3+ (GhostNet)	-	-	-	74.50	20.35	7.68
DeepLabv3+ (GhostNet)	√	-	-	75.29	14.61	7.68
DeepLabv3+ (GhostNet)	-	√	-	74.70	20.35	7.68
DeepLabv3+ (GhostNet)	-	-	√	75.73	16.11	4.14
DeepLabv3+ (GhostNet)	-	√	√	76.01	16.11	4.14
DeepLabv3+ (GhostNet)	√	-	√	76.29	10.38	4.13
DeepLabv3+ (GhostNet)	√	√	-	75.65	14.61	7.68
DeepLabv3+ (GhostNet)	√	√	√	76.46	10.38	4.13

First, using ECA+GhostNet increased the MIoU by 0.79% and reduced the parameter size by 5.74 MB compared to the base model. This improvement resulted in a lighter model with better detection performance without affecting the segmentation efficiency. Using the EAN module independently improved the MIoU by 0.2% without increasing the parameter size. The addition of the EGFI module to the model resulted in a 1.23% increase in MIoU, a 4.24 MB reduction in parameter size, and a 3.54 G reduction in GFLOPS. Using the EGFI module greatly increased detection performance, reduces parameter size, and improves computational efficiency.

Second, the combination of the ECA+GhostNet with the EAN module resulted in a 1.15% increase in MIoU and a 5.74 MB reduction in parameter size. The combination of the EAN module with the EGFI module not only increased the MIoU by 1.51% but also reduced parameter size by 4.24 MB and GFLOPS by 3.54 G. Additionally, the incorporation of the EGFI module with the ECA+GhostNet led to a 1.79% enhancement in MIoU, a 9.97 MB decrease in parameter size, and a 3.55 G reduction in GFLOPS. In summary, these findings suggest that the pairing of any two modules can improve the model's performance while simplifying the model structure.

Compared to the base model, the MIoU increased by 1.96%, the parameter size reduced by 9.97 MB, and the GFLOPS decreased by 3.55 G.

To better evaluate the impact of each module on the model, we systematically calculated the confusion matrix for each model as each module was integrated into the model, as shown in Figure 8.

Incorporating ECA+GhostNet led to a comprehensive enhancement in model performance, reducing the eddy misdetection. Using the EAN module significantly decreased the background misidetected as eddies, thereby achieving more balanced results. Notably, the addition of the EGFI module greatly mitigated the situation where eddies were misidetected as background. Compared to the base model, GAD-Net demonstrated improvements in various aspects, including the detection accuracy, the misdetection of eddies as background, and the misdetection of background as eddies. Consequently, the eddy detection performance of the model was significantly enhanced.

Therefore, integrating ECA+GhostNet, the EAN module, and the EGFI module enhanced the eddy detection performance while concurrently simplifying the model's architecture and improving its detection efficiency.

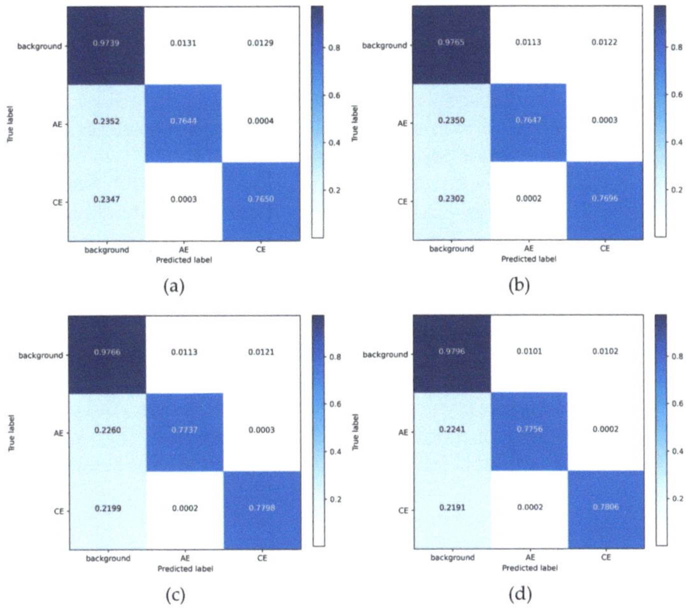

Figure 8. Confusion matrix calculations for the models. (**a**) Base model. (**b**) Base model integrating ECA+GhostNet. (**c**) Base model integrating ECA+GhostNet and EAN module. (**d**) GAD-Net.

In addition, spatial information loss, an issue that cannot be overlooked in deep learning models, prompted us to conduct ablation studies. In this experiment, we analyzed each step aimed at reducing spatial information loss, including the selection of the input size 640 × 640, the use of an ASPP module with small dilation rates, and the incorporation of the EGFI module. The results of this experiment are shown in Figure 9. In Figure 9, the red regions denote anticyclonic eddies, while the blue regions denote cyclonic eddies. The white box represents an eddy that is not detected relative to the ground truth, whereas the yellow box represents an eddy that is detected more than the ground truth.

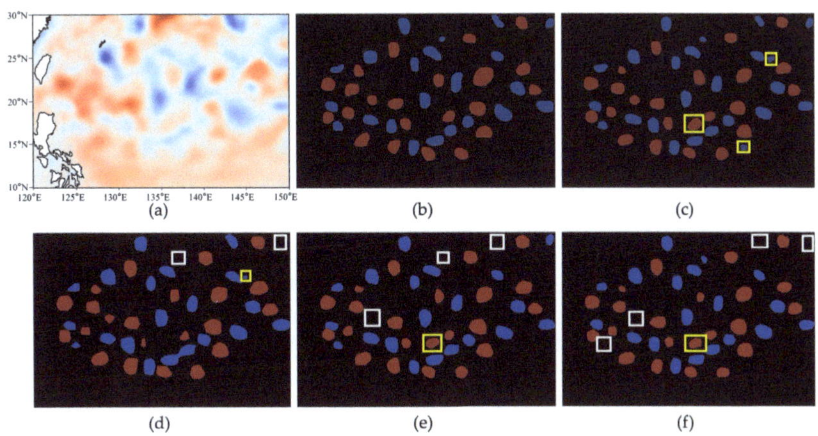

Figure 9. Results of ablation experiments with the spatial information loss. Red regions are anticyclonic eddies and blue regions are cyclonic eddies. White boxes are undetected eddies and yellow boxes are more detected eddies. (**a**) Input. (**b**) Ground truth. (**c**) GAD-Net with the input size of 640 × 640. (**d**) GAD-Net with the input size of 480 × 480. (**e**) GAD-Net without the ASPP module. (**f**) GAD-Net without the EGFI module.

Comparing detection results with different input sizes, it is observed that employing a large input size of 640 × 640 effectively mitigates spatial information loss in the model. Moreover, models with larger input sizes also exhibit smoother delineation of eddy boundaries. Subsequently, the models without the ASPP module and without the EGFI module both tend to miss the detection of small-size eddies, indicating a certain degree of spatial information loss. With the addition of these two modules, this problem is significantly improved; GAD-Net has no missed eddies and detects more eddies on the ground truth basis. Thus, GAD-Net has good spatial information extraction capability.

4.2. Comparison Experiment

To evaluate the eddy detection capabilities of GAD-Net, we conducted a comparison experiment. In this experiment, we trained a series of contemporary models and computed their performance indexes. In addition, the powerful ResNet and lightweight GhostNet were equipped with UNet, PSPNet, and Deeplabv3+ models, respectively, to emphasize the high performance and efficiency of the models. And the lightweight LR-ASPP [48], the high-performance HRNetv2 [49], and Segformer [50] were also included in the comparison experiment. Table 3 presents the results of this comparison experiment.

Table 3. Comparison experiment results.

Model	Recall (%)	Precision (%)	MIoU (%)	Params Size (MB)	GFLOPS (G)
UNet (ResNet)	83.24	86.56	74.71	167.31	226.61
UNet (GhostNet)	82.56	83.94	72.46	16.70	25.74
PSPNet (ResNet)	84.07	86.61	75.37	178.17	184.98
PSPNet (GhostNet)	83.77	84.02	73.40	10.07	4.46
DeepLabv3+ (ResNet)	84.63	87.07	76.09	154.53	176.78
DeepLabv3+ (GhostNet)	83.57	85.85	74.50	20.35	7.68
LR-ASPP	83.51	84.14	73.28	9.47	3.81
HRNetv2	85.07	86.02	75.69	37.56	58.33
Segformer	83.80	87.49	75.74	52.18	41.40
GAD-Net	85.40	86.77	76.46	10.38	4.13

In terms of Recall, GAD-Net reached the highest value of 85.40%. This is 0.33% higher than HRNetv2, 0.77% higher than Deeplabv3+ with ResNet, and 1.33% higher than PSPNet with ResNet. And the Recall of GAD-Net was also 1.89% higher than that of LR-ASPP. These comparisons show that the Recall of GAD-Net is significantly better than that of the high-performance and lightweight models. Therefore, compared with other models, GAD-Net can effectively alleviate missed eddy detection.

The Precision of GAD-Net was 86.77%, which was 0.72% lower than Segformer and 0.3% lower than Deeplabv3+ with ResNet. Notably, there was a 3.69% difference between Recall and Precision for Segfomer, and this imbalance may affect the model's performance. Nevertheless, GAD-Net maintained a good advantage in terms of Precision compared to other lightweight models. Specifically, GAD-Net's Precision was 2.63% higher than LR-ASPP, 0.92% higher than Deeplabv3+ with GhostNet, and 2.75% higher than PSPNet with GhostNet.

As for MIoU, GAD-Net was better than the other models. Specifically, the MIoU of GAD-Net exceeded that of Deeplabv3+ with ResNet by 0.37%, Segformer by 0.72%, and HRNetv2 by 0.77%. The advantage of GAD-Net over lightweight models was even more obvious, as its MIoU was 1.96% higher than that of Deeplabv3+ with GhostNet, 3.06% higher than that of PSPNet with GhostNet, and 3.18% higher than that of LR-ASPP. In conclusion, these comparisons show that GAD-Net has a strong comprehensive performance.

Regarding model complexity and computational efficiency, GAD-Net exhibited the characteristics of a lightweight model with a concise computational process. Although GAD-Net's parameter size and GFLOPS were slightly higher than those of LR-ASPP, its performance in eddy detection was considerably better. Moreover, compared with other high-performance models, the parameters and GFLOPS of GAD-Net were significantly reduced. These results demonstrate that GAD-Net effectively balances model complexity and detection performance.

Therefore, GAD-Net outperforms the other models in eddy detection performance, while also having a lighter model structure and a more efficient computational process. Figure 10 shows the eddy detection results of all models in the comparison experiment.

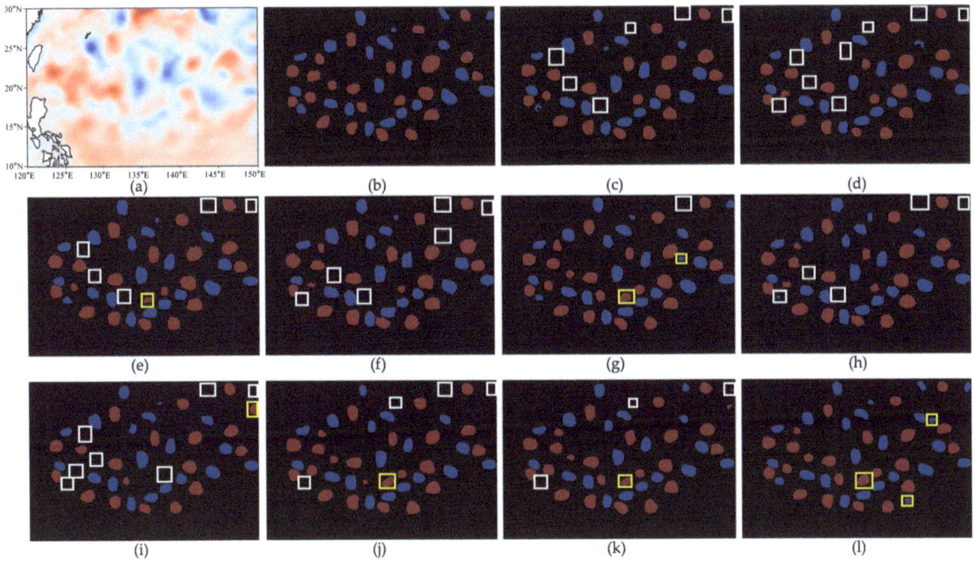

Figure 10. Eddy detection results of the comparison experiment. Red regions are anticyclonic eddies and blue regions are cyclonic eddies. White boxes are undetected eddies and yellow boxes are more detected eddies. (**a**) Input. (**b**) Ground truth. (**c**) UNet with ResNet. (**d**) UNet with GhostNet. (**e**) PSPNet with ResNet. (**f**) PSPNet with GhostNet. (**g**) Deeplabv3+ with ResNet. (**h**) Deeplabv3+ with GhostNet. (**i**) LR-ASPP. (**j**) HRNetv2. (**k**) Segformer. (**l**) GAD-Net.

As in Figure 9, in Figure 10, the red ones are anticyclonic eddies, the blue ones are cyclonic eddies, the white boxes are undetected eddies, and the yellow boxes are more detected eddies. We can see that GAD-Net has no white boxes and three yellow boxes, which indicates that the eddy detection of GAD-Net matches very well with the ground truth and can detect more eddies. Compared with other models, the actual eddy detection result of GAD-Net is better.

4.3. Detection Experiment

In this experiment, we aimed to assess the practicality of GAD-Net through eddy detection experiments. To achieve this, we selected four distinct model architectures for a comparative analysis: UNet with ResNet, DeepLabv3+ with ResNet, Segformer, and GAD-Net. By comparing the actual detection results of these models, we evaluated GAD-Net's detection capability.

Initially, we performed eddy detection on SLA images from different periods and seasons in our dataset, and the results are shown in Figure 11. Consistent with our comparison experiment, the white boxes in the figure represent the undetected eddies relative to the ground truth, while the yellow boxes indicate the extra detected eddies based on the ground truth. It is evident that GAD-Net has fewer white boxes compared to the other three models. This indicates a lower missed detection rate and better coincidence with the ground truth. Furthermore, GAD-Net has more yellow boxes than other models, indicating that it can detect more eddies based on the ground truth and performs better. And GAD-Net also exhibits a high accuracy and balance in eddy detection for different times and different sea conditions. Interestingly, the eddy boundaries determined by UNet and Segformer are smoother than the other models. This phenomenon may be related to the different structures between these models. Specifically, the symmetry structure of UNet and the multi-scale fusion module of Segformer may facilitate the smoothing of the eddy boundaries. But in general, GAD-Net still has better detection performance within the dataset.

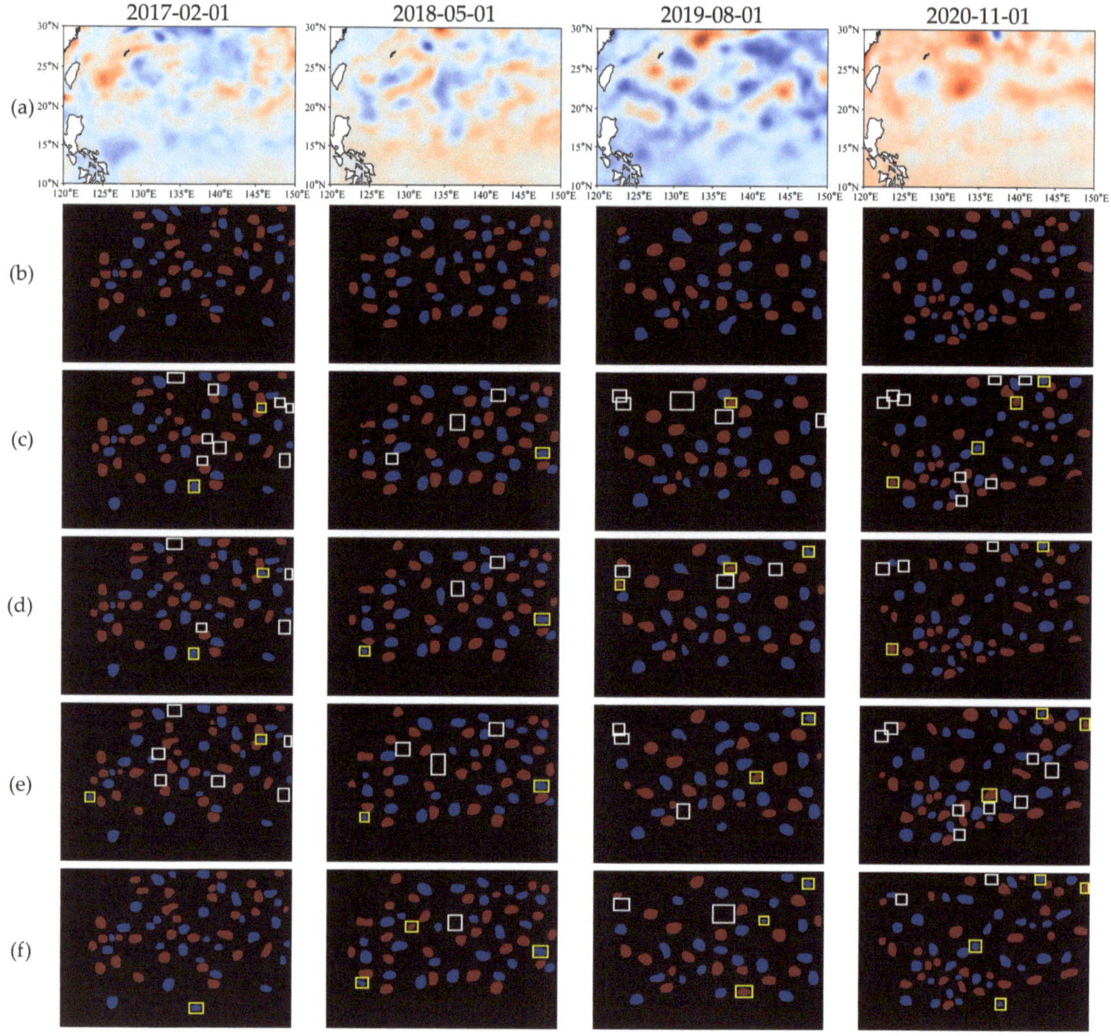

Figure 11. Eddy detection results within the dataset. Red regions are anticyclonic eddies and blue regions are cyclonic eddies. White boxes are undetected eddies and yellow boxes are more detected eddies. (**a**) Input. (**b**) Ground truth. (**c**) UNet with ResNet. (**d**) Deeplabv3+ with ResNet. (**e**) Segfomer. (**f**) GAD-Net.

To evaluate the performance of GAD-Net in detecting eddies on images of varying sparsity, we conducted eddy detection experiments on SLA images outside of the dataset. We assessed the eddy detection performance of the model based on the detection results from different sea areas and varying eddy sparsity levels. We performed eddy detection on SLA images in the Pacific Ocean (10°–30°N, 120°–150°W), Atlantic Ocean (20°–40°N,

10°–40°W), and Indian Ocean (10°–30°N, 50°–80°E) areas. Figure 12 shows the detection results for 1 January 2021.

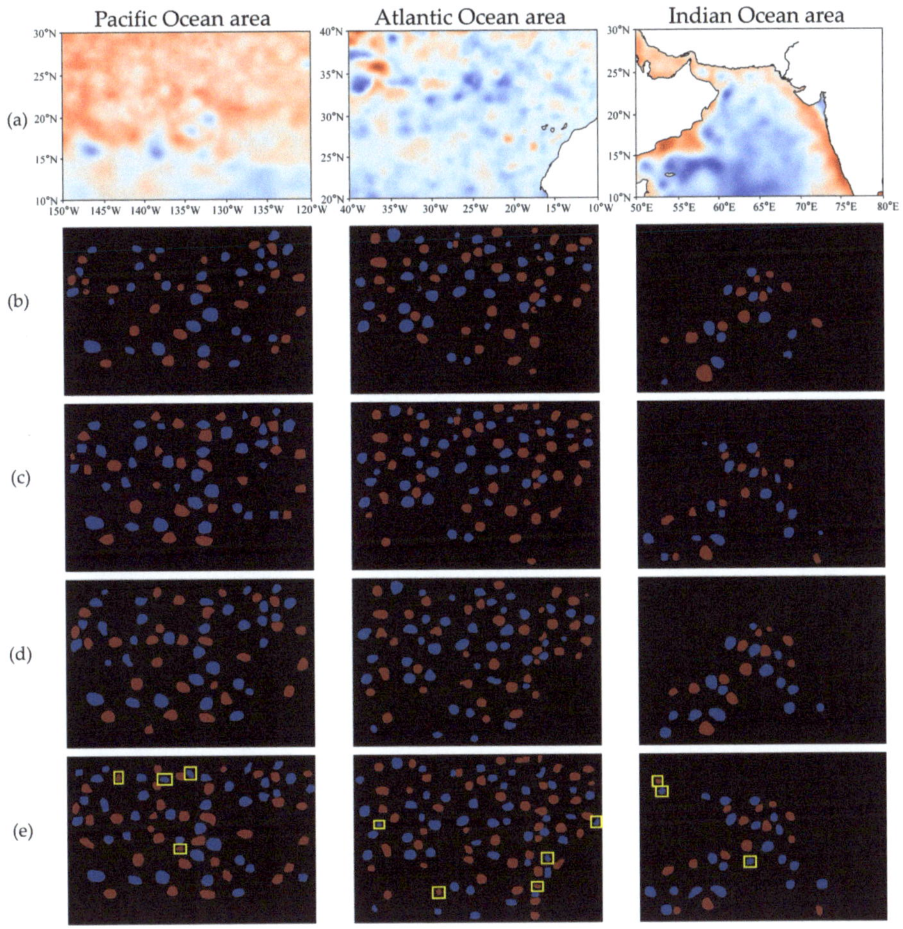

Figure 12. Eddy detection results outside the dataset. Red regions are anticyclonic eddies and blue regions are cyclonic eddies. Yellow boxes are more detected eddies. (**a**) Input. (**b**) UNet with ResNet. (**c**) Deeplabv3+ with ResNet. (**d**) Segfomer. (**e**) GAD-Net.

First and foremost, overall, GAD-Net has better eddy detection results, exhibiting good detection capabilities across images of varying sparsity. And the eddy boundaries detected by UNet and Segformer remain smoother. Compared to other models, GAD-Net is able to detect more small-size eddies, as shown in the yellow boxes in Figure 12. This indicates that GAD-Net has less spatial information loss and has good spatial feature extraction capability. Additionally, by observing the yellow boxes' positions in the Atlantic and Indian Ocean areas, it is found that GAD-Net's performance in detecting eddies around continents is superior to that of other models. It is noteworthy that the sparsity level around continental areas is significantly different from that of pure oceanic areas. Yet, GAD-Net still exhibits good detection performance, detecting more eddies than other models. Thus, GAD-Net possesses strong generalization capabilities, enabling it to detect ocean eddies in SLA data of varying sparsity levels.

Consequently, GAD-Net can reliably and accurately detect ocean eddies from SLA data under various sea conditions.

4.4. Validation Experiment

In this section, we conducted validation experiments to fully evaluate the eddy detection performance of GAD-Net. We performed eddy detection using the PET method [21] in areas outside the dataset and compared the results with those of GAD-Net. For areas where the two methods detected differently, we validated the method using manual discrimination of geostrophic flows. The PET method is a commonly used and advanced physical detection method for ocean eddies. This method employs SLA data and detects eddies by identifying closed sea surface height contour lines. Specifically, the PET method initially applies a Gaussian filter to the SLA data to remove large-scale noise and highlight the inherent features of eddies. Subsequently, the PET method scans the data to locate closed contour lines, which correspond to the boundaries of eddies. Furthermore, for each detected eddy, the PET method calculates a shape error to ensure that the detected features are consistent with those of eddies. The validation experiment results are shown in Figure 13.

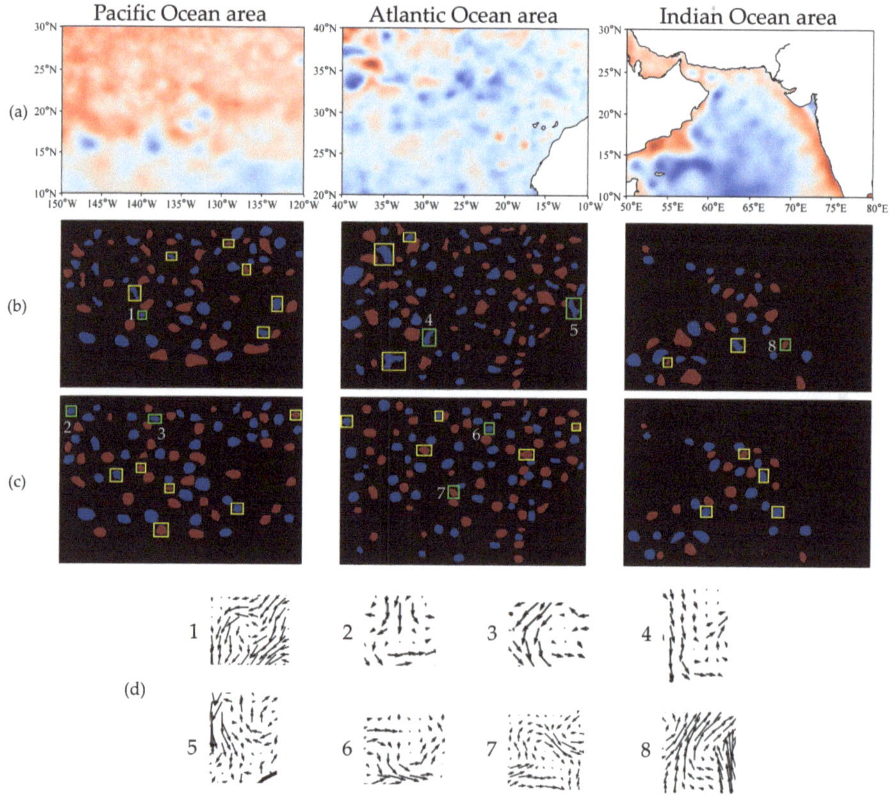

Figure 13. Validation experiment results. Red regions are anticyclonic eddies and blue regions are cyclonic eddies. Yellow boxes are more detected eddies, green boxes are incorrectly detected eddies, and the arrows indicate the regional geostrophic flow. (**a**) Input. (**b**) PET. (**c**) GAD-Net. (**d**) Regional geostrophic flow.

In Figure 13, the yellow boxes indicate more correct eddies detected than the other method, and the green boxes indicate incorrectly detected eddies. Comparing the detection results, it is observed that the eddies detected by GAD-Net exhibit smoother boundaries, and the eddies detected by PET are relatively larger in size. The count of yellow boxes for GAD-Net is slightly higher than that of the PET method, while the number of green boxes for both methods is equal. Overall, these observations indicate that GAD-Net and PET have comparable levels of performance in eddy detection, and both can detect ocean eddies stably.

Comparing the eddies in the yellow boxes, it is found that the eddies detected more by the PET method are mostly atypical elliptical eddies with larger sizes and small amplitude eddies. The lack of detection of atypical elliptical eddies by GAD-Net may be due to the fact that the manual labeling process is almost exclusively for typical elliptical eddies, leaving the model with a lack of data samples for atypical elliptical eddies. The reason for GAD-Net's missed detection of small-amplitude eddies is that the SLA data contours are not sufficiently plotted. These SLA images fail to plot closed contours that match the eddy features, resulting in the model's missed detection. In contrast, eddies detected more by GAD-Net are mostly elliptical eddies with small sizes. It can be seen that using the ASPP and EGFI modules improves the feature extraction capability of GAD-Net for small targets, giving the model better small-size eddy detection performance.

Comparing the incorrect eddies in the green boxes, it can be seen that the reason for these errors is similar for both methods, as sea surface anomalies that are not generated by eddies are detected as eddies. As shown in areas 1–8 in Figure 13, the complex sea conditions cause a strong opposite flow or other phenomena that create pseudo-eddies, producing eddy-like sea surface anomalies, which, in turn, lead to errors in eddy detection by both methods.

In addition, we applied the GAD-Net and PET methods to detect eddies in the dataset images from December 2020 and statistically counted the radii of the eddies, as shown in Figure 14. The comparison revealed that GAD-Net detected a higher proportion of eddies with a radius of less than 80 km than the ground truth and PET. Conversely, PET detected a higher proportion of eddies with a radius larger than 80 km. This observation, once again, demonstrates that GAD-Net exhibits good performance in detecting small-size eddies.

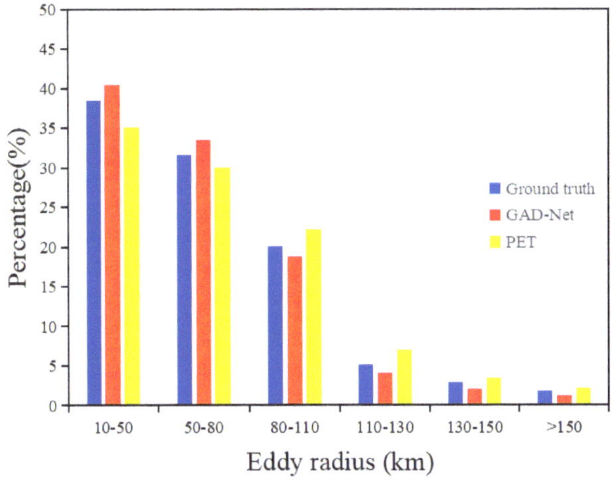

Figure 14. Distribution of eddy radius detected by GAD-Net and PET.

Regarding detection speed, GAD-Net detects one area in 0.04 s, which is much faster than the 2 s required by PET. Overall, GAD-Net's accuracy is comparable to that of the PET method, and it also has smaller eddy sizes with a faster detection speed.

5. Conclusions

This paper aims to detect ocean eddies in a lightweight and efficient way. To this end, we conducted in-depth research on constructing a lightweight and efficient deep learning model for ocean eddy detection. Initially, we established a novel ocean eddy dataset based on SLA data. This dataset encompasses ocean eddy information from 10°N to 30°N and 120°E to 150°E between 2017 and 2020. Subsequently, we developed a new lightweight eddy detection model, termed GAD-Net. For this model, we proposed and implemented ECA+GhostNet, the EAN module, and the EGFI module. The incorporation of these modules not only enhances the model's eddy detection performance, but also renders the model more lightweight and efficient. Thereafter, we employed GAD-Net to detect eddies in SLA images at different times and areas. Compared to other models, GAD-Net exhibits superior eddy detection capabilities, fewer model parameters, and more efficient computational processes. In addition, we compared GAD-Net with the traditional physical PET method. The results indicate that both methods have comparable eddy detection performance, although the eddy radius detected by GAD-Net is smaller, and the detection speed is faster. In general, the GAD-Net model is capable of detecting ocean eddies stably and efficiently, making it a lightweight and stable deep learning eddy detection model.

Although GAD-Net possesses accurate and efficient capabilities for eddy detection, the model is based on altimeter data for eddy detection. We did not conduct an in-depth exploration of the potential issues that may arise when detecting eddies using other types of data. Therefore, we will initiate research on ocean eddy detection using different datasets in the future. For instance, in L1-level SAR datasets, eddy samples are extremely scarce. We will investigate whether using our altimeter dataset as pre-trained weights can alleviate the issue of limited SAR data. Alternatively, we will consider what aspects need attention and improvement when constructing an eddy detection model specifically for L1-level SAR data. Additionally, we will explore whether we can develop a multimodal data eddy detection model to address the issue of pseudo-eddies in single altimeter data. These questions will be the focus of our future research.

Author Contributions: Conceptualization, H.S. and H.L.; methodology, H.S. and H.L.; software, H.S. and M.X.; validation, M.X. and T.X.; investigation, T.X. and H.Y.; data curation, T.X. and H.Y.; writing—original draft preparation, H.S.; writing—review and editing, M.X. and H.L. All authors have read and agreed to the published version of the manuscript.

Funding: This research was funded by the National Program on Global Change and Air–sea Interaction (Phase II)—Parameterization assessment for interactions of the ocean dynamic system.

Data Availability Statement: The model and the dataset are made public on https://github.com/Sun0532/GAD-Net/tree/master, accessed on 5 November 2024.

Acknowledgments: The authors would like to thank the Copernicus Marine Environment Monitoring Service for providing the free data.

Conflicts of Interest: The authors declare no conflicts of interest.

References

1. Chen, G.; Yang, J.; Han, G. Eddy morphology: Egg-like shape, overall spinning, and oceanographic implications. *Remote Sens. Environ.* **2021**, *257*, 112348. [CrossRef]
2. Chelton, D.B.; Schlax, M.G.; Samelson, R.M.; de Szoeke, R.A. Global observations of large oceanic eddies. *Geophys. Res. Lett.* **2007**, *34*, L15606. [CrossRef]
3. McWilliams, J.C. The nature and consequences of oceanic eddies. *Geophys. Monogr. Ser.* **2008**, *177*, 5–15.
4. Morrow, R.; Le Traon, P.-Y. Recent advances in observing mesoscale ocean dynamics with satellite altimetry. *Adv. Space Res.* **2012**, *50*, 1062–1076. [CrossRef]
5. Chelton, D.B.; Gaube, P.; Schlax, M.G.; Early, J.J.; Samelson, R.M. The influence of nonlinear mesoscale eddies on near-surface oceanic chlorophyll. *Science* **2011**, *334*, 328–332. [CrossRef] [PubMed]
6. Zhang, Z.; Wang, W.; Qiu, B. Oceanic mass transport by mesoscale eddies. *Science* **2014**, *345*, 322–324. [CrossRef] [PubMed]
7. Wunsch, C. Where do ocean eddy heat fluxes matter? *J. Geophys. Res. Ocean.* **1999**, *104*, 13235–13249. [CrossRef]
8. Roemmich, D.; Gilson, J. Eddy transport of heat and thermocline waters in the North Pacific: A key to interannual/decadal climate variability? *J. Phys. Oceanogr.* **2001**, *31*, 675–687. [CrossRef]
9. van Westen, R.; Dijkstra, H. Ocean eddies strongly affect global mean sea-level projections. *Sci. Adv.* **2021**, *7*, eabf1674. [CrossRef]
10. Faghmous, J.H.; Frenger, I.; Yao, Y.; Warmka, R.; Lindell, A.; Kumar, V. A daily global mesoscale ocean eddy dataset from satellite altimetry. *Sci. Data* **2015**, *2*, 1–16. [CrossRef] [PubMed]
11. Xu, M.E.; Li, H.; Yun, Y.; Yang, F.; Li, C. End-to-End Pixel-Wisely Detection of Oceanic Eddy on SAR Images With Stacked Attention Network. *IEEE J. Sel. Top. Appl. Earth Obs. Remote Sens.* **2023**, *16*, 9711–9724. [CrossRef]

12. Chen, G.; Yang, J.; Tian, F.; Chen, S.; Zhao, C.; Tang, J.; Liu, Y.; Wang, Y.; Yuan, Z.; He, Q. Remote sensing of oceanic eddies: Progresses and challenges. *Natl. Remote Sens. Bull* **2021**, *25*, 302–322. [CrossRef]
13. D'Alimonte, D. Detection of mesoscale eddy-related structures through iso-SST patterns. *IEEE Geosci. Remote Sens. Lett.* **2009**, *6*, 189–193. [CrossRef]
14. Dong, C.; Nencioli, F.; Liu, Y.; McWilliams, J.C. An automated approach to detect oceanic eddies from satellite remotely sensed sea surface temperature data. *IEEE Geosci. Remote Sens. Lett.* **2011**, *8*, 1055–1059. [CrossRef]
15. Karoui, I.; Chauris, H.; Garreau, P.; Craneguy, P. Multi-resolution eddy detection from ocean color and sea surface temperature images. In Proceedings of the OCEANS'10 IEEE, Sydney, NSW, Australia, 24–27 May 2010; pp. 1–6.
16. Xu, G.; Dong, C.; Liu, Y.; Gaube, P.; Yang, J. Chlorophyll rings around ocean eddies in the North Pacific. *IEEE Sci. Rep.* **2019**, *9*, 2056. [CrossRef] [PubMed]
17. Nencioli, F.; Dong, C.; Dickey, T.; Washburn, L.; McWilliams, J.C. A vector geometry–based eddy detection algorithm and its application to a high-resolution numerical model product and high-frequency radar surface velocities in the Southern California Bight. *J. Atmos. Ocean. Technol.* **2010**, *27*, 564–579. [CrossRef]
18. Chaigneau, A.; Gizolme, A.; Grados, C. Mesoscale eddies off Peru in altimeter records: Identification algorithms and eddy spatio-temporal patterns. *Prog. Oceanogr.* **2008**, *79*, 106–119. [CrossRef]
19. Chelton, D.B.; Schlax, M.G.; Samelson, R.M. Global observations of nonlinear mesoscale eddies. *Prog. Oceanogr.* **2011**, *91*, 167–216. [CrossRef]
20. Mason, E.; Pascual, A.; McWilliams, J.C. A new sea surface height–based code for oceanic mesoscale eddy tracking. *J. Atmos. Ocean. Technol.* **2014**, *31*, 1181–1188. [CrossRef]
21. Pegliasco, C.; Delepoulle, A.; Mason, E.; Morrow, R.; Faugère, Y.; Dibarboure, G. META3. 1exp: A new global mesoscale eddy trajectory atlas derived from altimetry. *Earth Syst. Sci. Data* **2022**, *14*, 1087–1107. [CrossRef]
22. Li, X.; Liu, B.; Zheng, G.; Ren, Y.; Zhang, S.; Liu, Y.; Gao, L.; Liu, Y.; Zhang, B.; Wang, F. Deep-learning-based information mining from ocean remote-sensing imagery. *Natl. Sci. Rev.* **2020**, *7*, 1584–1605. [CrossRef]
23. Lguensat, R.; Sun, M.; Fablet, R.; Tandeo, P.; Mason, E.; Chen, G. EddyNet: A deep neural network for pixel-wise classification of oceanic eddies. In Proceedings of the IGARSS 2018—2018 IEEE International Geoscience and Remote Sensing Symposium, Valencia, Spain, 22–27 July 2018; pp. 1764–1767.
24. Ronneberger, O.; Fischer, P.; Brox, T. U-net: Convolutional networks for biomedical image segmentation. In Proceedings of the Medical Image Computing and Computer-Assisted Intervention–MICCAI 2015: 18th International Conference, Munich, Germany, 5–9 October 2015; pp. 234–241.
25. Sun, X.; Zhang, M.; Dong, J.; Lguensat, R.; Yang, Y.; Lu, X. A deep framework for eddy detection and tracking from satellite sea surface height data. *IEEE Trans. Geosci. Remote Sens.* **2020**, *59*, 7224–7234. [CrossRef]
26. Chen, L.-C.; Zhu, Y.; Papandreou, G.; Schroff, F.; Adam, H. Encoder-decoder with atrous separable convolution for semantic image segmentation. In Proceedings of the European Conference on Computer Vision (ECCV), Munich, Germany, 8–14 September 2018; pp. 801–818.
27. Xu, G.; Xie, W.; Dong, C.; Gao, X. Application of three deep learning schemes into oceanic eddy detection. *Front. Mar. Sci.* **2021**, *8*, 672334. [CrossRef]
28. Zhao, H.; Shi, J.; Qi, X.; Wang, X.; Jia, J. Pyramid scene parsing network. In Proceedings of the IEEE Conference on Computer Vision and Pattern Recognition (CVPR), Hawaii, HI, USA, 21–26 July 2017; pp. 2881–2890.
29. Yu, C.; Wang, J.; Peng, C.; Gao, C.; Yu, G.; Sang, N. Bisenet: Bilateral segmentation network for real-time semantic segmentation. In Proceedings of the European Conference on Computer Vision (ECCV), Munich, Germany, 8–14 September 2018; pp. 325–341.
30. Santana, O.J.; Hernández-Sosa, D.; Smith, R.N. Oceanic mesoscale eddy detection and convolutional neural network complexity. *Int. J. Appl. Earth Obs. Geoinf.* **2022**, *113*, 102973. [CrossRef]
31. Santana, O.J.; Hernández-Sosa, D.; Martz, J.; Smith, R.N. Neural network training for the detection and classification of oceanic mesoscale eddies. *Remote Sens.* **2020**, *13*, 2625. [CrossRef]
32. Saida, S.J.; Sahoo, S.P.; Ari, S. Deep convolution neural network based semantic segmentation for ocean eddy detection. *Expert Syst. Appl.* **2023**, *219*, 119646. [CrossRef]
33. Hammoud, M.A.E.R.; Zhan, P.; Hakla, O.; Knio, O.; Hoteit, I. Semantic Segmentation of Mesoscale Eddies in the Arabian Sea: A Deep Learning Approach. *Remote Sens.* **2023**, *15*, 1525. [CrossRef]
34. Long, J.; Shelhamer, E.; Darrell, T.-A. Fully convolutional networks for semantic segmentation. In Proceedings of the IEEE Conference on Computer Vision and Pattern Recognition (CVPR), Boston, MA, USA, 7–12 June 2015; pp. 3431–3440.
35. Saida, S.J.; Ari, S. Automatic Detection of Ocean Eddy based on Deep Learning Technique with Attention Mechanism. In Proceedings of the National Conference on Communications (NCC), Mumbai, India, 24–27 May 2022; pp. 302–307.
36. Milletari, F.; Navab, N.; Ahmadi, S.-A. V-net: Fully convolutional neural networks for volumetric medical image segmentation. In Proceedings of the 2016 Fourth International Conference on 3D Vision (3DV), Stanford, CA, USA, 25–28 October 2016; pp. 565–571.
37. Lin, T.-Y.; Goyal, P.; Girshick, R.; He, K.; Dollár, P. Focal loss for dense object detection. In Proceedings of the IEEE International Conference on Computer Vision (ICCV), Venice, Italy, 22–29 October 2017; pp. 2980–2988.
38. Han, K.; Wang, Y.; Tian, Q.; Guo, J.; Xu, C.; Xu, C. Ghostnet: More features from cheap operations. In Proceedings of the IEEE Conference on Computer Vision and Pattern Recognition (CVPR), Seattle, WA, USA, 16–18 June 2020; pp. 1580–1589.

39. He, K.; Zhang, X.; Ren, S.; Sun, J. Deep residual learning for image recognition. In Proceedings of the IEEE Conference on Computer Vision and Pattern Recognition (CVPR), Las Vegas, NV, USA, 27–30 June 2016; pp. 770–778.
40. Hu, J.; Shen, L.; Sun, G. Squeeze-and-excitation networks. In Proceedings of the IEEE Conference on Computer Vision and Pattern Recognition (CVPR), Salt Lake City, UT, USA, 18–22 June 2018; pp. 7132–7141.
41. Sun, H.; Li, H.; Xu, M.; Yang, F.; Zhao, Q.; Li, C. A lightweight deep learning model for ocean eddy detection. *Front. Mar. Sci.* **2023**, *10*, 1266452. [CrossRef]
42. Wang, Q.; Wu, B.; Zhu, P.; Li, P.; Zuo, W.; Hu, Q. ECA-Net: Efficient channel attention for deep convolutional neural networks. In Proceedings of the IEEE Conference on Computer Vision and Pattern Recognition (CVPR), Seattle, WA, USA, 16–18 June 2020; pp. 11534–11542.
43. Woo, S.; Park, J.; Lee, J.-Y.; Kweon, I.S. Cbam: Convolutional block attention module. In Proceedings of the European Conference on Computer Vision (ECCV), Munich, Germany, 8–14 September 2018; pp. 3–19.
44. Park, J.; Woo, S.; Lee, J.-Y.; Kweon, I.S. Bam: Bottleneck attention module. *arXiv* **2018**, arXiv:1807.06514.
45. Loshchilov, I.; Hutter, F. Decoupled weight decay regularization. *arXiv* **2017**, arXiv:1711.05101.
46. Loshchilov, I.; Hutter, F. Sgdr: Stochastic gradient descent with warm restarts. *arXiv* **2016**, arXiv:1608.03983.
47. Keskar, N.S.; Mudigere, D.; Nocedal, J.; Smelyanskiy, M.; Tang, P.T.P. On large-batch training for deep learning: Generalization gap and sharp minima. *arXiv* **2016**, arXiv:1609.04836.
48. Howard, A.; Sandler, M.; Chu, G.; Chen, L.-C.; Chen, B.; Tan, M.; Wang, W.; Zhu, Y.; Pang, R.; Vasudevan, V. Searching for mobilenetv3. In Proceedings of the IEEE International Conference on Computer Vision (ICCV), Seoul, Republic of Korea, 27 October–2 November 2019; pp. 1314–1324.
49. Wang, J.; Sun, K.; Cheng, T.; Jiang, B.; Deng, C.; Zhao, Y.; Liu, D.; Mu, Y.; Tan, M.; Wang, X. Deep high-resolution representation learning for visual recognition. *IEEE Trans. Pattern Anal. Mach. Intell.* **2020**, *43*, 3349–3364. [CrossRef]
50. Xie, E.; Wang, W.; Yu, Z.; Anandkumar, A.; Alvarez, J.M.; Luo, P. SegFormer: Simple and efficient design for semantic segmentation with transformers. In Proceedings of the Advances in Neural Information Processing Systems 34 (NeurIPS), Online, 6–14 December 2021; pp. 12077–12090.

remote sensing

MDPI

Article

Enhanced Inversion of Sound Speed Profile Based on a Physics-Inspired Self-Organizing Map

Guojun Xu [1], Ke Qu [2], Zhanglong Li [1,*], Zixuan Zhang [2], Pan Xu [1], Dongbao Gao [1] and Xudong Dai [3]

[1] College of Meteorology and Oceanography, National University of Defense Technology, Changsha 410073, China; xuguojun@nudt.edu.cn (G.X.); xupan@nudt.edu.cn (P.X.); gaodongbao@nudt.edu.cn (D.G.)
[2] College of Electronics and Information Engineering, Guangdong Ocean University, Zhanjiang 524088, China; quke@gdou.edu.cn (K.Q.); zhangzixuan1@stu.gdou.edu.cn (Z.Z.)
[3] No. 92677 Troops of PLA, Qingdao 266000, China; daixudong10@163.com
* Correspondence: lizhanglong19@mails.ucas.ac.cn

Abstract: The remote sensing-based inversion of sound speed profile (SSP) enables the acquisition of high-spatial-resolution SSP without in situ measurements. The spatial division of the inversion grid is crucial for the accuracy of results, determining both the number of samples and the consistency of inversion relationships. The result of our research is the introduction of a physics-inspired self-organizing map (PISOM) that facilitates SSP inversion by clustering samples according to the physical perturbation law. The linear physical relationship between sea surface parameters and the SSP drives dimensionality reduction for the SOM, resulting in the clustering of samples exhibiting similar disturbance laws. Subsequently, samples within each cluster are generalized to construct the topology of the solution space for SSP reconstruction. The PISOM method significantly improves accuracy compared with the SOM method without clustering. The PISOM has an SSP reconstruction error of less than 2 m/s in 25% of cases, while the SOM method has none. The transmission loss calculation also shows promising results, with an error of only 0.5 dB at 30 km, 5.5 dB smaller than that of the SOM method. A physical interpretation of the neural network processing confirms that physics-inspired clustering can bring better precision gains than the previous spatial grid.

Keywords: sound speed profile; self-organizing map; physics-inspired clustering

Academic Editor: Massimiliano Pepe

Received: 1 November 2024
Revised: 17 December 2024
Accepted: 31 December 2024
Published: 2 January 2025

Citation: Xu, G.; Qu, K.; Li, Z.; Zhang, Z.; Xu, P.; Gao, D.; Dai, X. Enhanced Inversion of Sound Speed Profile Based on a Physics-Inspired Self-Organizing Map. *Remote Sens.* **2025**, *17*, 132. https://doi.org/10.3390/rs17010132

1. Introduction

Despite formidable challenges, the relentless pursuit of more precise information regarding the ocean's sound speed profile (SSP) remains unabated. On the one hand, the spatiotemporal distribution of SSPs plays a pivotal role in oceanic observations, enabling the exploration of phenomena ranging from global climate change to micro-scale turbulence through SSP inversion [1,2]. On the other hand, as a critical parameter in underwater waveguides, the SSP's distribution profoundly influences underwater sound propagation [3].

The most accurate way to obtain SSP is through direct measurement using sound speed profilers or conductivity–temperature–depth (CTD) instruments. However, this method only provides data for one measuring point and can be time-consuming and labor-intensive, making it economically impractical to acquire wide-area, synchronous SSPs. After the introduction of acoustic thermometry of ocean climate (ATOC) by Munk [4], various frameworks for acoustic inversion techniques have been developed, including matched field processing [5], compressed sensing [6], and deep learning [7]. By introducing environmental parameters into the optimization process, matching field processing can significantly enhance inversion accuracy. Compressed sensing enhances the real-time performance of SSP inversion by expressing it as an underdetermined linear problem and applying regularization through a least-squares cost function. This deep learning approach optimizes the inversion model via a data-driven method, substantially increasing the upper limit of inversion accuracy. As the acoustic signal is directly influenced by the SSP, the acoustic inversion method can provide precise and timely inversion results, and these acoustic inversions have enabled the retrieval of wide-area sound speed profiles. However, it is important to note that the acoustic signal is typically considered an integrated probe, reflecting the cumulative refraction effects along the sound propagation path. As a result, reconstructed SSPs reflect averaged effects along the propagation path. In typical application scenarios, the SSP obtained through acoustic inversion reflects the average SSP over a distance ranging from several to tens of kilometers between the transmitter and receiver, potentially limiting its spatial resolution.

To address the need for wide-area, high-resolution, quasi-real-time SSPs, satellite remote sensing stands out as the sole quasi-real-time global ocean observation platform, demonstrating considerable potential in SSP inversion. Theoretically, remote sensing-based inversion can achieve a spatial resolution for SSPs comparable to that of sea surface parameters, providing spatial variation information at scales finer than several hundred meters. Utilizing satellite remote sensing to acquire ocean surface parameters, a novel approach has emerged, inferring the correlation between surface parameters and SSP disturbances based on historical profiles and sea surface conditions. According to the principle of thermal expansion, Carnes initially validated a nearly linear correlation between sea level anomalies (SLAs) and the amplitudes of empirical orthogonal function (EOF) of temperature profiles through an analysis of extensive historical data [8]. Subsequently, experiments conducted in the northwestern Pacific and northwestern Atlantic utilized sea surface temperature anomalies (SSTAs) and SLAs as inputs to infer subsurface profiles based on a single empirical orthogonal function-based regression (sEOF-r) [9]. The effectiveness of this approximate linear physical relationship for subsurface profile inversion has been confirmed, and its applicability to global SSPs has been validated by Chen [10]. While it is impractical to describe complex air–sea dynamical systems using physical equations, a series of "physical" methods, such as sEOF-r, has demonstrated that reasonable results can be achieved through approximate physical expressions.

Contrastingly, recent advancements in machine learning algorithms have led to the emergence of data-driven approaches. Hjelmervik employed gradient descent algorithms to infer temperature and salinity profiles based on sea surface remote sensing data [11,12]. Chapman proposed self-organizing maps (SOMs) to reconstruct subsurface current profiles [13]. Su improved the inversion of global ocean salinity profiles by combining extreme gradient boosting with gradient-boosting decision trees [14]. Ali adopted artificial neural networks to invert SSP and discussed the error distribution at different depths [15]. Ou applied neural networks for multivariate regression to enhance inversion results for SSP using various sea surface parameters [16]. Due to their unrestricted nature in uncovering implicit nonlinear relationships between multiple parameters, data-driven methods have demonstrated clear advantages in precision over "physical" methods and have become mainstream in designing inversion methodologies. Based on the powerful data-mining capabilities of data-driven methods, the implicit relationships between input parameters and SSPs can be further utilized. Chen incorporated input data from an echo sounder and the maximum layer depth, in addition to remote sensing parameters, to estimate SSPs [17]. With the advantage of merging multi-source information, the inversion accuracy showed a significant improvement. Qu compared the applicability of SOM and sEOFr in the South

China Sea. By analyzing the entire solution process, they found that enhancing the consistency of the SSP basis function significantly improves the inversion accuracy [18]. Building on remote sensing parameters, Li introduced surface sound speed, measured by a surface velocimeter, as an additional solution condition [19]. Although this algorithm increased the cost of in situ measurements, it enabled the inversion of full-depth SSPs. Zhao employed long short-term memory neural networks to model the relationship between sea surface parameters and SSPs [20]. This algorithm capitalizes on the strong temporal correlation in SSP perturbations, achieving high-precision SSP predictions with limited samples within a confined spatial range.

The success of any inversion method, whether driven by physical equations or data, relies on the assumption that all samples adhere to the similar disturbance laws of the SSPs. Therefore, when conducting SSP inversion from remote sensing data, it is common to divide the ocean area into a 1° or 2° grid and process the samples within each grid cell separately. However, the ocean is a complex system influenced by various factors such as time, localization, monsoon, and circulation dynamics. These factors can hinder statistical consistency across the geographical grid and introduce errors in inversion results. Decreasing grid sizes may improve statistical consistency but will significantly reduce the sample size and pose challenges for applying machine learning algorithms. Conversely, expanding the grid can increase the sample size, but ensuring statistical consistency becomes challenging.

This study proposes an enhanced SSP inversion method from remote sensing data, utilizing a physics-inspired self-organizing map (PISOM). By employing the dimensionality reduction and generalization techniques of the SOM method, a PISOM can effectively cluster the physical relationships between sea surface parameters and profile parameters, overcoming spatial grid limitations and ensuring statistical consistency in the disturbance laws of SSP samples during inversion. Training is conducted using Argo data from the South China Sea spanning from 2007 to 2019, along with remote sensing data, to deduce the SSPs of 2020. This PISOM method significantly considers both sample statistics consistency and sample size, leading to improved effectiveness in inversion results. Furthermore, guided by physical mechanisms, neural network processing is analyzed. The findings demonstrate that the statistical consistency in inversion relationships strongly correlates with seasons rather than spatial positions, indicating a need for further improvement in conventional spatial grid processing. Our main contributions can be summarized as follows: (1) We introduce a new PISOM algorithm that enhances inversion accuracy by incorporating physical mechanism constraints into the neural network algorithm. This novel method clusters the inversion relationships of SSPs according to their physical expressions. Consequently, by utilizing the clustered training set, the neural network algorithm can more effectively delineate the perturbation features of SSPs and the relationship between the input–output parameters; (2) Recognizing that the spatial division of the inversion grid fails to ensure statistical consistency, clustering based on physical constraints reveals the significant influence of seasonal factors on inversion relationships. Consequently, we present a grid-free sample-clustering method that accounts for both the consistency of statistical rules and the adequacy of sample sizes; (3) We incorporate transmission loss to assess the validity of the inversion results at the application level, revealing that the improved method significantly enhances the accuracy of sound field calculations and can provide suitable SSP information for these applications.

2. Data

The key to using remote sensing data to obtain SSPs lies in establishing the inversion relationship between sea surface parameters and SSPs based on historical samples. By training historical SSPs alongside SLAs and SSTAs, we can establish this relationship.

During the final inversion process, only an SLA and an SSTA are required as inputs to obtain the corresponding SSP.

2.1. SSP Samples

Argo floats are the primary means of obtaining global ocean SSP samples. This study primarily focuses on addressing the challenge of initial grid division for inversion, which is particularly prominent in the South China Sea. The South China Sea is the largest marginal sea in the western Pacific, with an average depth of 1212 m. The central part has an average depth exceeding 4000 m, reaching a maximum of 5559 m. Due to the intricate topography and basin-scale circulation influences, SSPs within the South China Sea exhibit non-gradual spatial variations unlike those observed in open oceans. Profiles in a range of 8–24°N and 109–121°E were selected for testing inversion. All the Argo data were obtained from the Global Ocean Argo Collection [21].

Due to political and economic factors, the availability of SSP samples is severely limited, posing challenges in implementing conventional geographic grid division methods. To address this issue, our primary focus was on profiles within a depth range of 10 to 1000 m, which can account for both the main disturbance depth of SSPs and the number of retained samples. For training purposes, data from 2007 to 2019 were utilized, resulting in 7200 profiles. Validation was conducted using profiles from 2020 (132 in total). All profiles were linearly interpolated to conform with the standard depth levels defined by the World Ocean Atlas 2023 (WOA23). This approach ensures consistent sampling and facilitates error comparison with other methodologies.

During SSP processing using empirical orthogonal function (EOF) analysis, the background profile from WOA23 (https://www.ncei.noaa.gov/products/world-ocean-atlas, accessed on 23 July 2024.) was utilized. The WOA23 data were chosen as annual averages for the 2005–2017 period, with a spatial resolution of 0.25°. Similar to Argo, temperature and salinity profiles were transformed into sound speed profiles using Del Grosso's empirical formulas [22].

All the SSPs are illustrated in Figure 1, revealing significant differences between the South China Sea and the open ocean. In particular, disturbances in sound speed surpassing 20 m/s pose a significant challenge to SSP inversion. The disturbances primarily occur within the uppermost 300 m, gradually diminishing in intensity as they extend deeper. Notably, the presence of complex internal waves, fronts, and turbulence causes non-monotonic disturbances with large amplitudes at several depths, providing the main source of inversion errors in daily temporal resolution inversion.

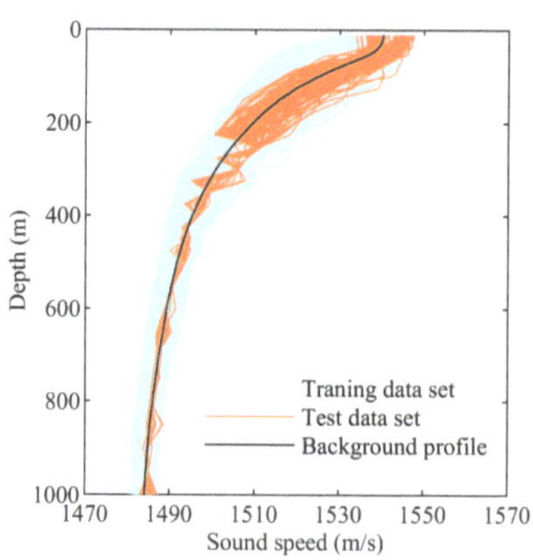

Figure 1. SSP samples and background profile.

2.2. Remote Sensing Data

SLA and SSTA data have been extensively validated as the most effective variables for SSP inversion, as demonstrated in numerous studies. All remote sensing data were obtained from the Copernicus Marine Environment Monitoring Service (CMEMS) [23,24]. The SLA data were derived by merging data from various altimetry missions and were processed using optimal interpolation, achieving a spatial resolution of 0.25°. The SSTA data were estimated by the Group for High-Resolution Sea Surface Temperature (GHRSST) project in conjunction with in situ observations, utilizing a spatial resolution of 0.05°. All remote sensing data had a daily temporal resolution. The establishment of a one-to-one correspondence between the remote sensing data and the same-day Argo SSPs was based on the spatial proximity principle.

3. Methods

Physics-driven methods involve specifying physical equations, providing valuable constraints on inversion results. However, due to the complex nature of the air–sea dynamical system, using approximate physical expressions inevitably introduces errors. While data-driven methods offer the advantage of avoiding approximate expressions, they often yield unrealistic results due to the absence of physical constraints. In this study, we propose a combination of physics-based expressions and statistical neural networks. Given the difficulty in precisely establishing inversion relationships between parameters using physical expressions and the challenge of ensuring the statistical consistency of all samples for inversion, we propose a clustering approach based on approximate physics formulas. This approach guarantees high statistical consistency among physics-inspired samples, enhancing their accuracy.

3.1. Dimensionality Reduction in SSPs

The disturbance of SSPs can be mathematically represented as a three-dimensional matrix. Here, each element in the column corresponds to a sampling point along the depth dimension, while the other two dimensions represent sequences of time and space. When addressing the inversion problem associated with an SSP, it is crucial to consider that the inversion's complexity is significantly impacted by the number of unknowns involved. Hence, it becomes necessary to reduce the dimensionality of SSPs. An SSP, $c(z)$, can be expressed as follows [25]:

$$c(z) = c_0 + \sum_{n=0}^{m} a_n \psi_n, \tag{1}$$

where z is the sampling point along the depth, ψ is the basis function describing the disturbance mode of SSPs, a is the projection coefficients of the basis function, and n is the order of the basis function. Due to the influence of the barotropic mode, the inversion of SSPs using remote sensing parameters includes a zeroth-order mode with the same amplitude across all depths. In inverse problems, a higher-order basis does not necessarily provide better approximations to the true values due to the presence of noise. Typically, using $m = 3$ strikes a balance between effectively capturing disturbances and avoiding noise introduced by higher-order modes. In the following experiment, $m = 3$ yields the best inversion accuracy. The EOF is the most classical basis function of SSPs, involving extracting principal components from samples. Assuming the SSP samples form an $m \times n$ matrix, where m represents the number of depth sampling points and n represents the number of samples, we can obtain the anomaly matrix, X, by subtracting the background profile from the sample matrix. The eigenvalues of this matrix can be used to calculate the disturbance modes in sound speed [26]:

$$\frac{XX^T}{N}K = KE, \tag{2}$$

where K is the eigenvector and E is the diagonal matrix of the eigenvalues. In practical applications, K is the modal function of each order of the EOF. The accuracy and effectiveness of EOFs are contingent on the consistency of the disturbance laws among the samples. A crucial objective in employing the EOF is to cluster profiles with consistent disturbance laws and inversion relationships, yielding more precise and effective EOFs, along with their projection coefficients.

3.2. *Physics-Driven SOM*

According to a regression analysis of many historical samples, an approximate linear relationship was observed between sea surface parameters and the EOF projection coefficients of SSPs. Based on this linear relationship, Carnes proposed the sEOF-r method, which can be expressed as follows [9]:

$$a_n = A_{n,0} + A_{n,1} \times SLA + A_{n,2} \times SSTA + A_{n,3} \times SLA \times SSTA, \tag{3}$$

where $A_{n,m}$ is the m-th linear fitting parameter for the n-th projection coefficients, a_n. In the training phase, based on historical samples, a_n and their corresponding SLA and $SSTA$ can be utilized to calculate three approximate linear fitting parameters A. During the solving phase, these coefficients and remote sensing parameters can be directly employed to derive projection coefficients for reconstructing the SSP. To avoid confusion, in the subsequent clustering process, the EOF coefficient derived directly from the SSP is denoted as a_n, whereas b_n represents the EOF coefficient calculated using the physical expression (3). Although Equation (3) does not provide an exact analytical representation, it exhibits consistent fitting coefficients when the disturbance law of the profiles is maintained, resulting in inversion results with a reasonable level of precision. These identical fitting parameters indicate consistency in heat and energy transfer, as well as sound speed disturbance modes resulting from air–sea interactions. Given the absence of precise physical formulations for air–sea interactions and SSP disturbances, conventional physics-informed neural networks (PINNs) incorporating partial differential equations to enforce physical constraints are not applicable in SSP inversion. To address this limitation, we propose a novel approach that combines physics-driven principles with data-driven techniques, integrating physical relationships into neural network inversion based on clustering statistically consistent samples according to Equation (3).

A flowchart of the proposed method is shown in Figure 2. Initially, clustering is conducted based on disturbance modes to process SSPs exhibiting consistent disturbances. Subsequently, the clustered samples are utilized to train an SOM to establish a generalized neural network structure. Then, by employing the input remote sensing parameters, the best-matching neuron (BMU) can be found in the neural network. Finally, extracting the projection coefficient elements from the BMU yields the inversion result. The SOM inversion process can be described as follows:

1. Clustering: To circumvent the conventional approach of employing spatial grids for classification, we propose a method utilizing an SOM network to cluster based on the correlation between remote sensing parameters and SSPs to achieve dimensionality reduction. Based on linear initialization, raw training samples are projected onto the linear subspace formed by three parameter types: remote sensing param-

eters (SLA; SSTA), EOF projection coefficients of the historical samples (a_n), and linearly reconstructed SSP projection coefficients (b_n) obtained from all sample data using the sEOF-r method. The first two parameter types rely on data-driven techniques to establish statistical relationships between surface and profile parameters. The third parameter type incorporates Equation (3) as a constraint to capture the approximate linear relationship, thereby clustering with the first and second parameter types based on deviations from this physical relationship. Dimensionality reduction for the samples is achieved by configuring a small number of neurons in the SOM network. Classification is performed on the clustering networks using the nearest neuron based on Euclidean distance. Training samples are classified according to their disturbance laws while retaining the first and second parameter types as clustering training samples. Importantly, after clustering, each cluster represents distinct disturbance laws. This requires a separate recalculation of EOFs and their coefficients for each cluster;

2. Generalization: Based on clustered samples, disturbance-consistent samples are re-input into the SOM network to generate a generalized network and form a solution topology. The cluster of samples closest to the solving profile's time is selected as the clustering training sample. Actual testing has shown that setting the number of neurons in the SOM network to three times the number of input samples during generalization maintains inversion accuracy. Training with the SOM network generates a generalized neural network, which is derived under Equation (3)'s near-linear relationship constraint. The ensemble of these neurons constitutes the network structure, which describes the SSP that may be formed under the disturbance law of the training sample;
3. Matching: Based on the input parameters, the BMU is determined on the generalized SOM network. The BMU is defined as the neuron that exhibits the minimum Euclidean distance to the input parameters within the generalized neural network. The input actually constitutes an incomplete neuron, and Chapman derived a function to calculate the truncated distance with the complete neuron on the neural network [27]:

$$d_p(x, u^p) = \sum\nolimits_{i \in avail} \left(1 + \sum\nolimits_{j \in missing} (cor_{ij}^c)^2\right) \times (x_i - u_i^p), \quad (4)$$

where d is the Euclidean distance, p represents the number of the neuron on the generalized network, u is the neuron vector, set $avail$ represents the existing elements of the incomplete neuron, and set $missing$ is the element to be solved for SSP reconstruction. The truncated distance of each neuron on the generalized network can be calculated through an exhaustive search. The smallest distance represents the BMU, indicating the closest match between the air–sea relationship and the input parameters;
4. Extraction: The inversion result can be obtained from the missing part of the BMU, i.e., the corresponding coefficients, a_n, of the SSP. By combining these coefficients with the EOFs derived from the principal component analysis of this cluster, Equation (1) can be utilized to reconstruct the sound SSP.

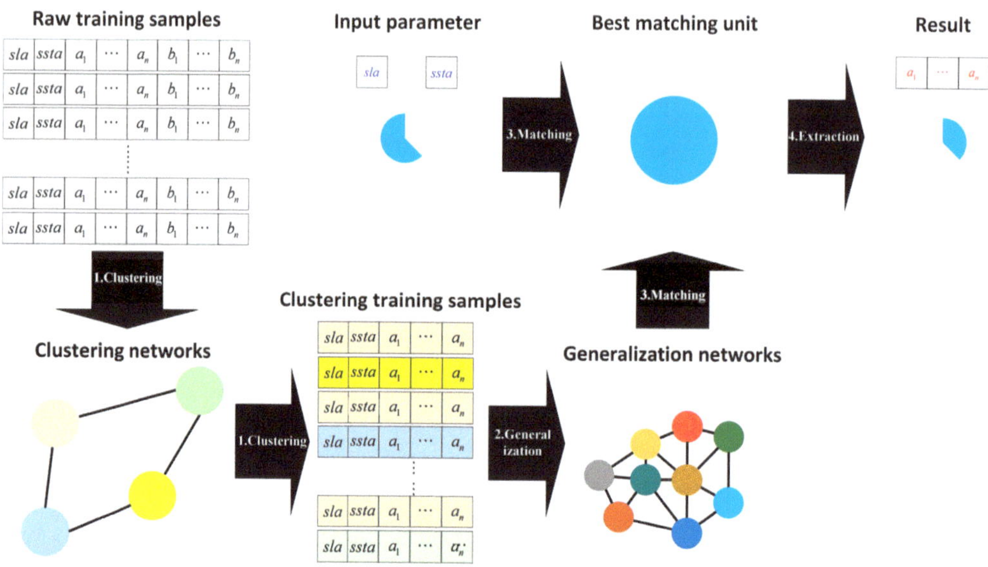

Figure 2. Flow chart of the physics-inspired SOM inversion. The black italicized variable represents the input parameters extracted from the training samples; the blue italicized variable denotes the input parameters utilized for solution information; and the red italicized variable signifies the output parameters serving as the reconstruction coefficients of the SSP.

4. Results

The proposed method involves pre-classification to ensure the statistical consistency of the samples. Thus, the number of clusters plays a crucial role in determining the results. Table 1 presents the inversion errors for different numbers of clusters. It is evident that classification significantly enhances inversion accuracy, with post-clustering inversion results outperforming non-clustered ones. The highest accuracy is achieved with two clusters, and as the number of clusters increases, there is only a minimal change in accuracy. This can be attributed to further clustering potentially amplifying noise during inversion and reducing the number of training samples, resulting in slightly lower accuracy than the optimal cluster numbers. For subsequent analysis, we focus on examining results obtained using two clusters, as this represents an optimal choice and provides evidence of physical mechanisms behind optimal clustering. The root-mean-squared error (RMSE) was used to quantify the SSP reconstruction error.

Table 1. Errors in different cluster numbers.

	1 Cluster	2 Clusters	3 Clusters	4 Clusters	5 Clusters
RMSE (m/s)	3.78	3.63	3.71	3.70	3.68

4.1. SSP Reconstruction Error

An error comparison between the SOM and PISOM inversion methods is illustrated in Figure 3. At most points, the SOM method with clustering has significantly higher accuracy than the simple SOM method. The average error of the SSPs obtained through the PISOM is 3.63 m/s, whereas for the SOM, it is 3.78 m/s. The SOM's performance is enhanced by incorporating a pre-clustering procedure. The classic SOM method encounters significant anomalies in the region, leading to limited accuracy with almost no cases of error below 2 m/s. By contrast, the PISOM demonstrates an error rate below 2 m/s for approximately one-fourth of the cases. The PISOM method can handle the features carried by disturbances more effectively by simply clustering the disturbance characteristics. Although the reduction in mean error between the two methods is not statistically significant, the inversion results can still be regarded as a great improvement. The main source of large errors arises from notable anomalies in SSPs caused by dynamic factors such as fronts and water masses, which pose challenges for accurate representation using EOFs. In cases where significant anomalies are absent in the samples, the reconstructed SSP more consistently aligns with the disturbance principal components, resulting in more pronounced improvements in error. Among the 47 samples exhibiting reconstruction errors below three, the PISOM method demonstrates an approximately 20% decrease in error compared with the SOM method.

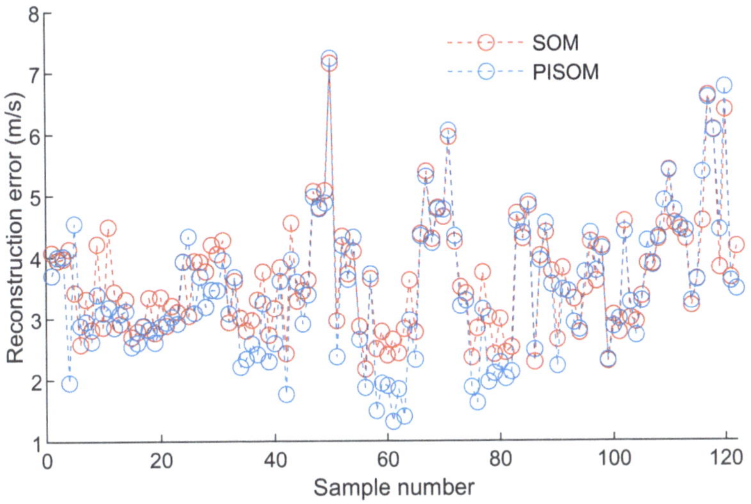

Figure 3. Errors in reconstruction for different sample numbers.

Figure 4 shows the errors at different depths. In most depths, the PISOM method outperforms the SOM method, with the surface layer showing the most significant improvement in inversion accuracy. This is because the core of the method lies in establishing a relationship between SSPs and surface remote sensing parameters, which are closely associated with the surface part of the SSP. The deep sea below 600 m is the least improved part, primarily because this section is less affected by surface remote sensing parameters and has relatively consistent sound speeds with minimal disturbances. Notably, the maximum error occurs around 300 m. Typically, due to the daily resolution of remote sensing inversion methods, diurnal variations in the mixed layer often become the primary source of inversion error. In the data presented here, significant errors can be observed at depths of approximately 300 m, 500 m, 750 m, and 1000 m. These errors mainly arise from random large gradient disturbances at specific depths in a small subset of the sample, resulting in a sharp increase in mean error. Since capturing such random large anomalies during inversion poses challenges, these data exhibit high error values specifically due to anomalies at certain depths, highlighting the difficulty in representing these conditions accurately through SSP inversion techniques. Furthermore, this reveals the highly challenging nature of SSP inversion in the South China Sea.

To explain the sources of error, Figure 5 presents two representative examples. In the example of high-precision reconstruction on the left, the PISOM method has a clear improvement in accuracy compared with the SOM method, particularly showcasing its performance advantage closer to the sea surface. Random disturbances caused by water masses change the smooth variation trend of the sound speed gradient at depths around 250 m and 500 m. While data-driven methods relying on the main statistical characteristics struggle to represent such random disturbances, the PISOM method still effectively describes these disruptions better than the SOM method. In the high-error example on the right, similar to the error distribution shown in Figure 4, random errors at depths around 300 m, 500 m, 750 m, and 1000 m constitute most of the mean reconstruction errors. Both methods suffer severe performance degradation in the presence of intense random disturbance. However, the PISOM method is still closer to the actual measured profile. The precise representation of these outlier points constitutes a significant challenge for nearly all SSP inversion techniques. This is primarily because the infrequent appearance of these outlier points in the samples does not constitute the primary component of the disturbance features in the training data; thus, the basis functions cannot reconstruct these outlier points. Additionally, the input parameters used for inversion are insufficient to deduce these outlier points. Sea surface parameters solely encompass sea surface information, while in acoustic inversion, acoustic propagation signals reflect the averaged effects along the propagation path. Neither of these inputs provides detailed structural information about the SSP at specific depths.

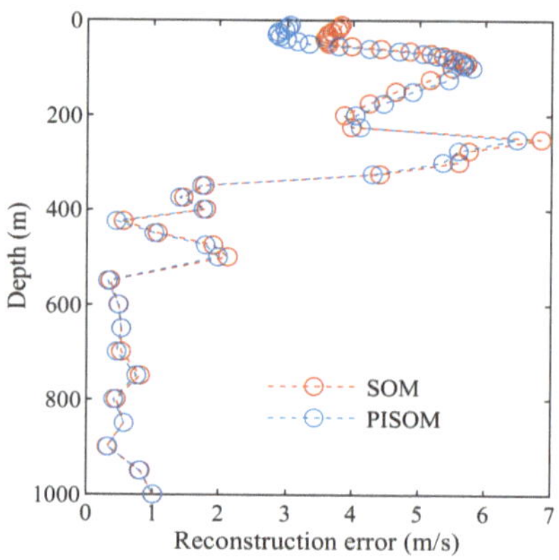

Figure 4. Errors in reconstruction for different depths.

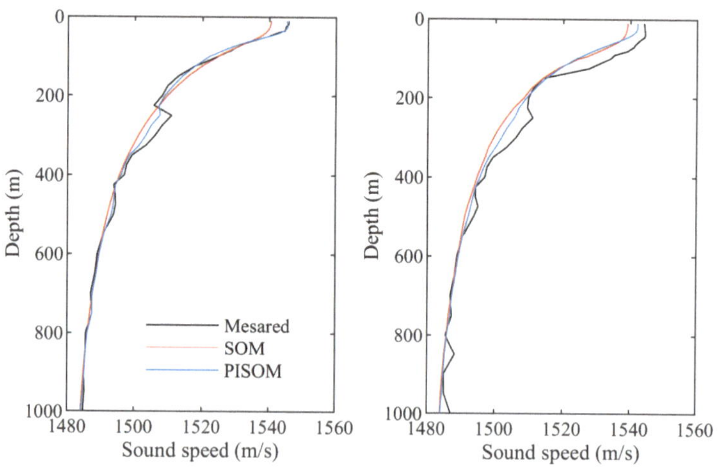

Figure 5. Examples of sound speed profile reconstruction.

4.2. Validation of Transmission Loss

The primary objective of SSP inversion is to conduct calculations on the sound field, and the most direct way to verify the effectiveness of the results is to perform sound field calculations. We performed validation to forecast transmission loss using the normal mode model KRAKENC based on the reconstructed SSP. Figure 6a shows that the errors in the PISOM and SOM, and the results were 2.62 m/s and 2.81 m/s, respectively. Considering the reconstructed depth of the SSP is 1000 m, characterized by minimal disturbances below this threshold, the inversion results below 600 m exhibit a high level of consistency with the measurements, and we extrapolated the depth profiles down to 4000 m based on WOA23 data. The sound source and receiver were positioned at a depth of 50 m. The frequency used was 100 Hz, and the seabed had a density of 1.73 g/cm^3 with a sound speed of 1541 m/s. Additionally, we considered an attenuation coefficient of 0.09 dB/λ and a water depth of 4000 m. A comparison of the calculated transmission loss is presented in Figure 6b. Significant transmission loss errors are caused by inaccuracies in the SSP reconstruction in the direct wave part before the first convergence zone. However, both inversion methods accurately capture the interference structure of the sound field in the first two convergence zones, indicating that the inversion method utilizing remote sensing parameters, which can achieve a globally high-spatial-resolution estimation of SSP, effectively meets the precision requirements for sound field calculations. With increasing propagation distance, the error in the sound field calculation accumulates due to the SSP inversion error. In the fifth convergence zone, the position predicted using the SOM method for the convergence zone deviates at longer distances, causing a significant deviation in the interference structure, with a transmission loss error of 6 dB at 30 km; conversely, the PISOM method can still predict the sound field's interference structure reasonably well, with a sound field calculation error of approximately 0.5 dB. From the perspective of sound field prediction, without conducting in situ measurements, the SSP inversion based on remote sensing parameters can indeed provide SSP information suitable for sound field calculation applications, and the PISOM method can effectively enhance the applicability of this method without introducing additional models or heavy computational burden.

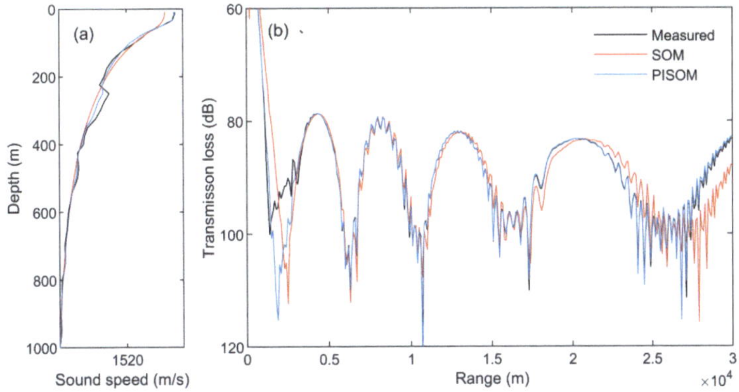

Figure 6. Transmission loss calculated using different SSPs, (**a**) SSPs, (**b**) Transmission loss.

4.3. Interpretation of Neural Network Processing

Due to the incorporation of physical linear relationship constraints, the neural network architecture becomes more interpretable, enabling us to gain insights into the underlying processes and mechanisms governing the SOM. Figures 7 and 8 depict the spatial and temporal distribution of all samples. The spatial distribution analysis reveals that the influence of spatial position on establishing the inversion relationship between surface parameters and SSPs is negligible. Within the South China Sea region, both clusters exhibit a uniform distribution, unaffected by their spatial positions. Despite the presence of intricate and intense mesoscale phenomena such as eddies, internal solitary waves, fronts, and Pacific exchange water masses that can significantly impact SSPs and complicate SSP inversion procedures, their effect on establishing the inversion relationship remains inconspicuous. The temporal distribution reveals a uniform spread of samples across all twelve months of the year, and subsequent clustering analysis demonstrates the significant influence of seasonal factors in establishing inversion relationships. Cluster 1 predominantly occurs from April to October, while Cluster 2 mainly occupies December to February, with March and November acting as transitional periods between these two clusters. During the summer, the strong influence of the southwest monsoon and intense sea surface irradiance result in a negative sound speed gradient at the surface. Conversely, in winter, the northeast monsoon leads to the formation of a robust mixed layer on the sea surface. These distinct sound speed distributions give rise to different barotropic and baroclinic modes within the ocean, characterized by varying relationships between the sea surface and the SSPs. Consequently, two inversion clusters are formed, demonstrating that previous spatial grid methods lack effectiveness in statistically clustering consistent samples.

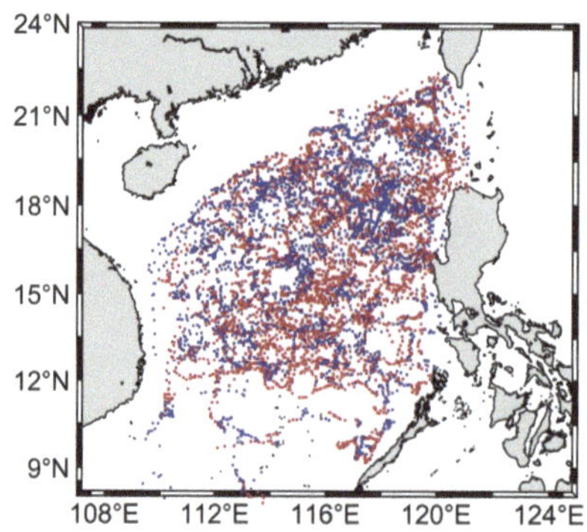

Figure 7. Spatial distribution of two sample clusters.

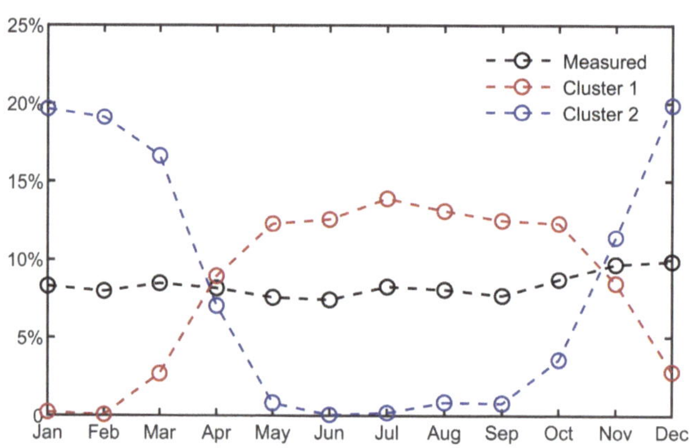

Figure 8. Temporal distribution of two sample clusters.

5. Conclusions

Sea surface parameters primarily influence SSPs through energy and matter transfers dominated by barotropic and baroclinic modes. This complex physical relationship can be approximated as a linear equation (Equation (3)). In this study, parameters adhering to this physical mechanism were incorporated as elements in the SOM clustering process, enhancing sample consistency in perturbation laws and improving inversion accuracy. The effectiveness of this PISOM was validated through SSP inversion experiments conducted in the South China Sea and compared with the simple SOM inversion.

The experiments confirmed that applying PISOM significantly enhanced the precision of SSP inversion. When samples from 2020 were used as the test set, many samples exhibited random large disturbances caused by water masses, posing a challenge for nearly all SSP inversion methods due to the difficulty in accurately representing such rare and intense disturbances with EOFs derived from a principal component analysis. When only profiles without abnormal disturbances were considered, the PISOM method demonstrated an approximate 20% reduction in error. Despite encountering notable errors near the depth between the mixed layer and thermocline due to limited information and temporal resolution, remote sensing-based SSP inversion methods validate their ability to provide globally high-spatial-resolution SSP information without requiring any in situ measurements, benefiting sonar system applications. The PISOM method enables reasonable transmission loss calculations while reducing errors by approximately 5.5 dB at 30 km compared with the SOM method.

After analyzing the clustering results inspired by the physical linear relationship, we discovered that seasonality plays a crucial role in determining the consistency of inversion relationships. In our South China Sea inversion experiment, samples mainly clustered around the summer and winter seasons. This finding demonstrates the limitations of previous spatial grid processing methods. Therefore, incorporating the proposed clustering process during preprocessing can effectively enhance inversion performance.

Author Contributions: G.X. and Z.L. provided research ideas and wrote the paper. K.Q. and Z.L. formulated the initial research questions and determined the research direction of this article. Z.Z., P.X., D.G. and X.D. provided support in data acquisition. G.X. and K.Q. coordinated the writing work. All authors have read and agreed to the published version of the manuscript.

Funding: This research was funded by the open fund of the National Key Laboratory of Science and Technology on Underwater Acoustic Antagonizing, grant number JCKY2024207CH07.

Data Availability Statement: The data generated in this study are not publicly available due to their use in an ongoing study by the authors but can be made available from the corresponding author upon reasonable request.

Conflicts of Interest: The authors declare no conflicts of interest.

References

1. Behringer, D.; Birdsall, T.; Brown, M.; Cornuelle, B.; Heinmiller, R.; Knox, R.; Metzger, K.; Munk, W.; Spiesberger, J.; Spindel, R.; et al. A demonstration of ocean acoustic tomography. *Nature* **1982**, *299*, 121–125. [CrossRef]
2. Colosi, J.A.; Duda, T.F.; Alford, M.; Lin, Y.-T.; Voet, G. On the feasibility of using short-range, high-frequency transmissions to characterize the vertical-spectra of small-scale internal waves and turbulence in a bottom boundary layer. *J. Acoust. Soc. Am.* **2024**, *155*, A278–A279. [CrossRef]
3. Uzhansky, E.; Lunkov, A.; Katsnelson, B. Effect of an internal Kelvin wave on sound propagation in a coastal wedgea. *J. Acoust. Soc. Am.* **2024**, *155*, 3357–3370. [CrossRef]
4. Munk, W.; Wunsch, C. Ocean acoustic tomography: A scheme for large scale monitoring. *Deep-Sea Res. Part A* **1979**, *26*, 123–161. [CrossRef]
5. Tolstoy, A.; Diachok, O.; Frazer, L.N. Acoustic tomography via matched field processing. *J. Acoust. Soc. Am.* **1991**, *89*, 1119–1127. [CrossRef]
6. Bianco, M.; Gerstoft, P. Compressive acoustic sound speed profile estimation. *J. Acoust. Soc. Am.* **2016**, *139*, EL90–EL94. [CrossRef]
7. Lu, J.; Zhang, H.; Wu, P.; Li, S.; Huang, W. Predictive modeling of future full-ocean depth SSPs utilizing hierarchical long short-term memory neural networks. *J. Mar. Sci. Eng.* **2024**, *12*, 943. [CrossRef]

8. Carnes, M.R.; Mitchell, J.L.; de Witt, P.W. Synthetic temperature profiles derived from Geosat altimetry: Comparison with air-dropped expendable bathythermograph profiles. *J. Geophys. Res.* **1990**, *95*, 17979–17992. [CrossRef]
9. Carnes, M.R.; Teague, W.J.; Mitchell, J.L. Inference of Subsurface Thermohaline Structure from Fields Measurable by Satellite. *J. Atmos. Ocean. Technol.* **1994**, *11*, 551–566. [CrossRef]
10. Chen, C.; Ma, Y.; Liu, Y. Reconstructing Sound speed profiles worldwide with Sea surface data. *Appl. Ocean Res.* **2018**, *77*, 26–33. [CrossRef]
11. Hjelmervik, K.T.; Hjelmervik, K. Estimating temperature and salinity profiles using empirical orthogonal functions and clustering on historical measurements. *Ocean Dynam.* **2013**, *63*, 809–821. [CrossRef]
12. Hjelmervik, K.; Hjelmervik, K.T. Time-calibrated estimates of oceanographic profiles using empirical orthogonal functions and clustering. *Ocean Dynam.* **2014**, *64*, 655–665. [CrossRef]
13. Chapman, C.; Charantonis, A.A. Reconstruction of Subsurface Velocities From Satellite Observations Using Iterative Self-Organizing Maps. *Geosci. Remote Sens. Lett.* **2017**, *14*, 617–620. [CrossRef]
14. Su, H.; Yang, X.; Lu, W.; Yan, X.-H. Estimating subsurface thermohaline structure of the global ocean using surface remote sensing observations. *Remote Sens.* **2019**, *11*, 1598. [CrossRef]
15. Jain, S.; Ali, M.M. Estimation of sound speed profiles using artificial neural networks. *IEEE Geosci. Remote Sens. Lett.* **2006**, *3*, 467–470. [CrossRef]
16. Ou, Z.; Qu, K.; Wang, Y.; Zhou, J. Estimating sound speed profile by combining satellite data with in situ sea surface observations. *Electronics* **2022**, *11*, 3271. [CrossRef]
17. Cheng, C.; Yan, F.; Gao, Y. Improving reconstruction of sound speed profiles using a self-organizing map method with multi-source observations. *Remote Sens. Lett.* **2020**, *11*, 572–580. [CrossRef]
18. Qu, K.; Zou, B.; Zhou, J. Rapid environmental assessment in the South China Sea: Improved inversion of sound speed profile using remote sensing data. *Acta Oceanol. Sin.* **2022**, *41*, 78–83. [CrossRef]
19. Li, Q.; Li, H.; Cao, S.; Yan, X.; Ma, Z. Inversion of the full-depth sound speed profile based on remote sensing data and surface sound speed. *Acta Oceanol. Sin.* **2022**, *44*, 84–94. [CrossRef]
20. Zhao, Y.; Xu, P.; Li, G.; Ou, Z.; Qu, K. Reconstructing the sound speed profile of South China Sea using remote sensing data and long short-term memory neural networks. *Front. Mar. Sci.* **2024**, *11*, 1375766. [CrossRef]
21. Li, Z.Q.; Liu, Z.H.; Lu, S.L. Global Argo data fast receiving and post-quality-control system. *IOP Conf. Ser. Earth Environ. Sci.* **2020**, *502*, 012012. [CrossRef]
22. Del Grosso, V.A. New equation for the speed of sound in natural waters (with comparisons to other equations). *J. Acoust. Soc. Am.* **1974**, *56*, 1084–1091. [CrossRef]
23. Good, S.; Fiedler, E.; Mao, C.; Martin, M.J.; Maycock, A.; Reid, R.; Roberts-Jones, J.; Searle, T.; Waters, J.; While, J.; et al. The Current Configuration of the OSTIA System for operational roduction of foundation sea surface temperature and ice concentration analyses. *Remote Sens.* **2020**, *12*, 720. [CrossRef]
24. Donlon, C.J.; Martin, M.; Stark, J.; Roberts-Jones, J.; Fiedler, E.; Wimmer, W. The Operational Sea Surface Temperature and Sea Ice Analysis (OSTIA) system. *Remote Sens. Environ.* **2012**, *116*, 140–158. [CrossRef]
25. Cheng, L.; Ji, X.; Zhao, H.; Li, J.; Xu, W. Tensor-based basis function learning for three-dimensional sound speed fields. *J. Acoust. Soc. Am.* **2022**, *151*, 269–285. [CrossRef]
26. Bianco, M.; Gerstoft, P. Dictionary learning of sound speed profiles. *J. Acoust. Soc. Am.* **2017**, *141*, 1749–1758. [CrossRef] [PubMed]
27. Charantonis, A.A.; Testor, P.; Mortier, L.; D'Ortenzio, F.; Thiria, S. Completion of a sparse glider database using multi-iterative self-organizing maps (ITCOMP SOM). *Proc. Comput. Sci.* **2015**, *51*, 2198–2206. [CrossRef]

remote sensing

MDPI

Article

AquaPile-YOLO: Pioneering Underwater Pile Foundation Detection with Forward-Looking Sonar Image Processing

Zhongwei Xu [1,2], Rui Wang [1,*], Tianyu Cao [3], Wenbo Guo [3], Bo Shi [3] and Qiqi Ge [3]

1 Department of Information and Communication Engineering, Tongji University, Shanghai 201804, China; 2180135@tongji.edu.cn
2 China State Shipbuilding Corporation Haiying Enterprise Group Co., Ltd., Wuxi 214061, China
3 Department of Automation, Shanghai Jiao Tong University, Shanghai 200240, China; caotianyu2023@sjtu.edu.cn (T.C.); wenbo.guo@sjtu.edu.cn (W.G.); tommy7080@sjtu.edu.cn (B.S.); gqq@sjtu.edu.cn (Q.G.)
* Correspondence: ruiwang@tongji.edu.cn

Abstract: Underwater pile foundation detection is crucial for environmental monitoring and marine engineering. Traditional methods for detecting underwater pile foundations are labor-intensive and inefficient. Deep learning-based image processing has revolutionized detection, enabling identification through sonar imagery analysis. This study proposes an innovative methodology, named the AquaPile-YOLO algorithm, for underwater pile foundation detection. Our approach significantly enhances detection accuracy and robustness by integrating multi-scale feature fusion, improved attention mechanisms, and advanced data augmentation techniques. Trained on 4000 sonar images, the model excels in delineating pile structures and effectively identifying underwater targets. Experimental data show that the model can achieve good target identification results in similar experimental scenarios, with a 96.89% accuracy rate for underwater target recognition.

Keywords: AquaPile-YOLO; multi-scale feature fusion; deep learning; sonar image; underwater target recognition; attention mechanism

Academic Editor: Jaroslaw Tegowski

Received: 12 December 2024
Revised: 13 January 2025
Accepted: 16 January 2025
Published: 22 January 2025

Citation: Xu, Z.; Wang, R.; Cao, T.; Guo, W.; Shi, B.; Ge, Q. AquaPile-YOLO: Pioneering Underwater Pile Foundation Detection with Forward-Looking Sonar Image Processing. *Remote Sens.* **2025**, *17*, 360. https://doi.org/10.3390/rs17030360

1. Introduction

The detection of underwater pile foundations is important for harbor channel operations and marine engineering [1]. Traditionally, visual examinations by divers have been the main method for identifying underwater pile foundations, but this has limitations including poor safety, high cost, and low efficiency [2,3]. The development of high-resolution sonar imaging technology has opened new possibilities for underwater target detection by offering advantages such as long-range detection capabilities and real-time imaging [4]. However, due to the imaging principles of sonar technology and the impact of underwater environments, sonar images often exhibit high noise, poor contrast, and structural distortions, making the accurate detection and identification of underwater targets difficult [5,6].

The development of underwater pile foundation detection technology has garnered significant attention in the realms of maritime engineering and environmental monitoring. Over the past few decades, underwater target detection using high-resolution sonar imaging has progressed significantly. Early methods focused on feature extraction and enhancement techniques, such as mathematical morphology and level-set methods, to address the inherent noise and resolution issues of sonar imagery. With the advent of deep learning, innovations like the Mask R-CNN and improved YOLO frameworks have emerged, offering enhanced accuracy and robustness. Despite these advancements, key challenges remain, including the detection of small and densely packed targets under varying environmental conditions. This study addresses these challenges by integrating multi-scale feature fusion and attention mechanisms into the AquaPile-YOLO framework. These enhancements are pivotal for the real-time detection and precise identification of underwater pile foundations, enabling significant improvements in sonar image analysis.

Early research on sonar image processing primarily focused on feature extraction and image enhancement [5,7]. For instance, Lu et al. provided a comprehensive review of feature extraction technology for underwater targets using active sonar technology, establishing theoretical foundations for sonar image processing [2]. Subsequently, Calder et al. presented a novel concept for underwater identification of side-scan sonar images—a Bayesian approach to target detection. These early investigations established foundations for understanding and interpreting sonar images [3]. The application of computer vision technologies has enhanced sonar image processing. Foresti et al. proposed an underwater image target recognition method based on a computer vision system, employing computer vision analysis of sonar data [4]. Liu et al. investigated the application of mathematical morphology in acoustic image processing, proving the utility of morphological approaches for image enhancement and edge identification [6].

Deep learning has revolutionized sonar image processing, replacing older methods such as level sets [8], Markov random fields (MRFs) [9], and Curvelet transform [10]. Intelligence in sonar image processing has emerged as the most significant development trend [11]. Intelligence has improved target identification accuracy and efficiency under complex underwater situations [12,13]. Advances in image resolution and quality have made forward-looking sonar broadly applicable in engineering applications [14] like seabed sediment classification [15] and mine target detection [16,17]. Valdenegro-Toro et al. applied convolutional neural networks to target detection and recognition in forward-looking sonar images, initiating deep learning applications in sonar image processing [18]. Zhu et al. addressed the challenge of limited sonar data by proposing a deep network classification algorithm for identifying small bottom targets in high-resolution underwater sonar images, demonstrating the effectiveness of deep learning in small target detection [19].

Most deep learning-based sonar image detection methods rely on sliding window feature extraction, employing various computer vision techniques such as boosted classifiers [20], machine learning classifiers [21–24], and template matching [25,26]. However, these methods often perform poorly outside of the training set, especially in challenging scenarios like underwater tiny target recognition [14–16,27]. Recent research has proposed numerous innovations to address these challenges. For instance, Fan et al. [28] introduced an improved Mask R-CNN method for underwater object detection in forward-looking sonar images, achieving high accuracy. Zhang et al. [29] emphasized the importance of sonar image registration and proposed an improved CNN for learning similarity functions, significantly enhancing model performance. Additionally, Xie et al. [30] released a multi-beam forward-looking sonar image dataset, providing a benchmark for target detection. By integrating traditional methods' strengths with deep learning advancements, ongoing research aims to address these challenges, focusing on improving model generalization, efficiency, and robustness for sonar image processing applications.

Building on previous research, Zhang et al. proposed an improved YOLOv5 network for forward-looking sonar images [31], incorporating transfer learning and optimized clustering algorithms. Gaspar et al. have developed unsupervised methods for feature-based place recognition in poor visibility conditions [32], while Jiao et al. proposed the PLUD (Push the right Logit Up and the wrong logit Down) approach to improve sonar image feature representation for open-set and long-tail recognition challenges [33]. Li et al.

introduced TransYOLO, a new forward-looking sonar image target detector based on a TFFN feature fusion network with a transformer stack structure [34].

Leveraging these advancements, this paper proposes an underwater pile foundation detection approach for forward-looking sonar images named AquaPile-YOLO, which is an enhancement of the YOLOv5 algorithm. The AquaPile-YOLO algorithm is designed to overcome the aforementioned challenges by integrating multi-scale feature fusion and attention mechanisms. These enhancements are particularly beneficial for detecting small targets within sonar images. Additionally, the application of data augmentation techniques serves to bolster the model's robustness and generalization capabilities. The training dataset, comprising 4000 sonar images, underwent a series of augmentations including random cropping, rotation, and the introduction of noise to improve the model's adaptability across diverse environmental conditions.

This study proposes AquaPile-YOLO, an advanced algorithm for detecting underwater pile foundations in forward-looking sonar images. By integrating multi-scale feature fusion and attention mechanisms, the proposed method aims to improve detection accuracy and robustness for real-time applications. The ultimate goal is to overcome existing limitations in sonar-based target detection, enabling more reliable and efficient underwater engineering and environmental monitoring applications.

2. Methods

2.1. Forward-Looking Sonar

A forward-looking sonar is an imaging sonar that uses transducers to emit and receive sound waves, forming images from the intensity of sound wave reflections off of underwater targets [26]. Like an optical camera, a forward-looking sonar generates images, but sonar images typically show an overhead view rather than the frontal perspective of an optical camera. Figure 1 illustrates the imaging principle, depicting the 2D reconstruction of a 3D underwater target by a forward-looking sonar.

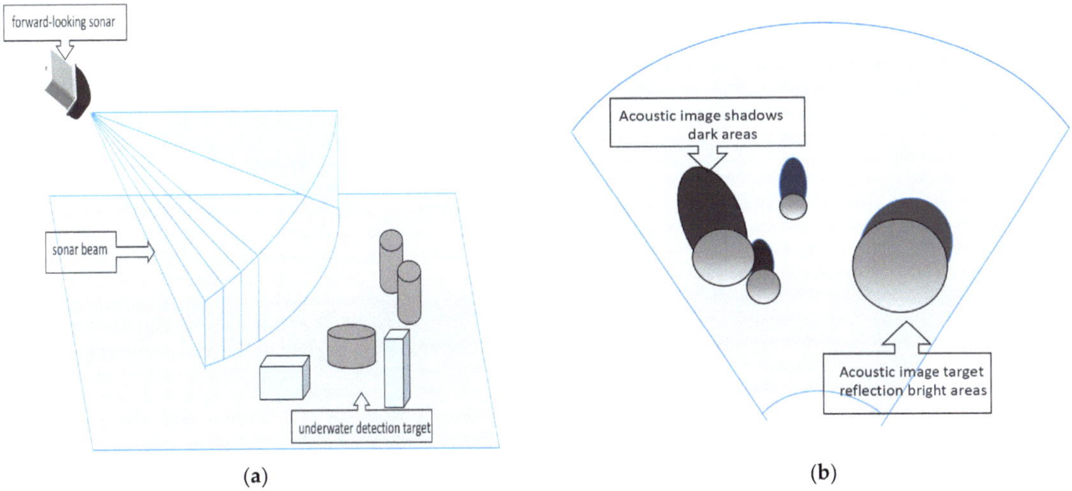

Figure 1. A diagram of a 2D reconstruction of an underwater 3D target using a forward-looking sonar: (**a**) A schematic of the FLS operational principle in an underwater environment, depicting the acoustic imaging process; (**b**) An illustration of the 2D reconstruction process, transforming 3D target data into a planar representation as captured by the sonar.

The equipment used in the experiments for this paper is the HY1645 model forward-looking sonar, manufactured by Haiying Marine in Wuxi, China [35]. The sonar utilizes two-dimensional acoustic imaging technology to obtain real-time, high-resolution images of underwater targets (including bearings and distances) for the autonomous recognition and transmission of information. It can meet the needs of autonomous detection in complex, low-visibility, shallow water environments. To meet engineering demands for portability, the device incorporates a novel sparse array design for multibeam imaging sonar systems, reducing the number of transducers while preserving imaging performance. This minimizes the number of transducers in the array while maintaining multibeam imaging performance [36]. A schematic diagram of the fan-scan function for detecting underwater pile foundation targets by a forward-looking sonar is shown in Figure 2.

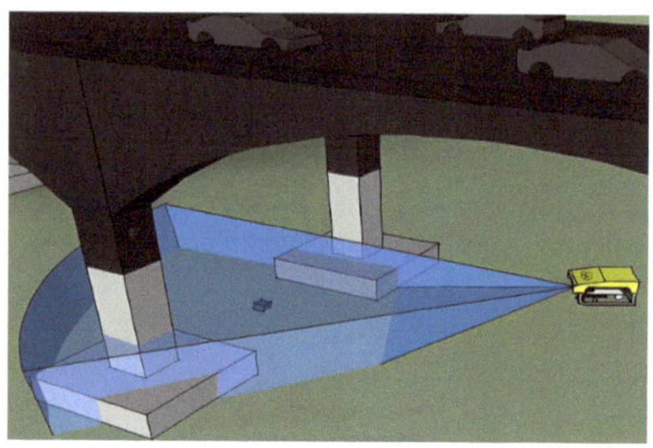

Figure 2. A schematic diagram of the underwater detection capabilities of a forward-looking sonar.

The system primarily consists of an underwater transducer, a transmitter/receiver module, a data acquisition processor, and acquisition software, among other components. Figure 3 illustrates a photo of the HY1645 imaging sonar transducer and its on-site installation. In the photo, the black part of the transducer is responsible for the reception and transmission of underwater acoustic signals, while the white part encloses the receiver and transmitter modules along with their associated circuitry. The entire assembly is encapsulated in waterproof housing for integrated packaging and communicates and receives power from the exterior through a single cable.

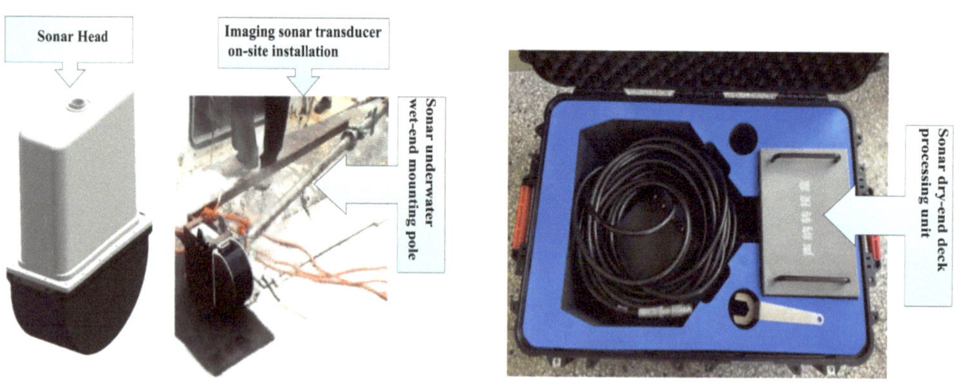

(a) (b)

Figure 3. The composition of the HY1645 forward-looking sonar system: (**a**) The wet end of the sonar, designed for underwater acoustic signal emission and reception; (**b**) The dry end components of the HY1645 sonar, including the data processing unit and associated cabling.

The main technical parameters and performance indicators of the HY1645 forward-looking sonar are presented in Table 1. A significant characteristic of forward-looking sonars is that, in most cases, the distance and bearing of objects can be directly read from the sonar data, but the elevation of underwater targets is lost. As a result, the image information from forward-looking sonars is typically challenging to interpret. For instance, during the detection of an underwater stepped structure at a hydropower station using the HY1645 forward-looking sonar, Figure 4a shows the surface photo of the stepped structure, while Figure 4b displays the corresponding underwater sonar data collected by the two-dimensional imaging sonar.

Table 1. Technical specifications of HY1645 forward-looking sonar.

Parameter	Value
Operating Frequency	450 kHz
Field of View	$90^\circ \times 20^\circ$
Maximum Range	100 m
Beam Width (Horizontal × Vertical)	$1^\circ \times 20^\circ$
Number of Beams	538
Beam Spacing	0.17°
Range Resolution	2.5 cm
Maximum Sampling Rate	15 Hz

(a)

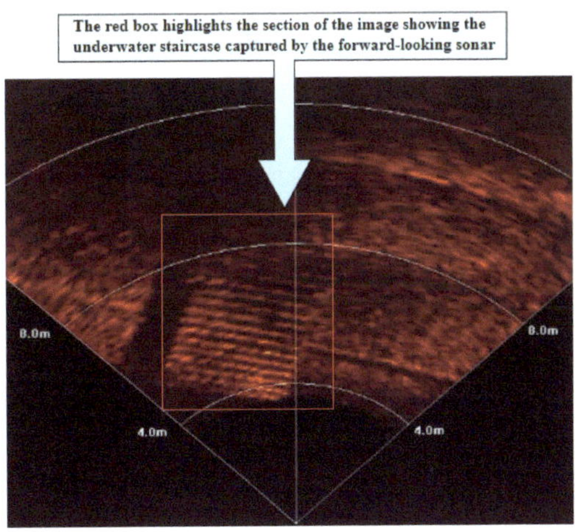

(b)

Figure 4. Sonar scanning experiment of the underwater ladder structure: (**a**) Photos of the above-water part of the ladder underwater structure sonar scanning experiment; (**b**) Forward-looking underwater ladder scanning experimental sonar data.

Acoustic imaging cannot capture the true color of detected objects, yielding purely grayscale initial data. Yellow sonar images are pseudo-colored, enhanced in contrast via software processing. The HY1645 imaging sonar can scan both static and dynamic underwater targets, like divers. Figure 5 illustrates the use of an imaging sonar to simulate the monitoring of a diver in a swimming pool.

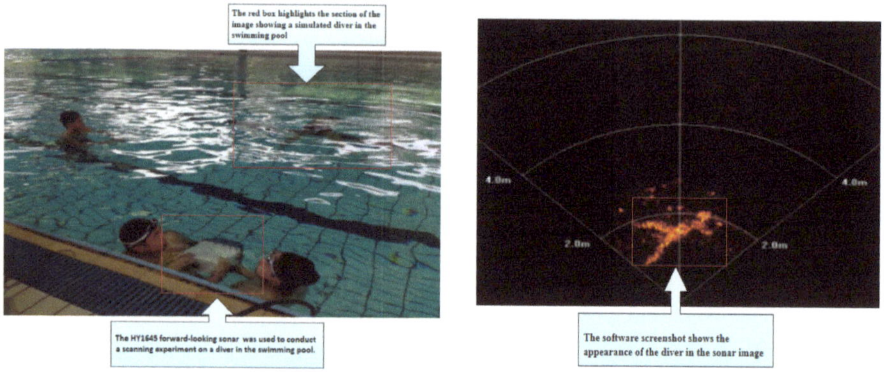

(a) (b)

Figure 5. Sonar scanning experiment of the diver's pool: (**a**) The forward-looking sonar scanning diver swimming pool experiment; (**b**) The forward-looking sonar scans the sonar image data of the diver's pool experiment.

2.2. *AquaPile-YOLO Network*

The YOLO (You Only Look Once) network [37] is a revolutionary real-time object detection system that can predict the positions and categories of objects in an image through a single forward propagation. YOLOv5 is an efficient object detection algorithm renowned for its speed and superior performance. Figure 6 illustrates the structure of the AquaPile-YOLO network. In recent years, through continuous updates and iterations [38,39], the YOLO network has been widely applied in engineering projects due to its stability. However, forward-looking sonar images present unique challenges, requiring adaptations for effective detection. This study aims to enhance the performance of AquaPile-YOLO in underwater pile foundation detection tasks by introducing a series of innovative improvements. These enhancements were designed to adapt to the particularities of forward-looking sonar images and increase the detection accuracy of underwater pile foundation targets.

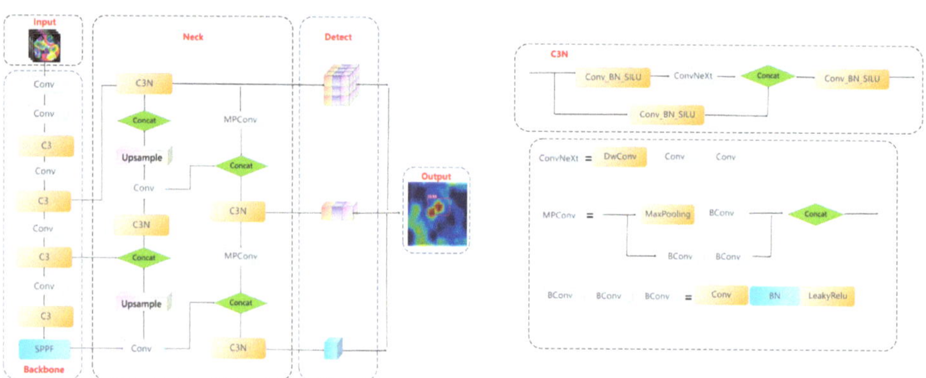

(a) (b)

Figure 6. The AquaPile-YOLO network structure pipeline diagram: (**a**) This panel presents an overview of the AquaPile-YOLO's architecture, illustrating the comprehensive workflow from input to output, and highlighting the integration of multi-scale feature fusion, attention mechanisms, and other key components that facilitate the detection of underwater pile foundations; (**b**) This panel zooms in on specific modules within the AquaPile-YOLO network, detailing the internal structure and connectivity of the components, such as the C3 Module with CBAM Attention, MPConv Module, and C3N Module, which are crucial for enhancing the network's performance in processing forward-looking sonar images.

2.2.1. Data Augmentation

To enhance the model's generalization and robustness, data augmentation techniques were employed. Operations such as rotation, scaling, flipping, and adding noise to the training images simulate the complexity of underwater environments, effectively increasing the diversity of the training data.

It was assumed that the sonar image dataset is divided into groups of four sub-images I1, I2, I3, and I4, each of a size of H × W. Through random cropping and flipping operations, each sub-image can generate a new sub-image I1′, I2′, I3′, and I4′. These sub-images were then concatenated into a larger image Imosaic of size 4H × 4W. The concatenation operation can be expressed as follows:

$$I_{mosaic}(x,y)=\begin{cases} I_1'(x-2H,y-2W) \text{ if } x\in[0,2H) \text{ and } y\in[0,2W) \\ I_2'(x-H,y-2W) \text{ if } x\in[2H,3H) \text{ and } y\in[0,2W) \\ I_3'(x-2H,y-H) \text{ if } x\in[0,2H) \text{ and } y\in[2W,3W) \\ I_4'(x-H,y-H) \text{ if } x\in[2H,3H) \text{ and } y\in[2W,3W) \end{cases} \tag{1}$$

where (x,y) represents the coordinate position in Imosaic.

2.2.2. Transfer Learning Strategy

Considering the scarcity of sonar image data, a transfer learning strategy was adopted. A pre-trained AquaPile-YOLO model, initially trained on a large dataset like ImageNet, served as the starting point. Then, by fine-tuning AquaPile-YOLO on the limited sonar image data, the model's performance was quickly enhanced. The transfer learning strategy by Huo et al. [40] for side-scan sonar image classification and target recognition is referenced.

Let the source domain be D_s = {χ,P(X)} and the target domain be D_t = {χ′,P(X′)}, where χ and χ′ represent the feature spaces, and P(X) and P(X′) represent the marginal probabilities. The task T is defined by the label space y and the target prediction function f(x). The goal of transfer learning is to improve the performance of the prediction function f_t for the target task T_t by discovering and transferring knowledge from D_s and T_s.

During the pre-training phase, a deep network F was trained on the source domain to learn general feature representations.

$$F^*=\text{argmin}_F L(F(X^s),Y^s) \tag{2}$$

where L is the loss function, X^s and Y^s are the input and label of the source domain, respectively, and F^* is the pre-trained network.

In the transfer phase, the pre-trained network F^* was transferred to the target domain and adapted to the target task through fine-tuning.

$$F'=\arg\min_F L\left(F\left(X^t\right),Y^t\right) \tag{3}$$

where X^t and Y^t are the inputs and labels for the target domain, respectively.

2.2.3. Multi-Scale Feature Fusion

Multi-scale feature fusion techniques were introduced into the AquaPile-YOLO to address the variability in target sizes within forward-looking sonar images. This strategy enhanced the model's ability to recognize targets of various scales by integrating feature maps at different resolutions. A Feature Pyramid Network (FPN) structure was employed to effectively combine deep semantic information with shallow detail information, thereby improving the detection accuracy of small targets. The feature fusion can be expressed as follows:

$$F_{fuse} = F_{upsample}(F_d) \oplus F_{downsample}(F_c) \tag{4}$$

where F_d represents the deep-layer feature map and F_c represents the shallow-layer feature map. $F_{upsample}$ and $F_{downsample}$ denote the upsampling and downsampling operations, respectively, while the symbol $\oplus$ signifies the operation of feature fusion.

The upper-level feature maps contain stronger semantic information due to the deeper network layers, while the lower-level features suffer less loss of positional information due to fewer convolutional layers. The FPN structure performs top–down upsampling to ensure that the bottom-level feature maps contain stronger semantic information (the backbone in Figure 6). Conversely, the PAN (Path Aggregation Network) structure performs bottom–up downsampling, enabling the top-level features to retain positional information (neck in Figure 6). The fusion of these two features ensures that feature maps of different scales contain both semantic and spatial information, thereby facilitating accurate predictions for images of various sizes.

Fine-tuning further trained the network on the target domain data, adjusting the network parameters to improve performance.

$$\theta' = \theta - \eta \nabla_\theta L(F(X^t;\theta), Y^t) \tag{5}$$

where θ is the network parameter, η is the learning rate, and ∇ stands for the gradient.

2.2.4. Attention Mechanism

The AquaPile-YOLO network incorporates advanced attention mechanisms to enhance the model's ability to focus on salient regions within the image, particularly in complex underwater environments characterized by noise and occlusions. This was achieved through the integration of the Convolutional Block Attention Module (CBAM) into the YOLOv5 network architecture. In this section, we will discuss the role of the C3 module (CSP Bottleneck with 3 convolutions) [41], MPConv, and the C3N module in enhancing the attention mechanisms of the AquaPile-YOLO network.

(1) C3 Module with CBAM Attention

An attention mechanism called CBAM was incorporated into the AquaPile-YOLO network to enhance the model's focus on key areas within the image. This mechanism comprises spatial and channel attention modules that adaptively adjust the weights of the feature maps, enhancing the model's response to target areas, especially in complex underwater environments with noise and occlusions.

For example, the channel attention for an input feature map Fattn is given by the following:

$$F_{attn} = \sum_c A_c \cdot F_c \tag{6}$$

where A_c is the attention weight of the c channel, typically calculated using learnable parameters and an activation function σ as follows:

$$A_c = \sigma(W \cdot F_c + b) \tag{7}$$

where W and b are the weight parameters and bias parameters in the deep network, respectively.

The C3 module comprises a main branch (primary path) and a shortcut branch (skip connection), which are merged at the output [41]. The main branch typically includes multiple Bottleneck layers, sequentially stacked to increase the network's depth and representational capacity. By replacing the default Bottleneck layers in the C3 module with

CBAM modules and iteratively creating multiple CBAM Bottleneck layers, the integration of CBAM attention mechanisms within the C3 module was achieved (C3-CBAM in Figure 6). The C3-CBAM module retains the advantages of the C3 module, such as efficient feature extraction and partial gradient flow sharing, while significantly enhancing feature representation through the CBAM's channel and spatial attention mechanisms. This synergistic combination endowed YOLOv5 with higher accuracy and robustness in object detection tasks, thereby improving the overall model performance on sonar objects.

(2) MPConv Module

The MPConv (Multi-Path Convolution) module is a novel architectural component introduced in the AquaPile-YOLO network to address the challenges posed by the diverse scales and orientations of underwater targets in sonar images. MPConv is designed to capture a rich set of features by processing the input data through multiple parallel convolutional paths with different kernel sizes and aspect ratios [42]. Each path is tailored to capture specific spatial hierarchies, allowing the network to represent a wide range of underwater structures effectively. The outputs from these parallel paths are then concatenated, forming a comprehensive feature representation that encapsulates both local and global contextual information. This multi-path processing approach enabled the AquaPile-YOLO network to achieve superior performance in detecting targets of varying sizes and complexities within sonar imagery.

(3) C3N Module

The C3N module, building on the strengths of the C3 module, introduces an innovative structure combining depth-separable convolution with a novel inverted Bottleneck design, inspired by the ConvNeXt architecture [43,44]. The C3N module consists of three convolutional layers followed by multiple ConvNeXt blocks, enabling efficient parameter utilization and enhanced feature correlation capture while mitigating information loss during dimensionality compression. The inverted Bottleneck structure of the C3N module, with a wider central section and narrower endpoint, empowers effective feature correlation capture and efficient feature space transformation processing. This results in robust feature extraction capability, particularly beneficial for detecting small, densely packed targets in sonar images, despite the imaging limitations of sonar technology.

By integrating these advanced modules—C3 with CBAM, MPConv, and C3N—the AquaPile-YOLO network achieved a heightened level of attention and discrimination, enabling it to excel in the detection of underwater pile foundation targets within forward-looking sonar images [45].

2.2.5. Loss Function Optimization

The loss function plays a crucial role in object detection tasks. The loss function for AquaPile-YOLO was optimized based on the characteristics of sonar image targets, employing a composite loss function that guides model training more comprehensively through classification loss L_{cls}, regression loss L_{reg}, and objectness loss L_{obj}.

The total loss function is given by the following:

$$\mathrm{L_{sonar}} = \frac{1}{\mathrm{N_{pos}}}\left(\mathrm{L_{cls}} + \mathrm{L_{reg}} + \mathrm{L_{obj}}\right) \quad (8)$$

where $\mathrm{N_{pos}}$ is the number of positive samples, $\mathrm{I}\{\cdot\}$ is the indicator function, $\mathrm{L_{Focal}}$ is the focal loss for classification, $\mathrm{L_{IoU}}$ is the IoU loss for regression, and $\mathrm{L_{BCE}}$ is the binary cross-entropy loss for objectness.

2.2.6. Soft-NMS (Soft-Non-Maximum Suppression)

In this study, the Soft-NMS algorithm [46] was used to improve the object detection process of AquaPile-YOLO. Soft-NMS adjusts the scores of detection boxes using a Gaussian function for continuous decay instead of simply setting the scores of overlapping detection boxes to zero, thereby improving the accuracy and robustness of small target detection.

In traditional NMS, given a set of detection boxes $B = \{b_1, \ldots, b_N\}$ and corresponding scores $S = \{s_1, \ldots, s_N\}$, the algorithm first selects the box with M the highest score, then removes all other boxes with an overlap higher than the threshold of N_t with M. This process is then recursively applied to the remaining boxes. Soft-NMS proposes a different approach by adjusting the score s_i of the detection box b_i using the following Formula (9):

$$s_i' = s_i \cdot e^{-IOU(M,b_i)^2/\sigma} \forall b_i \notin D \tag{9}$$

where IOU(M,b_i) represents the Intersection over Union between the detection boxes M and b_i, and σ is a parameter controlling the speed of score decay.

3. Experiments

3.1. *Experimental Design*

The purpose of this experiment was to validate the effectiveness of the proposed AquaPile-YOLO algorithm for underwater pile foundation detection using HY1645 forward-looking sonar images. The experimental environment was a designated section of a lake field test site, characterized by water depths ranging from 2 m to 20 m and a substrate primarily composed of sand and gravel, providing a controlled yet representative setting for underwater sonar testing.

The HY1645 forward-looking sonar was installed on a vessel using a lateral straddle mount, as shown in Figure 7, which illustrates the field experiment vessel with the sonar installed. Two devices were fixed onto an installation pole. Due to the weight of the equipment, the structure was designed to grip the vessel's edge from both sides beneath the bow. The installation pole was fixed to the side of the vessel, with the detection sonar located approximately 0.5 m below the water surface.

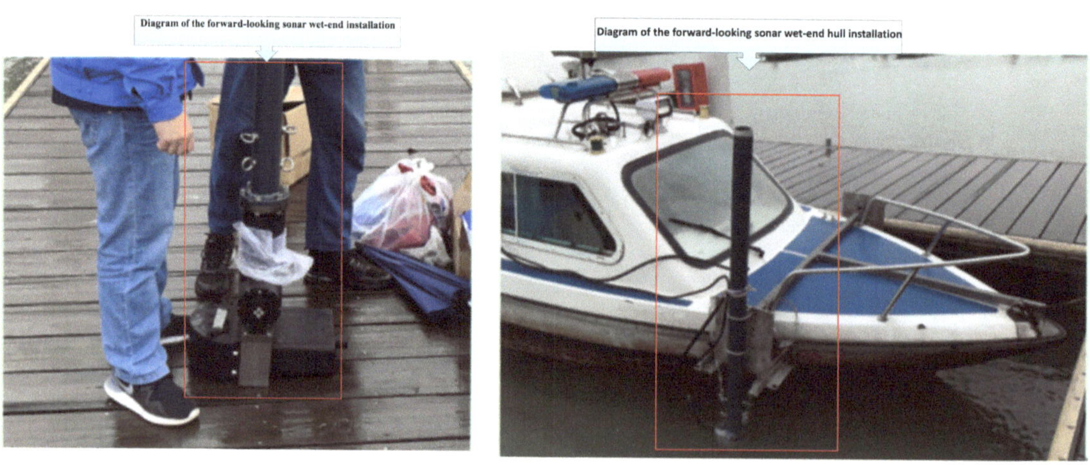

(**a**) (**b**)

Figure 7. HY1645 forward-looking sonar field test installation; Forward-looking sonar installation angle (**a**) and target (underwater pile foundation) distribution (**b**) diagram.

To prevent interference from the side lobes of the forward-looking sonar touching the water surface and causing noise, the emission direction of the detection sonar was set to a 30° downward tilt from the water surface, based on the scanning direction of the sonar beam opening angle. The sonar installation angle (left) and the distribution of the target (underwater pile foundation) (right) are shown in Figure 8.

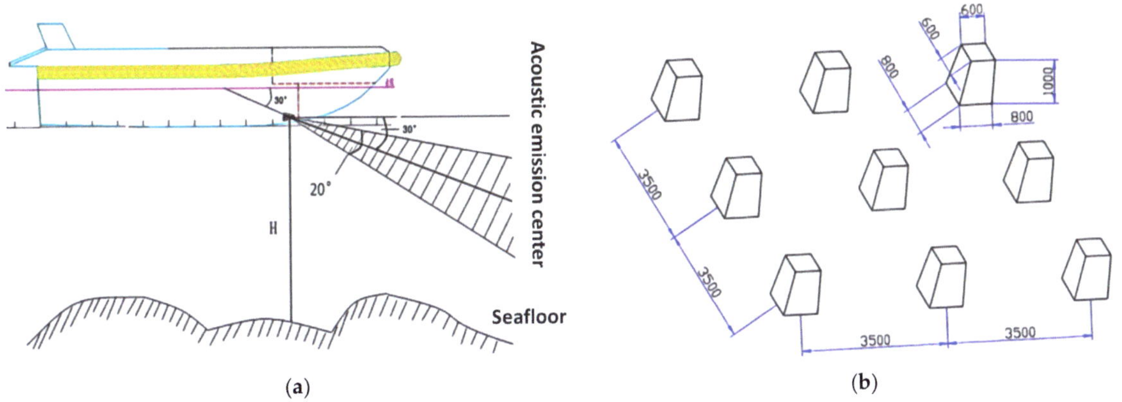

Figure 8. Forward-looking sonar installation angle (**a**) and target (underwater pile foundation) distribution (**b**) diagram.

3.2. Data Collection

During the data collection phase, we conducted field experiments at one lake in Wuxi City from November 26th to 27th, 2020. The trials involved multi-distance, multi-directional sonar detection of pre-set targets (track racks) at varying speeds (13 km/h, 15 km/h, and 18 km/h, equivalent to approximately 7, 8, and 10 knots, respectively). Utilizing a gimbaled mount, the sonar was adjusted to an optimal operational attitude to ensure the acquisition of target sonograms in real-time. Data were recorded in the AVI video format and were saved as JPEG/PNG/BMP snapshots for subsequent analysis and algorithm validation.

Data annotation was performed for underwater pile foundation targets in sonar images by conducting continuous long-term detection and comparing them with human observations and mapping charts. The dataset includes two category labels, "l" and "r", using the YOLO format. To construct the training dataset, 4000 sonar images were collected in the field experiment, covering various underwater environments and target conditions. The image data were preprocessed, including grayscaling, noise removal, and contrast enhancement, to improve the accuracy of subsequent target detection.

Due to the scarcity of sonar data, scholars in the field of sonar images have mostly used simulated datasets as the sample space, while actual measured datasets barely exceeded a few hundred images. This paper collected 4000 sonar images on-site as the dataset for deep learning training, which to some extent compensates for the lack of data in previous research in this field.

The original data collected by the HY1645 imaging sonar were in a custom format of acoustic signal data, with ".hca" and ".son" being the two formats. The HAICA.EXE executable program provided by the system is required to read them. The original acoustic signals were transformed into image data in the ".bmp" format. Using the original acoustic data collected by the HY1645 in the field experiment, 4000 two-dimensional sonar images with a pixel resolution of 848 × 600 were generated.

3.3. Experimental Procedure

This study employs a comprehensive dataset, containing a total of 4000 field-measured, forward-looking sonar data images, which were meticulously preprocessed to augment model detection capabilities. In the experiment, the dataset was divided into a training set (3000 images) and a validation set (1000 images). The experimental steps incorporate several key stages: data augmentation, model training, performance evaluation, and systematic recording of results. The data augmentation phase plays a crucial role in enhancing model adaptability across diverse environments, achieved through an array of techniques such as random cropping, rotation, and the strategic introduction of noise. The model training phase was executed within a strictly defined, controlled environment, where parameters including the learning rate and batch size were rigorously monitored.

The experiment was conducted on a system equipped with a high-performance CPU and GPU to ensure efficient operation. The CPU is Intel(R) Xeon(R) Gold 6130, and the GPU is Tesla V100-PCIE-32GB, with 32 GB of video memory. The operating system used is Ubuntu 18.04.5 LTS, and the deep learning framework is torch-2.0.0, as shown in Table 2 for the detailed experimental environment configuration.

Table 2. Experimental environment configuration.

Parameter	Setup
Ubuntu	18.04.5 LTS
Pytorch	2.0.0
Python	3.8
CUDA	11.8
GPU	Tesla V100-PCIE-32GB
CPU	Intel(R) Xeon(R) Gold 6130

In order to enhance the persuasiveness of the experiments, this study conducted a series of parameter adjustments based on the AquaPile-YOLO model and performed multiple experimental tests, ultimately selecting the hyperparameter settings as shown in Table 3.

Table 3. Hyperparameters during training.

Parameter	Setup
Epoch	300
Batch	32
NMS IoU	0.6
Initial Learning Rate	0.01
Final Learning Rate	0.01
Momentum	0.937
Weight Decay	0.0005

The formulas are as follows. Regular evaluations were undertaken using a validation set to ensure the model's performance was accurately gauged. Key metrics like precision, recall, and mAP were systematically recorded during this stage. The formulas are as follows.

$$Precision = \frac{TP}{TP + FP} \tag{10}$$

$$Recall = \frac{TP}{TP + FN} \tag{11}$$

$$AP = \int_0^1 P(R)dR \tag{12}$$

$$mAP = \frac{1}{N}\sum_{i=1}^{n} AP_i \quad (13)$$

where TP is the number of correctly predicted positive samples, FP is the number of negative samples incorrectly predicted as positive, and FN is the number of positive samples incorrectly predicted as negative. Moreover, average precision (AP) is the calculation of the area under the accuracy–response rate curve for a certain category. mAP is an auxiliary to the AP of all categories and can be used to evaluate the model's detection performance for all categories. In Formula (13), n is the number of categories; AP(j) is the AP of the jth category.

To ensure methodological rigor and reproducibility, all experimental settings and parameters were painstakingly documented. Furthermore, the entire experiment was repeated multiple times in order to confirm the consistency and reliability of the results obtained. Potential biases and errors that could arise during the course of the study were identified and discussed, along with the corresponding mitigation strategies proposed. This thorough experimental procedure aimed to provide a transparent and replicable guide for scholars seeking to replicate the study's setup, as well as to harness the enhanced capabilities of the AquaPile-YOLO model within their own research endeavors.

4. Results

4.1. Ablation Studies

In order to analyze the influence of different improvement strategies on the performance of model detection, three groups of experiments were designed to complete the training and testing under the premise of ensuring the same data set and training parameters and the experimental results are shown in Table 4.

Table 4. Results of ablation experiments.

Multi-Scale Feature Fusion	CBAM	Sonar Loss	Soft-NMS	Precision	Recall	mAP50	mAP50-95
×	×	×	×	0.886	0.76	0.789	0.517
✓	×	×	×	0.9	0.764	0.8	0.521
×	✓	×	×	0.912	0.764	0.808	0.524
✓	✓	×	×	0.919	0.771	0.811	0.525
✓	✓	✓	×	0.896	0.785	0.819	0.528
✓	✓	✓	✓	0.888	0.798	0.821	0.529

When only CBAM was enabled, the precision further increased to 0.912, while the recall remained at 0.764. The mAP50 improved to 0.808, and the mAP50-95 increased to 0.524. This demonstrates the significant effect of the CBAM on enhancing model performance. By combining MSFF and the CBAM, the performance continued to improve, with precision reaching 0.919, recall increasing to 0.771, mAP50 rising to 0.811, and mAP50-95 reaching 0.525. This combination clearly outperforms the use of MSFF or CBAM alone.

After introducing Sonar Loss, the precision slightly decreased to 0.896, but the recall improved to 0.785. The mAP50 increased to 0.819, and the mAP50-95 reached 0.528. This indicates that Sonar Loss is helpful in improving the recall rate and overall performance of the model, although it may slightly impact accuracy.

Finally, with all improvements (including Soft-NMS) enabled, the precision was 0.888, recall increased to 0.798, mAP50 reached 0.821, and mAP50-95 also increased to 0.529. Despite a slight decrease in accuracy, the improvements in recall and mAP values reflect the enhancement of overall detection performance.

In conclusion, by combining techniques such as multi-scale feature fusion, CBAM, Sonar Loss, and Soft-NMS, AquaPile-YOLO achieved improvements in various performance metrics, particularly in mAP, the most convincing indicators. These enhancements effectively enhance the detection capabilities of YOLOv5.

4.2. Comparisons

After ablation studies, the AquaPile-YOLO model was tested by comparative experiments. The test results showed that the model achieved an identification accuracy rate of 96.89% for underwater targets, confirming the effectiveness and reliability of the proposed method in actual underwater pile foundation detection.

This experiment compared the performance of five object detection algorithms, SSD300, YOLOv3, Faster R-CNN, Cascade R-CNN, and AquaPile-YOLO, on sonar images. The test results for each algorithm are shown in Table 5. Additionally, we compared our results with the recently published Underwater Acoustic Target Detection (UATD) dataset. This dataset includes identification results for underwater objects such as a ball, cube, tire, sc (square cage), and cc (circle cage). As shown in Table 6, the AquaPile-YOLO model performed superiorly across these categories, further validating its efficacy in various underwater detection scenarios.

Table 5. Comparison of AquaPile-YOLO with other models.

Model	Precision	Recall	mAP@50	Params/M	FPS
SSD300	0.238	0.403	0.670	23.88	9.1
YOLOv3	0.364	0.455	0.783	61.52	46.7
Faster R-CNN	0.328	0.429	0.760	41.35	19.4
Cascade R-CNN	0.333	0.438	0.752	69.15	15.5
AquaPile-YOLO	0.888	0.798	0.821	46.60	111.1

Table 6. Detection results of underwater targets with different scenarios.

Model	AP (Ball)	AP (Cube)	AP (Tyre)	AP (sc)	AP (cc)	AP (Pile)
Faster-RCNN (Resnet-18)	0.869	0.717	0.847	0.547	0.666	-
Faster-RCNN(Resnet-50)	0.870	0.686	0.889	0.621	0.538	0.328
Faster-RCNN(Resnet-101)	0.865	0.697	0.840	0.572	0.491	0.333
YOLOv3 (Darknet-53)	0.860	0.669	0.874	0.470	0.519	-
YOLOv3 (MobilenetV2)	0.868	0.573	0.738	0.518	0.498	0.364
AquaPile-YOLO	-	-	-	-	-	0.888

As shown in Figure 9a, the comparative analysis indicates that AquaPile-YOLO outperforms other state-of-the-art object detection models, including YOLOv3, Faster R-CNN, Cascade R-CNN, and SSD300, in terms of both precision and recall. Precision, which quantifies the proportion of true positive detections among all detected samples, and recall, which measures the model's ability to detect all actual positive instances, are critical metrics for object detection systems. AquaPile-YOLO achieves a precision of 0.888 and a recall of 0.798, with a mAP@50 score of 0.821, indicating its exceptional ability to identify underwater pile foundations while minimizing false positives accurately. This high level of precision and recall suggests that AquaPile-YOLO is particularly robust in scenarios requiring reliable underwater detection.

Figure 9b provides a detailed comparison of recall performance among the same set of object detection models, further emphasizing AquaPile-YOLO's superiority. With a recall value of 0.821, AquaPile-YOLO demonstrates its effectiveness in detecting all instances of underwater targets, outperforming YOLOv3 (0.783), Faster R-CNN (0.760), Cascade R-CNN (0.752), and SSD300 (0.670). This superior recall performance indicates that AquaPile-YOLO is more reliable in identifying underwater targets, making it highly suitable for applications where high recall is essential for operational success. The high recall rate is particularly crucial in underwater environments, where missing a target could have significant consequences, thus highlighting AquaPile-YOLO as the preferred model for critical detection tasks.

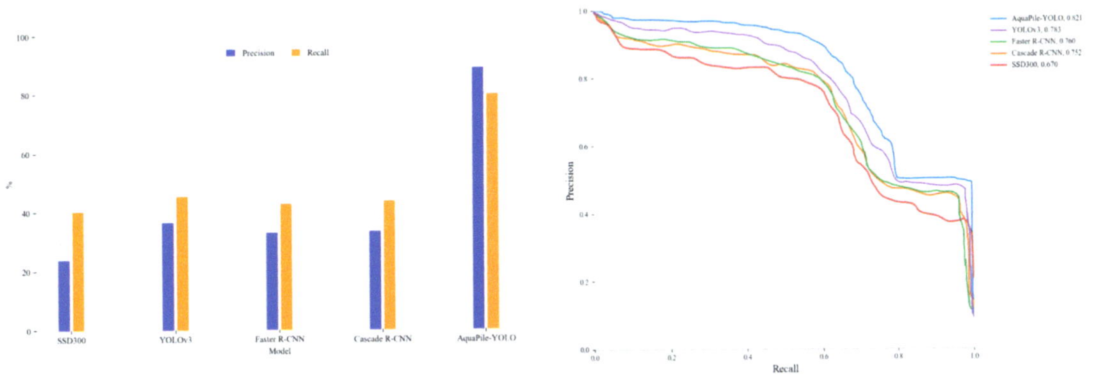

(a) (b)

Figure 9. (**a**) Bar chart comparison of precision–recall for different models; (**b**) Comparison of object detection models.

Figure 10a presents a compelling comparison of F1 performance for pile foundation detection using forward-looking sonar images across various algorithms. The F1 score, a balanced metric harmonizing both precision and recall, is depicted at varying confidence thresholds. This composite score provides a comprehensive measure of a model's exactness and completeness in detection. AquaPile-YOLO exhibits notably high F1 scores, signifying its ability to balance precision and recall. Notably, at a confidence threshold of 0.155, AquaPile-YOLO achieves an F1 score of 0.84, indicating robustness in accurately detecting pile foundations.

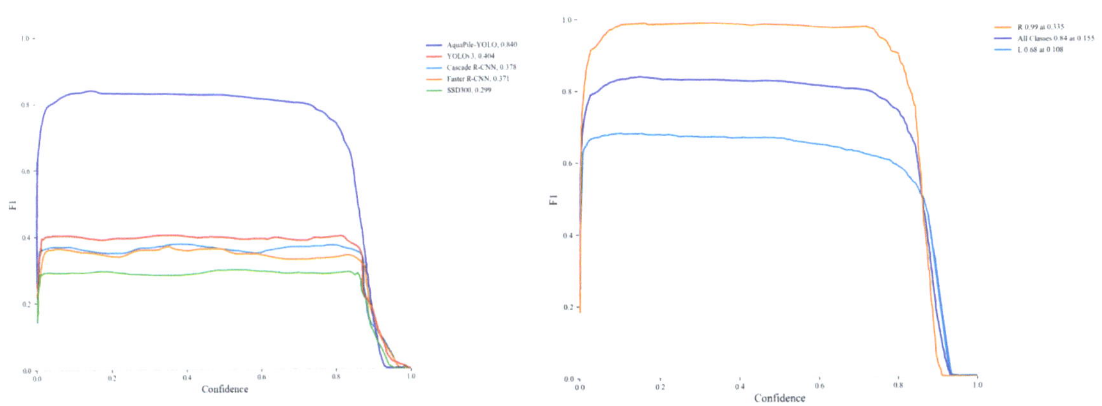

(a) (b)

Figure 10. (**a**) F1 performance comparison of different algorithms for pile foundation detection by forward-looking sonar images; (**b**) F1 performance curve for AquaPile-YOLO.

Figure 10b depicts the F1 performance curve for the AquaPile-YOLO model, illustrating how the model's F1 score fluctuates at different confidence thresholds. The curve represents the interplay between precision and recall, with each point reflecting the precision at various levels of recall. This visualization is instrumental in assessing the model's performance across the entire spectrum of detection confidence. The curve underscores

AquaPile-YOLO's consistently high performance, even at lower confidence thresholds, thereby validating its reliability and effectiveness in real-world applications. The "All classes" average F1 score encapsulates the model's overall efficacy in detecting a diverse range of underwater targets, further solidifying AquaPile-YOLO as a superior choice for sonar-based object detection tasks.

Figure 11 is a heatmap comparison, demonstrating the comprehensive performance of different algorithms on the target detection task. AquaPile-YOLO achieved a high score of 0.93 on this indicator. A comparison of the original image and detection results for each algorithm's heatmap indicates the model's strong comprehensive performance for detecting underwater pile foundation targets under various scenarios. Simultaneously, it shows the model performs well in sonar image target detection tasks, meeting real-time detection speed requirements and significantly improving accuracy. These results support the model in this paper as the preferred algorithm for sonar image target detection.

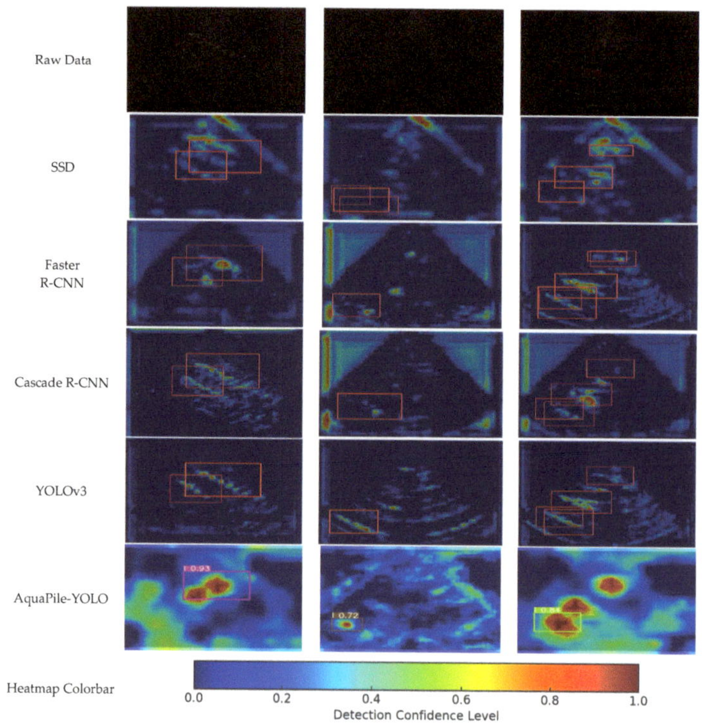

Figure 11. A heatmap comparison of different algorithms for pile foundation detection by forward-looking sonar images. The heatmap illustrates the performance comparison of various detection algorithms, with the red box highlighting the area of interest where the pile foundation targets are detected. Within this box, the intensity of the color indicates the confidence level of the detection, with warmer tones (reds and yellows) signifying higher confidence in the presence of a target.

5. Discussion

This study introduces AquaPile-YOLO, an advanced underwater pile foundation detection method utilizing forward-looking sonar imagery. The proposed method offers several advantages, including significantly improved detection accuracy achieved by the AquaPile-YOLO algorithm. The algorithm effectively captures underwater targets of varying sizes and enhances the detection of small targets, representing a critical advancement in the field.

The principal contributions of this study comprise the following: (1) the development and proposal of the AquaPile-YOLO algorithm, an innovative method for underwater pile foundation detection, which builds upon the foundational architecture of YOLOv5 and incorporates multi-scale feature fusion and attention mechanisms to achieve significantly improved detection accuracy; (2) the application of data augmentation techniques to improve model generalization and robustness; (3) the collation and use of 4000 sonar images as a training dataset, offering plentiful data for model training and validation; and (4) experimental results underscoring the considerable practical application value in detecting underwater pile foundation targets within sonar images.

Specifically, this study addresses the critical challenge of real-time, fast, and accurate template recognition and the detection of underwater pile foundations in sonar images. Key innovations include the following:

- Multi-scale Feature Fusion: By incorporating a multi-scale feature fusion scheme, this study effectively captures underwater targets of varying sizes, thereby improving small target detection accuracy.
- Enhanced Attention Mechanism: The attention mechanism is improved by combining Normalized Weighted Distance (NWD) and Intersection over Union (IOU), enhancing the model's ability to distinguish small targets and reducing scale sensitivity. This enhancement is complemented by structural modifications within the YOLOv5 network, allowing for a more nuanced focus on critical image regions.
- Application of Soft-NMS: Rather than traditional NMS, Soft-NMS better handles occlusions and overlapping targets, limiting missed and false detections in complex scenes.
- Data Augmentation Strategy: The model's generalization and adaptability to diverse environmental conditions are bolstered through data augmentation techniques like rotation, random cropping, and noise addition.

In addition to the aforementioned innovations, this study significantly contributed to the dataset by collecting 4000 real-measured sonar images from field experiments as a training dataset. This collection provides substantial data support for model training and validation and serves as a vital supplement to existing research datasets. This includes raw acoustic data from forward-looking sonar technology, sonar images, and video data, thereby facilitating further research and collaboration within the academic community.

Despite the promising results, our study has limitations. The AquaPile-YOLO algorithm has primarily been tested in controlled environments with specific water conditions, and its performance in more variable natural settings remains to be explored. Additionally, the model's computational requirements may pose challenges for real-time applications in resource-constrained environments.

The proposed AquaPile-YOLO method exhibits high applicability in marine engineering and environmental monitoring. Its ability to accurately detect underwater pile foundations can significantly enhance the efficiency and safety of harbor operations and underwater construction projects. Furthermore, the model's robustness to environmental variations makes it a promising tool for the long-term monitoring of underwater infrastructure.

Building on the foundation of the AquaPile-YOLO algorithm, future research will focus on refining and expanding capabilities for underwater pile foundation detection. The following five aspects outline the trajectory for future research and development:

- Algorithm Optimization: While the AquaPile-YOLO algorithm has demonstrated high accuracy, there is a need to continue optimizing the model structure. Reducing computational resource consumption and improving detection speed are essential to meet the demands of real-time detection, particularly in resource-constrained environments.
- Multimodal Data Fusion: To further improve detection accuracy and robustness, exploring the combination of sonar images with other sensor data, such as optical

images or LiDAR data, is a promising avenue. Multi-modal data fusion could provide a more comprehensive understanding of the underwater environment and enhance the algorithm's capabilities.

- Broader Environmental Adaptability: Assessing the model's performance across a broader range of underwater environments is crucial. Testing the algorithm in various water qualities, lighting conditions, and underwater structures will enhance the model's generality and adaptability, ensuring its effectiveness in diverse marine settings.
- Automation and Intelligence: The development of an automated sonar image collection system, integrated into underwater robots or autonomous underwater vehicles (AUV/USV/ROV/UUV), is essential for achieving fully autonomous underwater detection tasks. This advancement would increase the efficiency and safety of underwater operations.
- Engineering Application Deployment: Integrating the AquaPile-YOLO model into existing underwater monitoring systems for long-term deployment and performance evaluation is vital. Such integration will provide insights into the model's practical performance and longevity, facilitating its adoption in marine engineering and environmental monitoring projects.

Through these future directions, we expect to enhance the performance of underwater pile foundation detection technology and promote its application in the fields of marine engineering and environmental monitoring.

6. Conclusions

This paper proposes an underwater pile foundation detection method for forward-looking sonar images based on the AquaPile-YOLO algorithm. By introducing modules such as multi-scale feature fusion, attention mechanisms, and Soft-NMS, the model's detection accuracy for underwater pile foundation targets is significantly improved. The experimental results show that the AquaPile-YOLO model achieves an accuracy rate of 96.89% in underwater target identification tasks, demonstrating its efficiency and reliability in practical applications.

Author Contributions: Conceptualization, Z.X. and R.W.; methodology, Z.X., R.W. and T.C.; software, Z.X. and W.G.; validation, T.C. and Q.G.; formal analysis, R.W.; investigation, Q.G.; writing—original draft preparation, Z.X.; writing—review and editing, Z.X. and B.S.; visualization, Z.X. and W.G.; supervision, W.G.; project administration, Z.X.; funding acquisition, R.W. All authors have read and agreed to the published version of the manuscript.

Funding: This research was funded by the Hunan Province Key Laboratory of Credible Intelligent Navigation and Positioning (funding number CINP2024002).

Data Availability Statement: The original contributions presented in the study are included in the article; further inquiries can be directed to the corresponding author.

Acknowledgments: The authors express their sincere gratitude to China State Shipbuilding Corporation Haiying Enterprise Group Co., Ltd. for their invaluable support throughout this research endeavor. Special thanks are extended to Lei Dong, the sonar product designer, for his instrumental support in providing access to the advanced sonar equipment that was pivotal to our experimental work. We also acknowledge the contributions of Haiying Cui, whose expertise as a researcher significantly enhanced our field experiments. Additionally, we are thankful for the dedicated assistance of Hongyong Deng, whose technical prowess in equipment calibration and pool testing was crucial to the success of our research. The authors appreciate the commitment and efforts of these individuals, which played a vital role in the completion of this study.

Conflicts of Interest: Author Zhongwei Xu was employed by the company China State Shipbuilding Corporation Haiying Enterprise Group Co., Ltd. The remaining authors declare that the research was conducted in the absence of any commercial or financial relationships that could be construed as potential conflicts of interest.

References

1. Gan, J. Development of an Underwater Detection Robot for the Structures with Pile Foundation. *J. Mar. Sci. Eng.* **2024**, *12*, 1051. [CrossRef]
2. Lu, Y.; Sang, E. Feature extraction techniques of underwater objects based on active sonars—An overview. *J. Harbin Eng. Univ.* **1997**, *18*, 43–54. (In Chinese)
3. Calder, B.R.; Linnett, L.M.; Carmichael, D.R. Bayesian approach to object detection in sidescan sonar. *IEE Proc.-Vis. Image Signal Process.* **1998**, *145*, 221–228. [CrossRef]
4. Foresti, G.L.; Gentili, S. A Vision Based System for Object Detection in Underwater Images. *Int. J. Pattern Recognit. Artif. Intell.* **2000**, *14*, 167–188. [CrossRef]
5. Guo, H. Post-Image Processing of High-Resolution Imaging Sonar. Master's Thesis, Harbin Engineering University, Harbin, China, 2002. (In Chinese).
6. Liu, C.C.; Sang, E.F. Underwater Acoustic image processing based on mathematical morphology. *J. Jilin Univ. Inf. Sci. Ed.* **2003**, *21*, 52–57. (In Chinese)
7. Kelly, J.G.; Carpenter, R.N.; Tague, J.A. Object classification and acoustic imaging with active sonar. *J. Acoust. Soc. Am.* **1992**, *91 Pt 1*, 2073–2081. [CrossRef]
8. Ye, X.F.; Zhang, Z.H.; Liu, P.X.; Guan, H.L. Sonar image segmentation based on GMRF and level-set models. *Ocean. Eng.* **2010**, *37*, 891–901. [CrossRef]
9. Wang, X. Research on Underwater Sonar Images Objective Detection and Based Respectively on MRF and Level-Set. Ph.D. Thesis., Harbin Engineering University, Harbin, China, 2010. (In Chinese).
10. Sheng, H.; Meng, F.; Li, Q.; Ma, G.; Cao, Y. Enhancement Algorithm of Side-scan Sonar Image in Curvelet Transform Domain. *Ocean. Surv. Mapp.* **2012**, *32*, 8–17. (In Chinese)
11. Sheng, Z.; Huo, G. Detection of underwater mine target in sidescan sonar image based on sample simulation and transfer learning. *CAAI Trans. Intell. Syst.* **2021**, *16*, 385–392. (In Chinese)
12. Valdenegro-Toro, M. End-to-end object detection and recognition in forward-looking sonar images with convolutional neural networks. In Proceedings of the 2016 IEEE/OES Autonomous Underwater Vehicles (AUV), Tokyo, Japan, 6–9 November 2016.
13. Gong, W.; Tian, J.; Huang, H. Underwater sonar image small target recognition method based on shape features. *J. Appl. Acoust.* **2021**, *40*, 294–302. (In Chinese)
14. Bian, H.Y.; Sang, E.F.; Ji, X.C.; Zhao, J.Y. Simulation research on acoustic lens beamforming. *J. Harbin Eng. Univ.* **2004**, *25*, 43–45. (In Chinese)
15. Yang, C.Y.; Xu, F.; Wei, J.J. Seafloor sediments classification using a neighborhood gray level co-occurrence matrix. *J. Harbin Eng. Univ.* **2005**, *26*, 561–564. (In Chinese)
16. Gao, S.; Xu, J.; Zhang, P. Automatic target recognition of mine-like objects in sonar images. *Mine Warf. Ship Prot.* **2006**, *1*, 42–45. (In Chinese)
17. Fandos, R.; Zoubir, A.M.; Siantidis, K. Unified Design of a Feature-Based ADAC System for Mine Hunting Using Synthetic Aperture Sonar. *IEEE Trans. Geosci. Remote Sens.* **2014**, *52*, 2413–2426. [CrossRef]
18. Valdenegro-Toro, M. Objectness Scoring and Detection Proposals in Forward-Looking Sonar Images with Convolutional Neural Networks. In *Artificial Neural Networks in Pattern Recognition, Proceedings of the 7th IAPR TC3 Workshop, ANNPR 2016, Ulm, Germany, 28–30 September 2016*; Lecture Notes in Computer Science; Springer International Publishing: Cham, Switzerland, 2016; pp. 209–219.
19. Zhu, K.; Tian, J.; Huang, H. Underwater objects classification method in high-resolution sonar images using deep neural network. *Acta Acust.* **2019**, *44*, 595–603. (In Chinese)
20. Sawas, J.; Petillot, Y.; Pailhas, Y. Cascade of Boosted Classifiers for Rapid Detection of Underwater Objects. In Proceedings of the 10th European Conference on Underwater Acoustics, Istanbul, Turkey, 5–9 July 2010.
21. Reed, S.; Petillot, Y.; Bell, J. Automated approach to classification of mine-like objects in sidescan sonar using highlight and shadow information. *IEE Proc. Radar Sonar Navig.* **2004**, *151*, 48–56. [CrossRef]
22. Isaacs, J.C. Sonar automatic target recognition for underwater UXO remediation. In Proceedings of the 2015 IEEE Conference on Computer Vision and Pattern Recognition Workshops (CVPRW), Boston, MA, USA, 7–12 June 2015.
23. Williams, D.P.; Groen, J. A fast physics-based, environmentally adaptive underwater object detection algorithm. In Proceedings of the OCEANS 2011 IEEE—Spain, Santander, Spain, 6–9 June 2011.
24. Fandos, R.; Zoubir, A.M. Optimal Feature Set for Automatic Detection and Classification of Underwater Objects in SAS Images. *IEEE J. Sel. Top. Signal Process.* **2011**, *5*, 454–468. [CrossRef]

25. Myers, V.; Fawcett, J. A Template Matching Procedure for Automatic Target Recognition in Synthetic Aperture Sonar Imagery. *IEEE Signal Process. Lett.* **2010**, *17*, 683–686. [CrossRef]
26. Hurtós, N.; Palomeras, N.; Nagappa, S.; Salvi, J. Automatic detection of underwater chain links using a forward-looking sonar. In Proceedings of the 2013 MTS/IEEE OCEANS—Bergen, Bergen, Norway, 10–14 June 2013. [CrossRef]
27. Kocak, D.M.; Dalgleish, F.R.; Caimi, F.M.; Schechner, Y.Y. A focus on recent developments and trends in underwater imaging. *Mar. Technol. Soc. J.* **2008**, *42*, 52–67. [CrossRef]
28. Fan, Z.; Xia, W.; Liu, X.; Li, H. Detection and segmentation of underwater objects from forward-looking sonar based on a modified Mask RCNN. *Signal Image Video Process.* **2021**, *15*, 1135–1143. [CrossRef]
29. Zhang, P.; Tang, J.; Zhong, H.; Ning, M.; Liu, D.; Wu, K. Self-Trained Target Detection of Radar and Sonar Images Using Automatic Deep Learning. *IEEE Trans. Geosci. Remote Sens.* **2021**, *60*, 1–4. [CrossRef]
30. Xie, K.; Yang, J.; Qiu, K. A Dataset with Multibeam Forward-Looking Sonar for Underwater Object Detection. *Sci. Data* **2022**, *9*, 739. [CrossRef] [PubMed]
31. Zhang, H.; Tian, M.; Shao, G.; Cheng, J.; Liu, J. Target Detection of Forward-Looking Sonar Image Based on Improved YOLOv5. *IEEE Access* **2022**, *10*, 18023–18034. [CrossRef]
32. Gaspar, A.R.; Matos, A. Feature-Based Place Recognition Using Forward-Looking Sonar. *J. Mar. Sci. Eng.* **2023**, *11*, 2198. [CrossRef]
33. Jiao, W.; Zhang, J.; Zhang, C. Open-set recognition with long-tail sonar images. *Expert Syst. Appl.* **2024**, *249 Pt A*, 123495. [CrossRef]
34. Li, Y.; Ye, X.; Zhang, W. TransYOLO: High-Performance Object Detector for Forward Looking Sonar Images. *IEEE Signal Process. Lett.* **2022**, *29*, 2098–2102.
35. Haiying Marine. HY1645 Imaging Sonar. Available online: https://www.haiyingmarine.com/index.php?a=shows&catid=74&id=106 (accessed on 15 January 2025).
36. Xia, W.; Jin, X.; Dou, F. Thinned Array Design With Minimum Number of Transducers for Multibeam Imaging Sonar. *IEEE J. Ocean. Eng.* **2017**, *42*, 892–900. [CrossRef]
37. Redmon, J.; Divvala, S.; Girshick, R.; Farhadi, A. You Only Look Once: Unified, Real-Time Object Detection. In Proceedings of the 2016 IEEE Conference on Computer Vision and Pattern Recognition (CVPR), Las Vegas, NV, USA, 27–30 June 2016.
38. Redmon, J.; Farhadi, A. YOLOv3: An Incremental Improvement. *arXiv* **2018**, arXiv:1804.02767v1.
39. Bochkovsky, A.; Wang, C.Y.; Liao, H.Y. YOLOv4: Optimal Speed and Accuracy of Object Detection. *arXiv* **2020**, arXiv:2004.10934.
40. Huo, G.; Wu, Z.; Li, J. Underwater Object Classification in Sidescan Sonar Images Using Deep Transfer Learning and Semisynthetic Training Data. *IEEE Access* **2020**, *8*, 47407–47418. [CrossRef]
41. Wang, C.Y.; Liao, H.Y.M.; Wu, Y.H.; Chen, P.Y.; Hsieh, J.W.; Yeh, I.H. CSPNet: A New Backbone that can Enhance Learning Capability of CNN. In Proceedings of the IEEE/CVF Conference on Computer Vision and Pattern Recognition Workshops, Long Beach, CA, USA, 16–17 June 2019.
42. Wang, C.Y.; Bochkovskiy, A.; Liao, H.Y.M. YOLOv7: Trainable bag-of-freebies sets new state-of-the-art for real-time object detectors. In Proceedings of the IEEE/CVF Conference on Computer Vision and Pattern Recognition, New Orleans, LA, USA, 18–24 June 2022.
43. Liu, Z.; Mao, H.; Wu, C.Y.; Feichtenhofer, C.; Darrell, T.; Xie, S. A ConvNet for the 2020s. In Proceedings of the IEEE/CVF Conference on Computer Vision and Pattern Recognition, New Orleans, LA, USA, 18–24 June 2022.
44. Xing, B.; Sun, M.; Ding, M.; Han, C. Fish sonar image recognition algorithm based on improved YOLOv5. *Math. Biosci. Eng.* **2024**, *21*, 1321–1341. [CrossRef] [PubMed]
45. Qin, K.S.; Liu, D.; Wang, F.; Zhou, J.; Yang, J.; Zhang, W. Improved YOLOv7 model for underwater sonar image object detection. *J. Vis. Commun. Image Represent.* **2024**, *100*, 104124. [CrossRef]
46. Bodla, N.; Singh, B.; Chellappa, R.; Davis, L.S. Soft-NMS--Improving Object Detection With One Line of Code. In Proceedings of the IEEE International Conference on Computer Vision, Venice, Italy, 22–29 October 2017. [CrossRef]

MDPI AG
Grosspeteranlage 5
4052 Basel
Switzerland
Tel.: +41 61 683 77 34

Remote Sensing Editorial Office
E-mail: remotesensing@mdpi.com
www.mdpi.com/journal/remotesensing

www.ingramcontent.com/pod-product-compliance
Lightning Source LLC
Chambersburg PA
CBHW041415210926
51850CB00013B/3278
* 9 7 8 3 7 2 5 8 3 6 0 9 3 *